REAL TIME PROGRAMMING 1999
(WRTP'99)

A Proceedings volume from the 24^h IFAC/IFIP Workshop,
Schloss Dagstuhl, Wadern, Saarland, Germany,
30 May – 3 June 1999

Edited by

A.H. FRIGERI
Universidade Federal do Rio Grande do Sul,
Porto Alegre, RS, Brazil

W.A. HALANG
Faculty of Electrical Engineering, FernUniversität Hagen,
Hagen, Germany

and

S.H. SON
Department of Computer Science,
University of Virginia, USA

Published for the

INTERNATIONAL FEDERATION OF AUTOMATIC CONTROL

by

PERGAMON
An Imprint of Elsevier Science

UK	Elsevier Science Ltd, The Boulevard, Langford Lane, Kidlington, Oxford, OX5 1GB, UK
USA	Elsevier Science Inc., 660 White Plains Road, Tarrytown, New York 10591-5153, USA
JAPAN	Elsevier Science Japan, Tsunashima Building Annex, 3-20-12 Yushima, Bunkyo-ku, Tokyo 113, Japan

First edition 1999

Library of Congress Cataloging in Publication Data

A catalogue record for this book is available from the Library of Congress

British Library Cataloguing in Publication Data

A catalogue record for this book is available from the British Library

ISBN 0-08-043548 3

Transferred to DIgital Printing 2008.

IFAC WORKSHOP ON REAL TIME PROGRAMMING 1999

Sponsored by
International Federation of Automatic Control (IFAC)
Technical Committee on Real Time Software Engineering (CCR)
University of Skövde

Co-sponsored by
International Federation of Information Processing (IFIP)
Working Group 5.4 on Industrial Software Quality and Certification

Organized by
GI/GMA – Fachgruppe "Echtzeitsysteme"
FernUniversität, Hagen

Co-organized by
University of Skövde
Schloss Dagsthul

Workshop Chair

Alceu Heinke Frigeri
  Universidade Federal do Rio Grande do Sul, *Brazil*

International Programme Committee

WRTP co-Chair
  Wolfgang A. **Halang**
    FernUniversität, *Germany*
ARTDB Co-Chair
  Sang H. **Son**
    University of Virginia, *USA*
Members

Brad **Adelberg**	Northwestern University	*USA*
Sten F. **Andler**	University of Skövde	*Sweden*
Sven-Arne **Andréasson**	Chalmers University of Technology	*Sweden*
Azer **Bestavros**	Boston University	*USA*
Alan **Burns**	University of York	*UK*
Matjaz **Colnarič**	University of Maribor	*Slovenia*
Alfons **Crespo**	Universitat Politecnica de Valencia	*Spain*
Anindya **Datta**	University of Arizona	*USA*
Jörgen **Hansson**	University of Skövde	*Sweden*
Mike **Hinchey**	University of Nebraska-Omaha	*USA*
Jörg **Kaiser**	Universität Ulm	*Germany*
Tei-Wei **Kuo**	National Chung Cheng University	*Taiwan*
Swamy **Kutti**	Deakin University	*Australia*
Kam-Yiu **Lam**	City University of Hong Kong	*Hong Kong*
Phillip **Laplante**	Pennsylvania Institute of Technology	*USA*
Kwei-Jay **Lin**	University of California	*USA*
C. Douglas **Locke**	Lockheed Martin Corporation	*USA*
Kin-F. **Man**	City University of Hong Kong	*Hong Kong*
Karlotto **Mangold**	ATM Computer	*Germany*
Marga Marcos **Muñoz**	Universidad del País Vasco	*Spain*
Sias **Mostert**	University of Stellenbosch	*South Africa*
Leo **Motus**	Tallinn Technical University	*Estonia*
Carlos E. **Pereira**	Univ. Fed. do Rio Grande do Sul	*Brazil*
Juan A. **de la Puente**	Universidad Politecnica de Madrid	*Spain*
Ragunathan **Rajkumar**	Carnegie Mellon University	*USA*

Helmut **Rzehak**	Universität der Bundeswehr München	*Germany*
Abd-El-Kader **Sahraoui**	Ecole Nationale Superieure D'Ingenieur des Constructions Aeronautique	*France*
Jean-Jacques **Schwarz**	Institut National des Sciences Appliquees de Lyon	*France*
Bran V. **Selic**	ObjecTime Limited	*Canada*
Alan **Shaw**	University of Washington	*USA*
Jacques **Skubich**	Institut National des Sciences Appliquees de Lyon	*France*
Jack A. **Stankovic**	University of Virginia	*USA*
Tomasz **Szmuc**	Stanislaw Staszic University of Mining and Metallurgy	*Poland*
Theodor **Tempelmeier**	Fachhochschule Rosenheim	*Germany*
Paulo Jorge **Veríssimo**	Universidade de Lisboa	*Portugal*
Horst **Wedde**	Universität Dortmund	*Germany*
Janusz **Zalewski**	University of Central Florida	*USA*
Lichen **Zhang**	Shantou University	*P.R. China*
Wei **Zhao**	Texas A&M University	*USA*

National Organising Committee

Chair

Mohammad A. Livani

University of Ulm, *Germany*

Members

Sten F. Andler	University of Skövde	*Sweden*
Joakim Eriksson	University of Skövde	*Sweden*
Wolfgang A. Halang	FernUniversität	*Germany*
Alceu Heinke Frigeri	Univ. Fed. do Rio Grande do Sul	*Brazil*
Janine Magnussen	FernUniversität	*Germany*
Nicole Probst	Schloß Dagstuhl	*Germany*

Preface

With its tradition of almost three decades, in 1999 the IFAC/IFIP Workshop on Real Time Programming (WRTP) joined forces with the Workshop on Active and Real-Time Database Systems (ARTDB). Both series of workshops have established themselves as excellent fora for exchanging information on recent scientific and technological advances and practices in real time computing, a field that is becoming an essential enabling discipline of both control engineering and computer science and engineering. As there is an accelerated growth of demands for functionality and dependability of real time systems, our intellectual and engineering abilities are being challenged to come up with practical solutions to the problems faced in the design and development of complex real time systems.

The now annual Workshop on Real Time Programming and the bi-annual Workshop on Active and Real-Time Database Systems are intended as meetings of relatively small numbers of experts in their fields, and to take place as truly international events. Therefore, each time they are held in different parts of the world. The 1999 Workshop maintained the outstanding quality of both series. It provided an opportunity to assess the state of the art, to present new results, and to discuss possible lines of future developments. Primarily, it focused on software development for real time systems, real time operating systems and active and real time database systems.

In particular, the technical programme of the Workshop covered latest research and developments in requirements engineering, software engineering, active and real time database systems, communication and clock synchronisation, embedded systems, formal methods, operating systems and scheduling. Out of 58 submissions from 19 countries (Australia, Brazil, Britain, Canada, P.R. China, France, Germany, Greece, Israel, Italy, the Netherlands, Poland, Romania, Singapore, Slovenia, Spain, Sweden, Switzerland, and the U.S.A.), the International Programme Committee selected 26 regular papers and 8 reserve papers for presentation at the Workshop. These contributions come from Europe, North America, Australia, and the Far East. In addition to these high quality technical papers, the programme also featured two world renowned keynote speakers, and a discussion panel about the state of the art in the field of active real time database systems.

As all scientific conferences to be successful, the 1999 Workshop required the dedicated effort of many individuals and organisations before the participants departed for Schloß Dagstuhl. The International Programme Committee did an outstanding job selecting the best candidates out of many high quality submissions. We are indebted to IFAC with its Technical Committee on Real Time Software Engineering for sponsoring the Workshop. The support of the IFAC Secretariat is also gratefully acknowledged. Formal co-sponsorship was provided by IFIP's Working Group on Industrial Software Quality and Certification. The financial support provided to Central European and overseas participants of the Workshop by Deutsche Forschungsgemeinschaft and by the University of Skövde is particularly appreciated.

All attendees appreciated the excellent conference facilities of *International Conference and Research Centre for Computer Science* — Schloß Dagstuhl, and the highly professional service provided by its staff. They greatly enjoyed the warm hospitality of this extraordinary site located in the small German state Saarland. It is its ambience which fostered fruitful technical exchanges between the participants, both in formal sessions and in informal discussions, in a very friendly and relaxed atmosphere.

Alceu Heinke Frigeri   Wolfgang A. Halang   Sang H. Son
Universidade Federal do Rio Grande do Sul   FernUniversität   University of Virginia

CONTENTS

EMBEDDED SYSTEMS AND CASE STUDIES

FORMAL METHODS

OPERATING SYSTEMS AND SCHEDULING

Real-Time Software Architectures and Design Patterns: Fundamental Concepts and Their Consequences

Janusz Zalewski

Dept. of Electrical & Computer Engineering

University of Central Florida

Orlando, FL 32816-2450, USA

`jza@ece.engr.ucf.edu`

Abstract. This paper discusses the principles of software architectures for real-time systems. The fundamental idea of a real-time architecture is based on the concept of feedback used in control engineering. A generic architecture is derived for three major categories of real-time systems. Then a fundamental design pattern is presented, valid for all major architectures. This is followed by a discussion of variations in the basic architecture for distributed systems and safety related systems. Finally, tool support for architectural design and a case study are discussed. *Copyright © 1999 IFAC*

Keywords. Real-time systems, real-time computing, software architecture, design patterns, safety related systems, software tools, history of engineering.

1 Introduction

In recent years, a new area of research and engineering has emerged called software architecture (Buschmann *et al.* 1996, Donohoe 1999, Rechtin and Maier 1997, Shaw and Garlan 1996, Witt *et al.* 1994). It is not clear, however, what are the principles of software architectures, what are good and bad examples or practices, and how to build high quality software according to certain principles and following good practices. The state of the art is such that almost every block diagram with interconnected boxes and a couple of arrows pointing from one box to another is claimed to be a description of a software architecture.

By any measure, this is far from being acceptable. Just like not every combination of construction materials can lead to a good house structure and not every combination of structural elements can lead to a good house architecture, there are only certain selective arrangements of software elements that can constitute a good software architecture. To find out what are these arrangements of elements and structures, which are crucial to software architectures, we have to look at the basic principles. We have to determine what constitutes an architecture of software. In particular, in this paper, we look at the issue of what constitutes an architecture of a real-time system software.

Architectures specific to real-time and related systems have been studied or proposed in a number of papers, in the last one and a half decade, for example (Atkinson *et al.* 1996, Baker and Scallon 1986, Boasson 1993, Emery et al. 1996, Levenson 1984, Schoch and Laplante 1995, Shaw 1995). Some of these papers propose interesting ideas to approach the real-time software architecture problem from the perspective of control engineering. This approach seems to be very promising and is pursued by us, but to be fully fruitful and understood it has to be tied to some more fundamental concepts.

There are certain fundamental laws of nature, which have to be taken into account by engineers. For example, for a house architecture, such a law of nature is the force of gravity. To start building a house from the roof would not be a good idea, even though such houses standing on a roof do exist, one in the author's hometown (http ref. n.d.).

In the field of real-time systems, such a fundamental law exists as well, although its significance has not been sufficiently articulated. It is the feedback principle, one of the most fundamental laws, which both the nature and the society rely upon. Basically, every human being and every society can exist because of this fundamental feedback property, to mention only the most obvious examples: regulation of temperature of a human body and self-regulating principles of economy.

In this paper, the author claims that the fundamentals of real-time computing and real-time systems architectures are deeply rooted in the feedback principle, the basic law of control engineering, and date back to the B.C. era; at least that's

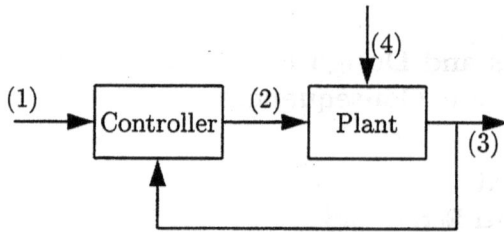

Fig. 1. Abstract view of the feedback principle (Watt's speed governor). (1) Desired value (speed set point); (2) Command signal (error value); (3) Controlled variable (actual speed); (4) Disturbances (load and steam pressure).

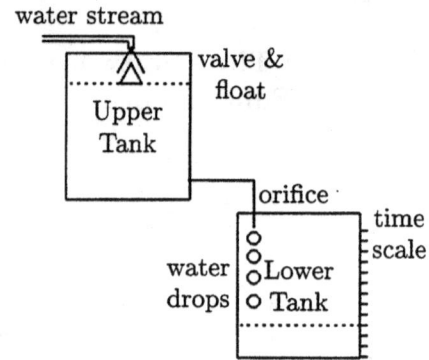

Fig. 2. Sketch of a water clock of Ktesibios (water level is marked by dotted lines).

how far they can be traced. The rest of the paper is structured as follows. Section 2 outlines these basic roots in the history of engineering. Section 3 derives the basic architecture of a real-time system, and Section 4 discusses the basic design patterns. Distributed real-time architectures and safety-related architectures are treated in Sections 5 and 6, respectively. A case study is presented in Section 7 and principles of design tool support for real-time architectures are discussed in Section 8. The paper ends with a conclusion summarizing the basic issues.

2 Ancient Roots of Real-Time Computing

Historically, the most familiar example of an engineering application of the feedback principle is probably the device known as the centrifugal speed governor (Bennett 1994, Mayr 1969). In 1767, James Watt invented a flyball speed governor to control the speed of steam engines. In this device, two spinning flyballs fastened to a shaft rise as a rotational speed increases, causing a mechanically connected steam valve to close, thus reducing the steam flow to the engine and regulating the speed. The control feedback principle applied in this device is illustrated in Figure 1.

Intuitively, we would agree that the Watt's speed governor operates in real time: engine's action (that is, that of the 'plant') to increase or decrease the speed causes an immediate (that is, real-time) response of the controller to close or open the valve. Moreover, we can calculate how "immediate" are these actions, because the exact timing relationships do exist between the controller and the plant. This is described by the equations of the system dynamics. All this means that the engine speed governor can be used as a model of a simple real-time system. A straightfor-

ward analysis reveals that the major elements of a controller are:

- plant interface (sensor and actuator)
- user interface (set point)
- processing power (subtraction operation and other transformations within the controller).

Studying the history of engineering, in particular, control engineering (Mayr 1969), reveals that the oldest written record of an application of the feedback principle was found to be that of a water clock, that is, a device to measure time. Presumably in the first half of the third century B.C., an inventive Greek craftsman, Ktesibios, invented a device which used a constant flow of water to measure the passage of time (Fig. 2). Water drops falling from the orifice of an upper tank accumulate in the lower tank and the increasing water level indicates how much time expired. However, the accuracy of the clock depends on the constant flow rate of water into the low-level container. This flow rate is being disturbed by variation in the upper-tank water level, that is, its water pressure. To solve the problem of variations in water flow rate, Ktesibios maintained a constant water level in the upper holding tank by applying a valve with a float. The role of the valve is to open and allow water flow in, when the water level in the upper tank goes down, and to close and cut the water stream off, when the water level in the upper tank reaches the uppermost level. The water level in the upper tank is determined (measured) by a float (a sensor) which operates as a part of the valve (an actuator).

This simple device invented in the third century B.C. fits very well into the model of feedback principle and system dynamics, and can be regarded as the first real-time device. In fact, it would be hard to find a better example for our purposes, because this device not only operates in real time

2

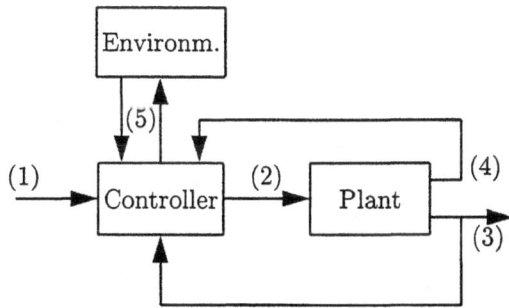

Fig. 3. Illustration of a control system.
(1) Desired value; (2) Controller commands;
(3) Controlled variables; (4) Other measured
variables; (5) Environment interface.

but also serves the purpose of measuring (computing) time. Interestingly, the oldest preserved written source on Ktesibios' water clock is a book on architecture, by Vitruvius (Vitruvius 1999).

3   Generic Architecture

In modern real-time systems, which in principle, are control systems, in one way or another, a number of elements has been added due to the introduction of computers. The computational power changed the nature of the controller and the environment, but not the nature of the feedback principle. Building on the examples presented thus far, we can introduce computing power to the feedback principle diagram. This issue has been discussed by several authors studying relationships between computation and control (Auslander 1993, Benveniste and Astron 1993, Stout and Williams 1995).

An illustrative example is presented in Fig. 3. The roles of interface variables are described in more details below:

(1) Desired value; a reference for the Controller to make necessary adjustments of controlled variables.

(2) Controller commands; signals applied to the Plant (outputs from the Controller) in order to achieve its desired behavior.

(3) Controlled variables; signals received from the Plant (inputs to the Controller), whose values are being controlled.

(4) Other measured variables; auxiliary signals received from the plant (inputs to the Controller) which are not controlled but used in the determination of the best values of Controller commands.

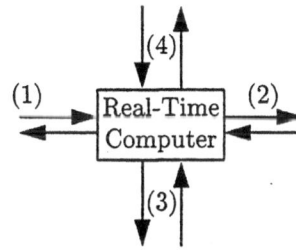

Fig. 4. Real-time computer system. (1) User interface; (2) Process interface; (3) Mass storage interface; (4) Communication link.

(5) Environment interfaces – user interface, mass storage interface, communication link to computer network.

It is evident that the modern controller became a digital processor or a real-time computer, and its interaction with the environment, in addition to that with sensors and actuators, includes interfaces with the following:

- plant operator
- computer network
- mass storage (database).

A unified diagram including explicitly all these elements is presented in Fig. 4.

In practice, a number of real-time systems exist which do not represent a complete system in a sense of Fig. 3, but nevertheless fit very well into this concept. Respective examples are discussed in the following three sections.

3.1   Data Acquisition System

If the connection (2) is broken, in Fig. 3, one has a plain data acquisition system (Fernández et al. 1998, Zalewski 1993). There are virtually no controller commands sent to the plant, which results in a system shown in Fig. 5.

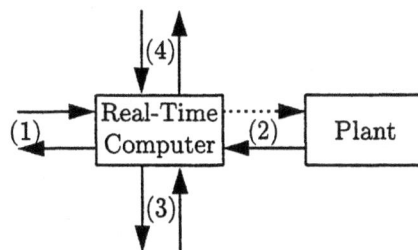

Fig. 5. Illustration of a data acquisition system
(notation same as in Fig. 4).

The most prominent example of a data acquisition system is an air traffic control system (Pozesky and Mann 1989). Surprisingly, it is not, in fact, an automatic control system at all, because there is no direct connection between the real-time computer and the plant (airspace). All commands are executed by the pilot who is receiving respective messages from an air-traffic controller (Fig. 6).

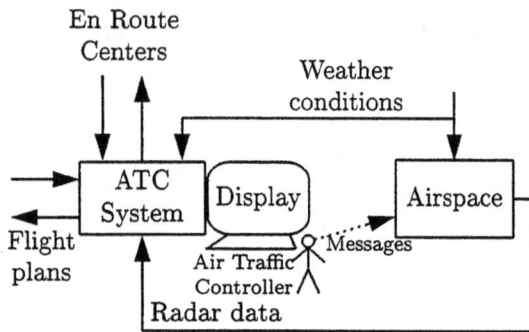

Fig. 6. Air traffic control system as a data acquisition system.

3.2 Programmed Controllers

If the feedback connection in Fig. 3, from the plant to the controller, is removed, with connection from the controller to the plant remaining intact, then we have a programmed control system (Fig. 7), for example, a traffic light control system, which in principle does not receive much feedback information. Simpler and more familiar examples of programmed control systems include a microwave oven controller and a washing machine controller. In principle, their control signals are precomputed in advance. All of these devices, however, continue to operate in real time.

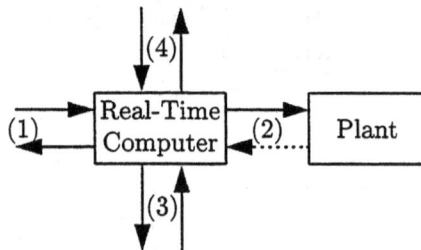

Fig. 7. Illustration of a programmed control (notation same as in Fig. 4).

For illustrative purposes, we can make a variation of a water clock which uses programmed control. This occurs, if we abondon the feedback

principle but still require the water level to change with a constant rate to indicate proportionally the time elapsed. For this purpose we just need one tank. To find a solution to this problem, one has to calculate an appropriate shape of a single tank, in which the water level decreases constantly over time (Edwards and Penney 1989). In the Cartesian coordinates (x, y), in ideal conditions, the shape of a tank can be described using Torricelli's law by the following differential equation relating the tank volume, V, and the shape of tank's walls in the vertical dimension, y:

$$\frac{dV}{dt} = -k\sqrt{y}$$

where a constant coefficient k depends on the exit area, discharge coefficient and gravitational acceleration.

On the other hand, the rate of change in water volume in the container, V, can be related to the shape of walls by the following equation:

$$\frac{dV}{dt} = A(y)\frac{dy}{dt}$$

where $A(y) = \pi x^2$ is the area of the water surface in the tank.

Comparing the right-hand sides and solving the resulting equation for $dy/dt = const$ we obtain the solution as the following formula:

$$y = f(x) = ax^4$$

for the shape of container walls, assuring us of the constant decrease in water level. The value of coefficient a can be obtained from other constants involved.

3.3 Reduced Architecture

Even if both connections between a plant and a real-time computer are broken, we can still claim that what remains is a reduced architecture for a valid real-time system (Fig. 8).

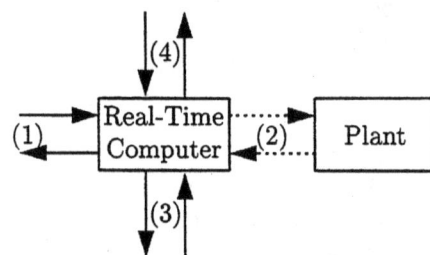

Fig. 8. Illustration of a reduced real-time architecture (notation same as in Fig. 4).

There are several practical examples of that kind of a real-time system. Putting emphasis on the distribution and communication, with relatively less interest in GUI and database access, brings us to a typical case of real-time simulation. With a slightly different emphasis, more on the database use and GUI, one has a real-time multimedia system.

4 Design Patterns

Once we understand the nature of an architecture of a real-time system, we can focus on developing its design and shaping its software architecture. It is at this point, when the concept of design patterns comes into play. What this means in practice is that our ability to apply engineering principles is significantly enhanced, because we can justify reliance on reusing proven existing solutions, commonly called design patterns.

The concept of design patterns becomes very clear if we take a look at other engineering disciplines. For example, there exists a clear pattern in designing radio receivers (Fig. 9). A radio receiver is always built out of certain components, such as an antenna, mixer and oscillator, detector, amplifier and speaker, connected in a predetermined way. It has been that way since radio receivers were conceived. What has changed and keeps changing is the technology and the implementations.

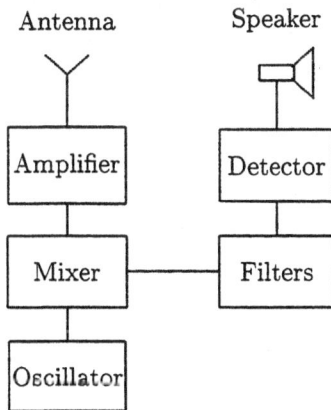

Fig. 9. Typical block diagram architecture of a radio receiver.

However, in real-time computing, somebody has to tell us what these major architectural components are and how are they related. Despite some well established concepts and principles of real-time computing (Halang *et al.* 1997, Stankovic 1988), thus far, there has been very little guidance on selecting real-time architectures,

either in the engineering literature or in practice (Clements 1997, Stuurman and Katwijk 1997). However, when one takes a closer look at Figures 3 and 4, with explanations (1)-(5) in the previous section, there is little doubt that one can start and should start designing the controller from the context diagram similar to that in Fig. 10.

The role of a context diagram cannot be overestimated. Even though it is a relatively old notational vehicle, it's been well established in real-time software design as the basis for architectural development (Hatley and Pirbhai 1988). It is at the context diagram level, where the interfaces between the software and the external world need to be defined and developed. For this very reason, the concept of a context diagram is indispensible as a starting point in designing a software architecture.

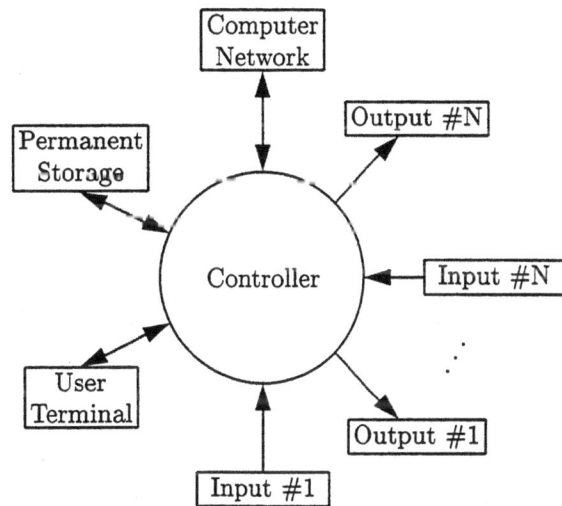

Fig. 10. The top level context diagram.

An example of the context diagram for the Air Traffic Control System (Fig. 6) is presented in Fig. 11.

It is compatible with the generic context diagram shown above and includes:

- two functionally different kinds of mass storage interfaces (for flight plans and flight services)

- two functionally different kinds of network interfaces (for *en route* centers and weather services)

- two sources of data (radars and time source), as well as

- a user interface to controller displays.

¿From Figures 10 and 11, it becomes immediately clear that the software components must

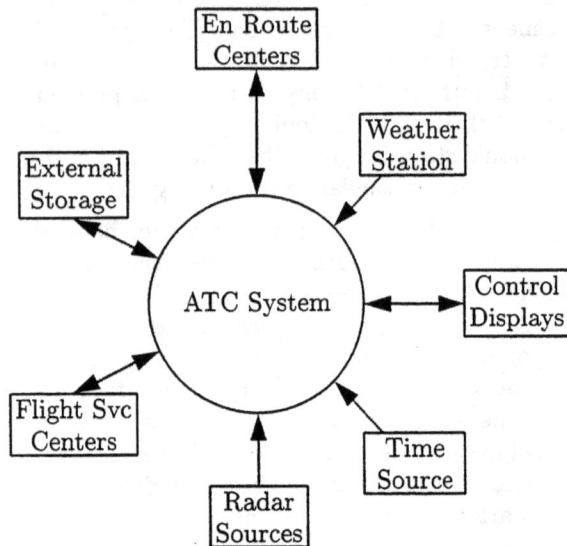

Fig. 11. The top level context diagram for the air traffic control system.

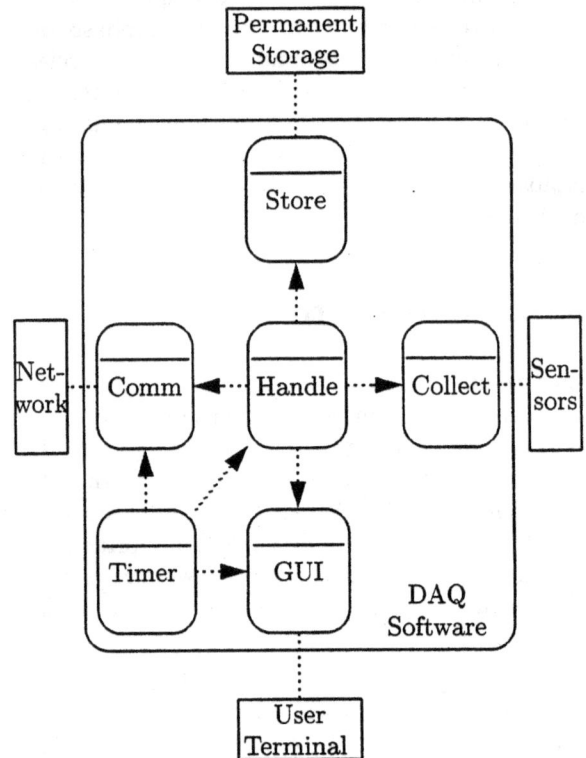

Fig. 12. Outline of a generic data acquisition system.

include parties responsible for the following interactions with all external elements:

- inputs from and outputs to the plant
- interaction with a user
- possible communication with other controllers/processors
- interaction with storage devices

enhanced by the processing (computational) capability. The time source can be internal or external depending on circumstances.

An example of such design, which can be considered a basic and generic design pattern for real-time systems software, is presented in Fig. 12, for a data acquisition application.

Respective software components need to comply with the principle of separation of concerns. They can be considered as sequential modules or individual concurrent tasks, and can run, respectively, on a single processor, on multiple processors, or even on a distributed system or network.

This basic design pattern can be expanded further into more comprehensive architectures, depending on the focus of particular applications. In the next sections, we consider the following additional issues, all derived from the basic concept of an architecture and design patterns:

- distributed real-time architectures (Gaspar *et al.* 1999)
- safety-related architectures (Anderson *et al.* 1999, Levenson 1984, Sha *et al.* 1995)
- real-time design tools for architectural support.

5 Software Architecture of a Distributed System

Depending on what is the focus of a distributed real-time architecture, there may be a variety of its particular instances. One such example (Fig. 13) comes from the area of high-energy physics, where multiple data collection and control facilities are spread over a large area surrounding an elementary particle accelerator (Gaspar *et al.* 1999).

Multiple software units can be created to access various (maybe the same) sources and destinations of data and to exchange information among themselves. Any single unit can perform individual functions and communicate with every other unit.

This leads to the concept of a dynamic architecture, where the organization of components and their interconnections may change during execution (Magee and Kramer 1996, Polze *et al.* 1999). Adding or deleting new components should have no impact or minimal impact on the operation, in a sense that no degradation of functionality should occur due to such dynamic changes.

An interesting observation is that this type of architecture forms a tree, with a root at the experiment hardware layer. Only this root cannot change, everything else is flexible. Since only the

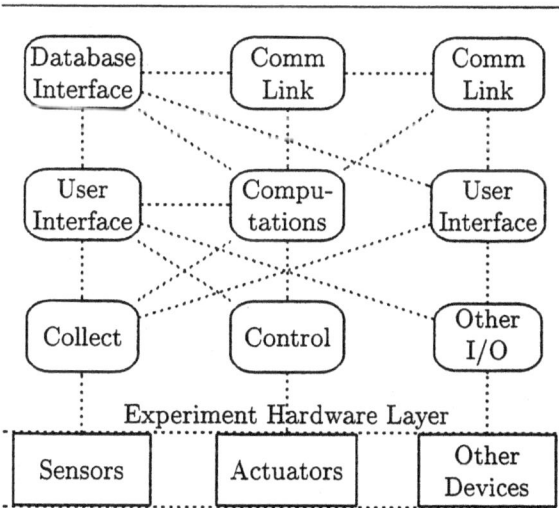

Fig. 13. Generic architecture of a distributed
real-time system.

root part is fixed, what happens above it is a matter of imagination of designers.

Communication links can operate individually or be lumped into a middleware layer (Muñoz 1999), as in Fig. 14, with program units communicating partially or exclusively via this layer. Examples of both variants exist, in applications as different as satellite on-board embedded real-time systems (Vardanega 1999) or large container terminal control systems in a big harbor (Katwijk et al. 1999), to name only a few published most recently.

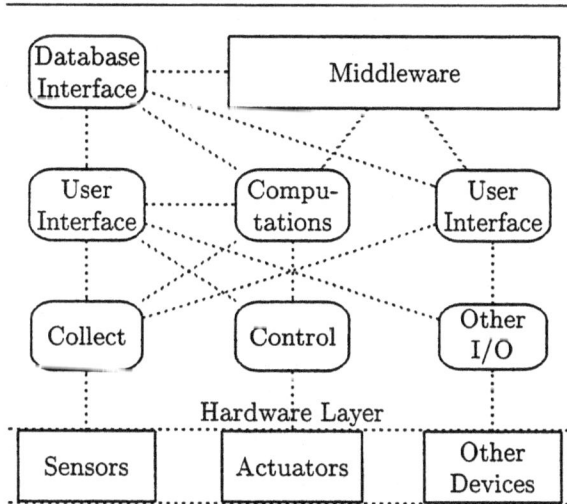

Fig. 14. Role of middleware in the architecture
of a distributed real-time system.

The primary advantage of having such a flexibility is that a number of new components can be created and the architecture expanded during the run of an experiment or operation of a pro-

cess. One such example is a dynamic GUI creation (Pedroza 1999). If new experiments are conceived which require including additional features to the GUI, or new characteristics are explored which need GUI reorganization, this can be done on-the-fly without jeopardizing the operation of an ongoing experiment or process.

6   Safety-Related   Software Architecture

Another important category of real-time systems is that including mission critical or safety-related systems (Anderson et al. 1999, Krämer and Völker 1997, Levenson 1984, Sha et al. 1995). Before discussing safety-related real-time architectures, it is worthwhile to take a closer look at the notion of safety and its relation to other critical system properties: reliability and security.

Considering critical system properties, we usually require guarantees on system behavior, requesting specifically that "nothing bad will happen" or that the risk of "something bad may happen" is low. The risk is usually analyzed in terms of potential hazards that are related to computer failures. If we ask the question, what is the risk, in terms of harm done to the environment or the computer system itself due to a computer (or software) failure, we may have the following answers:

1. Failure does not lead to severe consequences (high risk) to the environment or a computer system, nevertheless improving the failure rate is of principal concern (the notion of *reliability*).

2. Failure leads to severe consequences (high risk) to the environment, and later, maybe, to the computer system (the notion of *safety*).

3. Failure leads to severe consequences (high risk) to the computer system itself, and later, possibly, also to the environment (the notion of *security*).

In other words, reliability means minimizing undesired situations and their effects (to keep the system running), and safety and security mean preventing the environment or computer system, respectively, from undesired situations and their effects, because of the high risk involved (Fig. 15).

Probably the most common mission-critical system every researcher is familiar with, although rarely considered in these categories, is a car. In addition to meeting the required level of performance (for example, regarding speed, acceleration, fuel consumption, etc.), an embedded mi-

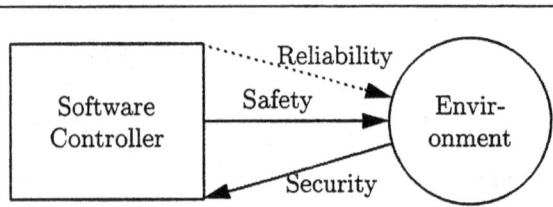

Fig. 15. Illustration of critical system properties.

croprocessor control software must meet several critical requirements, including:

- reliability, related to ignition control, cruise control, fuel gauge, odometer, etc.

- safety, related to air bag, seat belts control, anti-lock brakes, etc., and

- security, related to door locks, alarm, etc.

Stringent requirements on reaction times of air bag, anti-lock breaks, or alarms, make safety and security considerations a true real-time issue. In particular, a sporadic task handling the air bag release has one of the strictest requirements seen in practical applications.

The basic principle which should be observed when building software architectures for safety critical systems is "safety first". This comes from examples very common in practice, although not necessarily related to software, such as a lawn mower safety device, which is stopping operation immediately when an operator releases the handle. What is important to realize is that the actions of safety-related part of the system have precedence over the regular control procedure. This is confirmed for technologically more advanced systems, such as a nuclear reactor protection system (with a rod falling down into a reactor to absorb neutrons, in case of a danger).

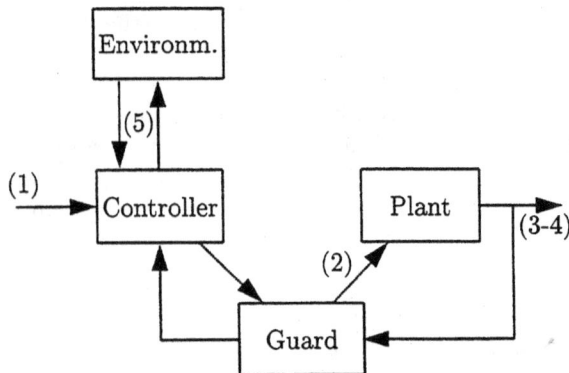

Fig. 16. Adding a safety guard to the structure of the control problem (notation as in Fig. 3).

Such observations suggest the use of a low-level construct, which acts as a guard detecting the danger first and then stopping the system and/or passing this information to a higher layer of control (Fig. 16). This is the role of a mower handle as well as that of the rods in a nuclear reactor protection system.

The assumption is to keep a guard as simple as possible, focusing only on the safety aspect. The safety component must first take care of all signals, before the controller. These factors limit the guard's basic functionality to reading the signals from the environment and determining whether they meet specifications, such as value ranges not exceeded, trends of variables within prescribed limits, validity of commands, etc. Thus the guard is filtering information passed to the control system. Particular algorithms what to do in case of detecting unsafe states depend on the application.

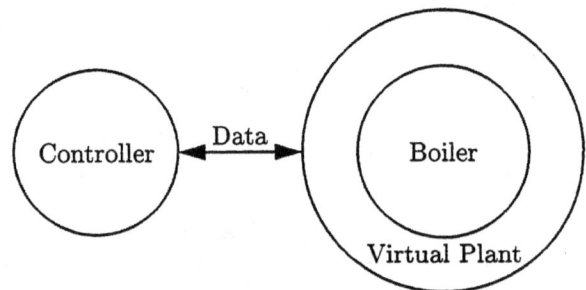

Fig. 17. Isolation provided by a guard as seen by the controller.

A guard conceived that way occupies virtual layers from the point of view of both the plant and the controller, by isolating the plant from the controller (Zalewski 1994). The isolation of a plant provided by a guard for the controller is illustrated in Fig. 17. The advantage of this situation is that the virtual plant is always in the safe state and the controller doesn't have to deal with errors in the plant at all.

Because safety is a property that affects the environment (Fig. 15), the computer software must be watched too, not to contribute to the violation of safety. The advantage of the situation presented in Fig. 18, due to the use of a guard, is that the ideal controller takes care of controller errors and the plant is never affected by them. The only difficulty is when the guard itself is faulty.

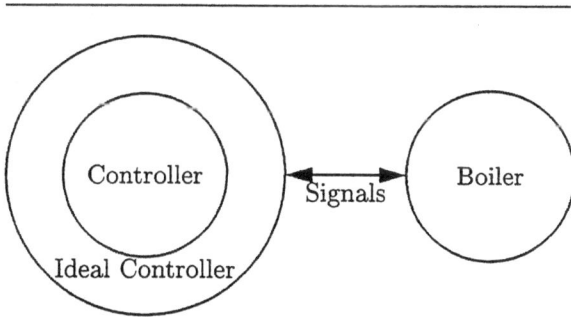

Fig. 18. Isolation provided by a guard
as seen by the plant.

7 Case Study

As a case study to illustrate how all these concepts worked in practice historically, we chose a water level control for a toilet tank (Coury 1997). An abstract view of the feedback control principle applied to a toilet tank is presented in Fig. 19.

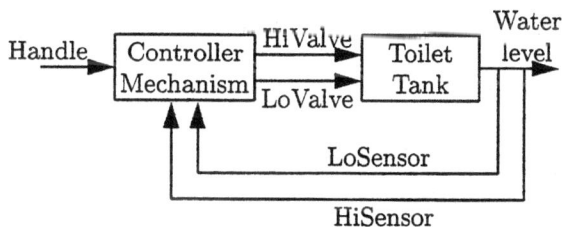

Fig. 19. Abstract view of the feedback principle
in a toilet tank control.

In terms of a generic architecture (Fig. 3 & 4), it is the water level in the tank which is controlled, although via two different preset values:

- low level value, which is determined by a *LoSensor*, a float sensing whether all water left the tank

- high level value, which is determined by a *HiSensor*, another float sensing that water in the tank reached the highest allowed level.

Value of the water level is regulated via two actuators being the valves cutting flush water flow and input water flow, respectively:

- *LoValve* actuator is the valve controlling the flow of water leaving the tank

- *HiValve* actuator is the valve controlling the flow of water entering the tank.

In normal operation, the *LoSensor* float is responsible for detecting the water level at which

the *LoValve* has to be closed. The *HiSensor*, in turn, is a float responsible for detecting the water level at which the *HiValve* has to be closed. Each sensor may actually be a part of a corresponding valve assembly, so it may be hard to separate them physically.

In addition to a plant interface via sensors and actuators, we also have a user interface, which is represented by a handle or button on the tank. A separation of a handle from the respective valve means in fact a careful user interface design.

It is interesting to see what is inside the toilet tank controller, in terms of its processing capabilities. Surprisingly, taking a closer look into this issue reveals a limited computational capability. Computation is in fact accomplished by the lever connecting the handle with the flush valve: pressing the handle causes lever to move and open the valve. This means clearly that the lever transforms an input signal into an output action and therefore performs analog computation, so in fact it is a computational device. Computation is based on mechanical principles and rather trivial but it is a computation.

By the same token, an earlier discussed water clock is a computational device, because it measures (computes) time. To achieve this, it performs arithmetic operation of summation by accumulating water. Being a hydraulic system, it also has storage, which changes its state when the tank is discharged. So does a toilet tank.

Why is it hard to observe computational capabilities or even believe that the computations take place? As pointed out in (Auslander 1993), analog computations suffer from the tyranny of impedance. Neither the toilet tank level, nor the Watt's speed governor valve, considered as primitive amplifiers have a significant input impedance. Small input impedance of transformational (computational) elements means that a significant power has to be applied to the input and go through the computational device to cause any visible effects on the output. This is generally true for all non-electronic devices. High input impedance, that is, isolation of input from output was only achieved with the advent of electronic, in particular, digital technology.

Furthermore, since there are two control loops (one for each water level), this is a multivariable control system. Each loop performs its individual computational task and synchronizes with another loop via sequencing. The tank controller responds with an action closing the upper valve, only after the lower valve has been closed. This means we have a synchronization in a concurrent system. Moreover, even though there are no distributed computations *per se*, because the tank itself does not exchange any signals with external tanks, a toilet tank is a part of a very large distributed sanitation system, including sewege. The effects of distribution and interconnection are visible, if the sewege is clogged.

In this sense, the toilet tank control is also safety related. In particular, the design of the tank always includes provision for a failure mode operation. A fault tolerant element (such as a hollow tube in a tank) is used to drain excess of water directly into the toilet, to respond properly in case of an upper valve failure, which otherwise would cause water overflow and severe damage to property.

In summary, the toilet tank controller complies with all modern requirements for a control system, in terms of Fig. 4.

8 Real-Time Design Tools

Developing software architectures for contemporary real-time applications, such as those described in (Gaspar *et al.* 1999) is too complex to be done manually by a single individual. Therefore automatic tools are needed to assist in the development process. One such category of tools are architecture design languages.

¿From the developer's perspective, architecture design languages are similar to specification languages, in that they try to help formalize the design process. For this to be effective, however, a certain view of a design methodology should be followed. This view assumes that a complete design methodology must include the following three components (Calvez 1993, Ludwig 1981):

- method, which is primarily a graphical notation applied to express important properties of an architecture

- techniques, that is, transformations which when applied to the notation allow for a successful completion of the development process (a set of such techniques including management aspects is often called a process)

- automatic software tools supporting the transformation techniques.

To this end, architecture design languages have not been very successful, mostly because of the lack of respective tool support and the lack of standardization. The importance of these two factors becomes more evident if we look at the success of design languages in other areas of computing. For example, VHDL is useful in hardware design, because it is standardized and supported by automatic software tools.

To be fully useful in the development of software architectures, however, the automatic tools have to provide support in the following four dimensions (Fig. 20):

- internal, related to all aspects of expressing real-time models via the specific notation and respective transformations

- horizontal, related to means of communication with other models and other tools (for example, via TCP/IP protocol suite)

- vertical, related to the next and previous phases of the development process, with respect to

  - support for code generation and prototyping, in particular, for specific programming languages and real-time kernels, and

  - support for design verification against the requirements (for example, to assure one-to-one correspondence of design components with the requirements)

- diagonal, related to the use of architectural models in different projects and different processes.

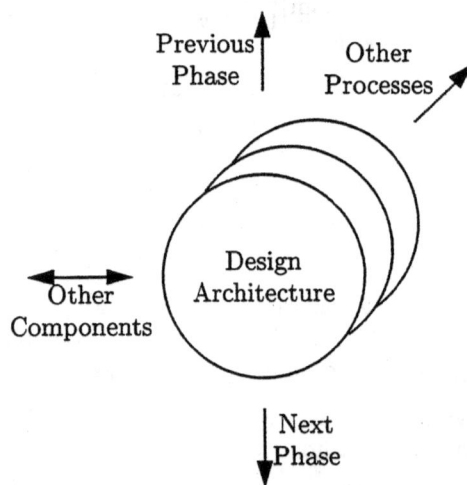

Fig. 20. Issues to consider when evaluating or selecting real-time design tools.

The advantage of this view, from the perspective of tool functionality, is that having the tools operational in all four dimensions, one gets the following properties covered for the software architecture:

- completeness, correctness, and consistency in the internal dimension

- interoperability and connectivity in the horizontal dimension

- testability and traceability in the vertical dimensions

- reusability and portability in the diagonal dimension.

Several software tools with this concept in mind operate already in the commercial marketplace. Since it is not our intention to promote any specific commercial software, we list only a few books which describe methodologies on which some of these tools are based (Douglass 1999, Harel and Politi 1999, Selic *et al.* 1994).

9 Conclusion

As one book states it (Selic *et al.* 1994), it is evident that *"The lack of a clear and accurate model of the actual system architecture has a severe impact on system understanding and is a prime contributor to system evolution and maintenance costs."* Therefore looking into the software architectures for real-time systems, that is, systems that must react within strictly defined timing constraints in response to external stimuli, should be of primary concern to software engineers and researchers.

In this paper, we tried to provide evidence that there is a clear template for real-time software architectures and design patterns, historically rooted in control engineering. From most likely the first engineered real-time device, a Greek water clock, to toilet tank controller, to steam engine speed governor, to air traffic control systems and distributed high-energy physics data acquisition systems, the same principles can be applied in designing real-time software architectures, forming a set of invariants that engineers can use further on in their design practice.

The most critical issue is to build a set of supporting software engineering tools to put these invariants into practice and assist designers in constructing real-time systems. All this can be resolved as long as the software architects will follow the advice given over 2000 years ago by Vitruvius, in the first sentence of his monumental work: *The architect should be equipped with knowledge of many branches of study and varied kinds of learning, for it is by his judgement that all work done by the other arts is put to test.* (Vitruvius 1999)

References

Anderson E., J. van Katwijk, J. Zalewski, New Method of Improving Software Safety in Mission-Critical Real-Time Systems, *Proc. 1999 International System Safety Conference*, Orlando, Florida, August 16–21, 1999

Atkinson C., C.W. McKay, A Generic Architecture for Distributed Non-Stop Mission and Safety Critical Systems, pp. 175-180, *Proc. 2nd IFAC Workshop on Safety and Reliability in Emerging Control Technologies*, Pergamon, Oxford, 1996

Auslander D.M., The Computer as Liberator: The Rise of Mechanical System Control, *Trans. of ASME: J. of Dynamic Systems, Measurement and Control*, Vol. 115, pp. 234–238, June 1993

Baker T.P., G.M. Scallon, An Architecture for Real-Time Software Systems, *IEEE Software*, Vol. 3, No. 3, pp. 50-58, May 1986

Bennett S., *Real-Time Computer Control: An Introduction.* Second Edition, Prentice Hall, 1994

Benveniste A., K.J. Astrom, Meeting the Challenge of Computer Science in the Industrial Applications of Control, *Automatica*, Vol. 29, No. 5, pp. 1169–1175, 1993

Boasson M., Control Systems Software, *IEEE Trans. on Automatic Control*, Vol. 38, No. 7, pp. 1094–1106, July 1993

Buschmann F. et al., *Pattern-Oriented Software Architecture: A System of Patterns*, John Wiley and Sons, 1996

Calvez J.P., *Embedded Real-Time Systems: A Specification and Design Methodology*, John Wiley and Sons, New York, 1993

Clements P.C., Coming Attractions in Software Architecture, *Proc. Joint Workshop on Parallel and Distributed Real-Time Systems*, Geneva, Switzerland, 1–3 April 1997, pp. 2–9, IEEE Computer Society Press, 1997

Coury B.G., Water Level Control for the Toilet Tank: A Historical Perspective, pp. 1179-1190, *The Control Handbook*, W.S. Levine (Ed.), CRC Press/IEEE Press, 1997

Donohoe P. (Ed.), *Software Architecture. Proc. TC2 First Working IFIP Conference*, Kluwer, Boston, Mass., 1999

Douglass B.P., *Doing Hard Time: Developing Real-Time Systems with UML, Objects, Frameworks, and Patters*, Addison-Wesley, Reading, Mass., 1999

Edwards C.H., D.E. Penney, *Elementary Differential Equations with Applications*, Prentice Hall, Englewood Cliffs, NJ, 1989

Emery D.E., R.F. Hilliard II, T.B. Rice, Experiences Applying a Practical Architectural Method, *Proc. Ada-Europe '96*, A. Strohmeier (Ed.), Notreux, Switzerland, 10–14 June, 1996, Springer-Verlag, Berlin, 1996

Fernández J.L. et al., A Case Study in Quantitative Evaluation of Real-Time Software Architectures, *Reliable Software Technologies*, L. Asplund (Ed.), Springer-Verlag, Berlin, 1998

Gaspar K., B. Franek, J. Schwarz, Architecture of a Distributed Real-Time System to Control Large High-Energy Physics Experiments, *Parallel and Distributed Computing Practices*, Vol. 2, No. 1, March 1999

Halang W. et al., Real-Time Computing Education: Responding to a Challenge of the Next Century, pp. 121–125, *Real-Time Programming 1997*, Pergamon, Oxford, 1997

Harel D., M. Politi, *Modeling Reactive Systems with Statecharts*, McGraw-Hill, New York, 1999

Hatley D.J., I.A. Pirbhai, *Strategies for Real-Time System Specification*, Dorset House, New York, 1988

van Katwijk J. et al., Software Development and Verification of Dynamic Real-Time Distributed Systems Based on the Radio Broadcast Paradigm, *Parallel and Distributed Computing Practices*, Vol. 2, No. 1, March 1999

Krämer B., N. Völker, A Highly Dependable Computing Architecture for Safety-Critical Control Applications, *Real-Time Systems*, Vol. 13, pp. 237-251, 1997

Leveson N., Software Safety in Computer Controlled Systems, *IEEE Computer*, Vol. 17, No. 2, pp. 48–55 , February 1984

Ludewig J., *Zur Erstellung der Spezifikation von Prozessrechner-Software*. Doctoral Dissertation, Report KfK 3060, Kernforschungszentrum Karlruhe, 1981

Magee J., J. Kramer, Dynamic Structure in Software Architectures, *ACM Software Engineering Notes*, Vol. 21, No. 6, pp. 3-14, November 1996

Mayr O., *Zur Frühgeschichte der technischen Regelungen*, Oldenburg Verlag, München, 1969 (English translation: *The Origins of Feedback Control*, MIT Press, Cambridge, Mass., 1970)

Muñoz C., J. Zalewski, Architecture and Performance of Java-Based Distributed Object Models, *Real-Time Systems Journal*, 1999 (submitted)

Pedroza H., *Dynamic Specification and Creation of Real-Time Graphical User Interfaces*, M.Sc. Thesis, University of Central Florida, Orlando, Fla., 1999 (in preparation)

Polze A. et al., Real-Time Computing with Off-the-Shelf Components: The Case for CORBA, *Parallel and Distributed Computing Practices*, Vol. 2, No. 1, March 1999

Pozesky M.T., M.K. Mann, The US Air Traffic Control System Architecture, *Proceedings of the IEEE*, Vol. 77, No. 11, pp. 1605-1617, November 1989

Rechtin E., M.W. Maier, *The Art of Systems Architecting*, CRC Press, Boca Raton, Fla., 1997

Schoch D.J., P.A. Laplante, A Real-Time Systems Context for the Framework for Information Systems Architecture, *IBM Systems Journal*, Vol. 34, No. 1, pp. 20–38, 1995

Selic B., G. Gullekson, P.T. Ward, *Real-Time Object-Oriented Modeling*, John Wiley and Sons, New York, 1994

Sha L., R. Rajkumar, M. Gagliardi, *A Software Architecture for Dependable and Evolvable Industrial Computing Systems*, Technical Report CMU/SEI-95-TR-005, Software Engineering Institute, Pittsburgh, Penn., July 1995

Shaw M., Beyond Objects: A Software Design Paradigm Based on Process Control, *ACM Software Engineering Notes*, Vol. 20, No. 1, pp. 27-39, January 1995

Shaw M., D. Garlan, *Software Architecture: Perspectives on an Emerging Discipline*, Prentice Hall, 1996

Stankovic J., Misconceptions about Real-Time Computing, *IEEE Computer*, Vol. 21, No. 10, pp. 10-19, October 1988

Stout T.M., T.J. Williams, Pioneering Work in the Field of Computer Process Control, *IEEE Annals of the History of Computing*, Vol. 17, No. 1, pp. 6–18, 1995

Stuurman S., J. van Katwijk, Evaluation of Software Architectures for a Control System: A Case Study, pp. 157–171, *Proc. 2nd Int'l Conf on Coordination Languages and Models*, D. Garlan, D. LeMetayer (Eds.), Springer Verlag, 1997

Vardanega T., On the Distribution of Control Functions in New-Generation On-Board Embedded Real-Time Systems, *Parallel and Distributed Computing Practices*, Vol. 2, No. 1, March 1999

Vitruvius Pollio, Marcus, *De architectura libri decem*, Rome, around 27–13 B.C. (Latest English translation: *Vitruvius: Ten Books of Architecture*, Cambridge University Press, New York, 1999)

Witt B., F.T. Baker, E.W. Merritt, *Software Architecture and Design: Principles, Models and Methods*, Van Nostrand Reinhold, New York, 1994

Zalewski J., Real-Time Data Acquisition in High-Energy Physics Experiments, pp. 112–115, *Proc. RTAW'93, IEEE Workshop on Real-Time Applications*, IEEE Computer Society Press, 1993

Zalewski J., Boiler Water Controller Based on DARTS and EWICS Safety Model, pp. 223-231, *Software Safety: Everybody's Business*, D. Del Bel Belluz, H.C. Ratz (Eds.), Institute for Risk Research, Waterloo, Ont., 1994

Http reference to Author's hometown building:
`http://www-ece.engr.ucf.edu/~jza/bld.jpg`

DISTRIBUTED, OBJECT-ORIENTED, ACTIVE, REAL-TIME DBMSS: WE WANT IT ALL - DO WE NEED THEM (AT) ALL?

Alejandro P. Buchmann, Christoph Liebig

*Department of Computer Science, Darmstadt University of
Technology, Wilhelminenstr. 7, 64283 Darmstadt, Germany*

Abstract: Whenever technologies converge there exists the potential for huge benefits
but also the risk of failure. The main pitfall when combining technologies that evolved
independently consists in attempting to provide the union of features without properly
considering the often incompatible assumptions and the crosseffects. In this paper
real-time databases, active databases, and distributed object systems are analyzed
together with some of the basic assumptions underlying previous work in these core
technologies. Crosseffects and potential incompatibilities are discussed in an attempt
to provide a better foundation for a configurable middleware platform that realistically
combines selected features of active, real-time and distributed object systems.
Copyright © 1999 IFAC

1. INTRODUCTION

The rapid evolution of today's event-based computing environments and the increased use of distributed systems in time-critical applications combined with the desire of reaping the benefits of objet-orientation, has lead researchers to investigate ever more complex systems. At the same time, the combination of these features requires a thorough understanding of each of the underlying technologies, and more important yet, of the possible cross-effects. The simple concatenation of buzz-words will lead to ill-understood systems whose behavior will be unpredictable and may pose a danger to life and/or property.

If the semantics of complex system software are not well understood the users will prefer to implement themselves a minimalistic version of the required functionality. The resulting systems are ad-hoc solutions that are expensive and can neither be extended nor exploited in a different context. Therefore, a middleware platform with clear semantics is needed that can satisfy the requirements of a variety of applications. The research community has tried to take up this challenge by combining (at least on paper) active databases, real-time databases, and distributed object systems in a single platform. Unfortunately, the full set of features of each technology is incompatible with the other technologies that are being combined, and cross effects between technologies have not been sufficiently studied and considered.

In this paper we try to highlight some of the problems that may arise when combining technologies with divergent goals and requirements. For example, real-time systems require predictability of resource consumption and execution time in order to give performance guarantees; active databases react to events and trigger rule executions that dynamically alter the work load and thereby make any form of predictability rather difficult. The same is true for object-orientation with its tendency to encapsulate the behavior of objects and to hide their internal implementation while real-time systems need the implementation details to predict worst case execution times. Finally, the lack of a global clock in distributed systems and the uncertainty caused by varying communication delays introduce many additional factors of unpredictability to both the active and real-time behavior of a system.

When faced with the seemingly insurmountable difficulties and apparent contradictions outlined in the previous paragraph one might be tempted to conclude that a combination of active, real-time, object-oriented and distributed functionality is impossible to achieve. Yet there are applications that do require several of these features. The question, therefore, is not whether we do need these features but rather: how can we combine meaningful subsets of them into generic software platforms with clear semantics that are modular and capable of satisfying the requirements of a variety of applications.

2. APPLICATIONS AND THEIR REQUIREMENTS

Applications that are often mentioned as motivation for research in the area of active, real-time, distributed databases fall typically into the domains of control, navigation, mobile systems, telecommunications, simulation, e-commerce auctions and profiling, and certain aspects of complex workflows.

Air traffic control systems cover a variety of aspects. While active real-time systems are typically mentioned to support the air traffic controller by filtering the information glut and alerting the controller of dangerous situations, there are several other interesting aspects [Liebig, Boesling and Buchmann 1999]. For once, the portion dealing with take-off preparation and the gate-to-runway movement are less time dependent but represent more complex workflows that are triggered by events. The new ATC systems will also depend more on electronically transmitted data delivered to the cockpit instead of voice traffic. To avoid swamping the crew with irrelevant information, event-driven publish-subscribe mechanisms that must fulfill reliability and timeliness constraints have been proposed. The new generation of ATC systems has been specified to use object-oriented middleware wherever feasible.

In [Locke 1997] aircraft mission control and space-craft control are described. They have similar properties and are characterized by small main memory databases with extremely short mean latency requirements (0.05 ms mean latency and 1 ms maximum latency). Databases for these environments are typically used to keep track of the air- or spacecraft infrastructure in addition to sensor data. These systems are typically built around cyclic executives. The resource limitations that are typical for aircraft and space vehicles make complex, multifunction software impractical in many cases.

On-board navigation systems cover a wide range of support systems with quite different demands. However, they all must obey strict timing constraints since the time available for a reaction is dependent on the movement of the vehicle. While the constraints may vary based on the type of vehicle, they depend on the determination of the current position (the event) to issue the pertinent instructions, either to a human operator or to an automatic controller. Many on-board navigation systems are built as cooperative distributed systems [Purimetla et al. 1995] in which the front end agents perform sensing and filtering operations while the back-end controller handles the events that the front-end controllers cannot handle because of their limited capabilities. Typical of on-board navigation systems is a fairly small portion of dynamic data that must be combined with a large amount of static data, e.g. maps and landmark information. The most demanding on-board navigation and control systems are those used for control of rather unstable fighter planes that are essentially flown by the on-board computers. While the signals detected by the sensors can be regarded as events, one must be careful not to imply that this means the kind of event handling associated with active databases. Instead, sensor signals are used to invoke specific methods and persistence of the sensor data is only required for auditability purposes. It is interesting to observe that much of the on-board navigation and control software for the new generation of fighter aircraft is being built in an object-oriented manner and that the TAO Object Request Broker and its Event Service have been modified to support hard timing constraints for use in this context [Harrison et al. 1998]. Industrial interest in real-time distributed object systems is demonstrated by the recently approved real-time CORBA specification [Object Management Group (OMG) 1999].

In the automotive industry manufacturers are exploring emerging technologies for spontaneous networking, such as Jini [Arnold et al. 1999], as a mechanism for integration of navigation systems, embedded systems and the infotainment that is expected to be part of the next generation of automobiles. In the latter case one has, in addition to the distribution, the typical timing problems associated with mobility and spontaneous networking.

In telecommunication applications [Raatikainen 1997] the notion of timing constraints is interpreted mostly statistically, i.e., a predetermined percentage of the transactions must meet its deadline. This approach is more a high throughput rather than a typical real-time approach. The functionality that is frequently mentioned encompasses Intelligent Network functionality, such as verification of PINs, call forwarding, user management actions to update, for example, the call forwarding option, televoting, mass calling, etc. The highest requirements are expected from mobile networks with extremely high transaction rates and requirements that 96% of the queries have a response time of 150 ms or less and the updates be performed in about 1 sec. with reliability of just a few seconds/year downtime. Many proposed telecommunication architectures assume an object-oriented paradigm and try to include standards, such as CORBA.

Simulations provide another class of demanding applications. Training simulation tries to present the trainee with a realistic environment and typical situations to which he must react. These systems often consist of two databases, a static read-only database for the environment and a dynamic database for the simulated situation.

Virtual testbenches are combinations of simulated systems with hardware in the loop. Such a virtual testbench may invoke very complex simulation routines that have timing constraints imposed by the hardware in the loop. For example, portions of a car might be simulated while others are actually running on the testbench and are being monitored and measured. The timing constraints imposed by the hardware in the loop are hard timing constraints and the volume of data may be extremely large since data gathering and analysis is the prime purpose of these systems. Test sequences and the simulation itself may be governed through plans or workflows that are conveniently expressed as event-condition-action rules.

In the area of e-commerce many interesting applications are just emerging, and the requirements of these applications with respect to activity, real-time and distribution have not been well analyzed. There are situations in e-commerce that appear to call for many of the features commonly associated with active and real-time systems. For example, profiling mechanisms must respond in real-time to the actions of a visitor to the site in order to retain the visitor and enhance the probability of a sale [Datta 1999]. This is an application in which no transaction processing is required but fast search and retrieval of profile data is required. The timing requirements are soft since missing a deadline does not invalidate completely the result of the profiling but the value of it may be diminished. In electronic auctions the proper timestamping and ordering of events may be critical.

As can be seen from the applications briefly described, there is no set of requirements common to all of them. However, some interesting generalizations may be possible.

Some form of monitoring and event-driven processing is common to all the analyzed applications.

The applications with tight deadlines often handle only a small data volume that fits into main memory, especially with current trends in memory availability. In many of these applications a small

volatile portion may have to be combined with larger volumes of static data. Long term permanence of the data is often restricted to whatever is needed for auditability. Predictability is very important for these applications.

Distribution is clearly needed in several applications, such as Air Traffic Control and mobile telephony and this will increase as we move into the realm of ubiquitous computing and spontaneous networking. The trend towards object oriented middleware platforms with real-time capabilities can be observed in some of the most demanding applications. Pertinent examples are Air Traffic Control and on-board navigation and control systems. The emerging platforms for spontaneous networking, such as Jini, are also built on the object paradigm. These systems will handle considerable amounts of data, although not all the data handled is structured. Multimedia data is usually streamed and the quality of service requirements are quite different from the transaction processing typically assumed by the active and real-time database community.

3. INTEGRATING THE KEY TECHNOLOGIES

From the previous section we must conclude that some combination of real-time, active, distributed and object-oriented functionality is required by a variety of applications. The question, therefore, is not whether we should combine them, but how to do it right. We should be concerned with the possible interference of the various features and ask ourselves:

- What are the key features in each base technology: real-time databases, active databases, and distributed OO-middleware?
- What subset of active features could be compatible with real-time requirements?
- How can predictability be ensured?
- How can active capabilities be provided in a distributed environment?
- What is the effect of the temporal fuzziness of distributed systems on active and real-time capabilities?
- How can active and real-time features be integrated into distributed object-oriented middleware?
- What are suitable correctness criteria for event composition and rule execution in distributed environments?

3.1 Real-time database issues

According to the definition of [Locke 1997] a real-time database is a data store whose operations execute with predictable response, and with application-acceptable levels of external consistency, temporal consistency, logical consistency, permanence and atomicity. In this definition, two key requirements that set an RTDB apart from a non-RT database system are the temporal consistency and the predictable response. Temporal consistency as defined in [Ramamritham 1993, Purimetla et al. 1995] has two aspects. Global temporal consistency refers to the absolute age of the data and characterizes its staleness. In real-time environments the data quality decreases as time progresses between the time data was acquired, e.g. from a sensor, and the time it is consumed. Global temporal consistency is referred to as external consistency in [Locke 1997]. Mutual temporal consistency refers to the relative age of a set of data values, for example, data used in a calculation should come from the same sensor cycle or, alternatively, the user may specify an acceptable maximum age difference that can be tolerated by the application.

Predictability in real-time database systems can be analyzed by looking at the various contributions to the total execution time of a transaction [Buchmann et al. 1989]. These are the time required for performing the database operations once data is in the buffer, the I/O time, the time lost due to transaction interference, the time used by the non-database portion of application processing, and the communication time. To determine worst case execution times in a deterministic way, each of these contributing elements must have an upper bound.

An upper bound for database operations can be achieved by limiting the size of data, for example, by setting an upper bound for the number of tuples in a relation. This approach is commonly used in real-time environments since it is the basis also for the calculation of the application dependent (non-database) contribution. Determining an upper bound for I/O operations is difficult, if not impossible, for disk resident databases. If one assumes a worst case of one page fault per access, nothing will be scheduled due to the extremely pessimistic worst case execution time that will be dominated by the high disk access time. Preexecution, an approach proposed in [O´Neil et al. 1996], in which a transaction is executed once without acquiring locks, just to determine what tuples are needed and to set the buffer, followed by the real execution in a second step, just shifts the problem but doesn't guarantee end to end predictability. Any disk-based approach can at best provide statistical values for the expected execution time. In addition, many real-time applications require mean read and write latencies in the range of 0.05 ms, something that is not achievable with disk-resident data. Therefore, the only feasible solution for real-time systems that

must give guarantees is avoiding I/O altogether by using main memory database systems. The drastic drop in memory costs and the availability of 64-bit address spaces make it possible to accommodate without problem the structured data required by typical real-time systems. For large-volume static and multimedia data that is streamed without concurrency control and transactional semantics a hybrid approach with large buffers for data staging is usually enough.

The time of interference is the time a transaction is either blocked waiting for resources held by another transaction or the time required for roll-back and restart due to a transaction abort, for example, because of deadlock. Conventional database systems use aggressive schedulers that acquire data resources dynamically and may hold these until committing. Conflicts between transactions are detected and resolved according to some resolution criterion. The reasons for doing so are simple and derived from the large disk access times: To avoid idling resources, such as the CPU, control passes to another transaction whenever an I/O operation is required. The overhead of passing control to another transaction is justified by the slow response of the disk. Therefore, a central objective of conventional DBMSs is to provide high intertransaction parallelism. To achieve high intertransaction parallelism it is necessary to lock as few data as possible, i.e., to provide small locking granules, for example, at the tuple level. Since the tuples that will be accessed by a transaction may depend on previous operations of the transaction, dynamic lock acquisition has become the method of choice to guarantee small locking units and thus high intertransaction parallelism. Dynamic lock acquisition results necessarily in scheduling policies based on conflict detection and conflict resolution. The bulk of the real-time database research has followed this approach because the assumption of the disk as the basic medium was never challenged. However, in main memory databases inter-transaction parallelism is less important since no long I/O waiting periods exist. Since the data will now be available in memory, no need for passing control to another transaction due to I/O is required and transactions will execute with fewer interruptions. Therefore, they can also lock larger units, for example, whole relations, which in turn makes it possible to determine the resources needed by a transaction ahead of time through simple syntactic analysis. This syntactic analysis can be done off-line for canned queries, a situation that is typical of real-time environments. Once the data resources are known, a whole new class of conflict avoiding schedulers becomes feasible [Ulusoy and Buchmann 1998]. The main advantage of these schedulers is, that the time of interference is eliminated since a transaction starts execution only when its resources are available. While this approach guarantees the execution time once a transaction is ready and scheduled for execution, the end-to-end execution time still depends on the waiting time in the queue.

The application-dependent portion of the worst case execution time can be determined with the help of tools by analyzing all the possible paths the application program may take.

The communication portion of the execution time depends on the degree of distribution of the system and the kind of network [Verissimo 1993]. There is a clear difference between a distributed system running in a LAN with a token ring network, in which an upper bound can be given, or a distributed system running over a wide area network or the Internet. It is also important to consider in this context the fuzziness derived from the lack of one central clock. This fuzziness will have an effect on all time stamps and everything dependent on them, from deadlines and temporal consistency to event composition and event consumption modes. Therefore, we will return to this problem later.

In addition to the predictability and concurrency control issues addressed above, real-time applications have special recovery needs. These are often less stringent than those typical of operational databases used in commercial applications. For example, since sensor data will arrive periodically, if a sensor reading is lost, the value can be captured in the next cycle. In general, no undo/redo is needed. Logging in real-time systems often has an archival function for long-term durability and auditability and less so for transaction roll-back or redo. In main memory databases alternate logging approaches based on messaging and asynchronous logging take the place of write-ahead logging assumed in conventional database systems. A summary of RTDBMS characteristics can be found in [Stankovic et al. 1999] and a good compendium of research results in [Bestavros et al. 1997, Bestavros and Fay-Wolfe 1997].

3.2 Active database issues

Active databases include Event-Condition-Action (ECA) rules as first class objects in the database [Dayal et al. 1988]. ECA rules have an explicit event part that determines, when a rule is to be executed, a condition that acts as a filter, and an action that may be any database-internal or external action. The events determine to a large extent the expressive power of an active database's rule mechanism. In their most general form they include database events, such as insert, delete, update; control events such as begin of transaction, commit, abort; temporal events that maybe absolute or relative, periodic or aperiodic; and user defined events. Primitive events may be combined through an event algebra that may include operators for sequence, disjunction, conjunction, negation, history, closure, etc. [Gehani et al. 1992, Gatziu and Dittrich 1993, Chakravarthy et al. 1994]. An event typically may trigger more than one rule, and an event may also participate in many different event compositions. This raises the question of event consumption, i.e., if a stream of events contains more than one event instance of a certain type that participates in a composite event definition, which event instance will be selected for the composition? This problem was solved for the centralized case in Snoop [Chakravarthy et al. 1994] through the definition of contexts. Contexts specify the consumption mode of events, i.e. they are policy definitions that define whether events should be consumed chronologically (chronicle), or whether the most recent instances should be used (recent). Besides these two obvious policies, two more have been defined. The cumulative policy specifies that all instances of an event that is part of a composite event will be accumulated until the composite event is fully composed. At that point all instances of the participating events will be removed. The continuous policy defines a sliding window, which is opened for each new primitive event arrival that initiates a new composition. The details of event composition and consumption are beyond the scope of this discussion, but we must observe the basic assumptions. All the event compositions and event consumption modes implicitly assume a central clock and a total order on events. These assumptions are quite reasonable for the centralized systems for which they were defined but cannot be sustained in a distributed environment. Again, we raise the issue here and defer the discussion.

ECA rules are triggered by events that may originate in a user transaction or they may be triggered by temporal or external events. Once a rule is triggered its condition should be evaluated and in case the condition is true, the action should be executed. The coupling mode determines how an ECA rule should be processed relative to the triggering transaction. This may be as a subtransaction of the user transaction either immediately following the event detection (immediate coupling mode) or it may be done at the end of the user transaction (deferred coupling mode). However, rules that are triggered by temporal events or by composite events that are based on primitive events that were generated in more than one user transaction cannot identify a single transaction to which they should be attached and must be executed in a separate transaction (detached coupling mode). Since events that are generated in a transaction may require bindings that reflect the state of the database and/or the transaction at the instant the event occurred, the participation in an event composition and the triggering of a separate transaction may imply a relaxation of the isolation property. This fact has not received the necessary attention so far, even for centralized systems.

For a comprehensive compendium of relevant work on active databases the reader is referred to [Paton 1998].

3.3 *Distribution issues*

Distribution affects all aspects of an active, distributed, real-time system. The discussion in this section focuses on event processing. Event-based computing is emerging as the paradigm of choice for composing applications in open distributed environments.

One of the core problems of distributed systems is the synchronization of processes running on physically and logically distributed nodes with communications that may exhibit unpredictable delays and without a central clock. The lack of a central clock implies that temporal order cannot unequivocally be established. Time, as provided by a distributed time service is imprecise with respect to clock readings at different nodes and inaccurate with respect to physical time.

Inaccuracies with respect to time have a major impact on timestamping which in turn is the basis for the determination of temporal consistency, both global and relative, the composition of events, and the consumption of events. The fundamental problem, then, is the characterization of the quality of the available time service and the proper consideration of its limitations.

A variety of approximations for modeling the time imprecision in distributed systems have been proposed. A thorough discussion of this issue is beyond the scope of this paper. Therefore, we will only address three approaches that have been adopted in the literature for use in distributed and active systems. For an excellent review of these issues the reader is referred to [Kopetz 1997].

Logical (Lamport) clocks [Lamport 1978] and vector time [Schwarz and Mattern 1994] make it possible to establish a partial order based on causality. Unfortunately, they are not appropriate for the open systems with external inputs that are the subject of this discussion.

When a sparse time base is assumed, the points at which events can be generated are discretized and predetermined. Only if events are at least two time granules apart, the sequence of these events

can be determined unequivocally. This is known as the 2g precedence model [Kopetz 1992]. In the 2g precedence model an upper bound to the precision is assumed and a virtual clock granularity with granularity g is defined. The 2g precedence model is very useful for dealing with embedded systems and other small and tightly controlled environments. However, since the granularity depends on the assumed precision, it is not a feasible approach for wide area networks and open distributed systems.

The Network Time Protocol (NTP) offers a standardized time service with a reliable error bound. NTP is based on the notion of strata that can guarantee the accuracy within an accuracy interval. By using NTP and injecting an external reference time, e.g. GPS time, it becomes possible to provide timestamping with accuracy intervals and partial ordering of events in large scale distributed systems [Liebig, Cilia and Buchmann 1999].

3.4 *Cross-effects*

Even a condensed discussion of the base technologies illustrates the magnitude of the problems. If two or more base technologies are combined, a whole new set of problems must be considered.

The time notion in active databases is quite different from the time notion in real-time systems. Temporal events in active databases determine when a rule is fired but no information on execution time or deadlines is provided. Real-time systems are primarily concerned with execution time and meeting deadlines, and the temporal consistency of the data.

Real-time requirements impose serious limitations on the active capabilities that can be provided. The first and most obvious is the triggering of new rules. Active databases in general do not limit the triggering of rules by the action part of another rule. Especially the object-oriented aDBMSs may, by the very nature of object-oriented systems, execute any method of arbitrary complexity in the action part. Emphasis in the active database community has been placed on ensuring termination, i.e., the avoidance of cycles. However, if timing constraints must be obeyed, the size of a task may not dynamically expand. The strongest limitation on the execution model of a real-time active database system consists in disallowing immediate and deferred coupling modes, thus allowing only detached execution of triggered transactions. Given the possible violation of the isolation property because of parameter transfer, this approach may not be acceptable. Furthermore, if a rule is triggered by a transaction that is aborted, the atomicity property is violated since a detached transaction is an independent transaction that cannot be rolled back. Therefore, sequential causally dependent detached transactions must be used. An alternative consists in limiting the depth of triggering, for example, to one rule. From the point of view of determining the execution time of such a transaction this is equivalent to the evaluation of a conditional statement. Limiting the depth of triggering raises, however, a serious problem, since the action part of the rules now must be strictly controlled not to produce any legal event, or we are faced with the problem of distinguishing between triggering and non-triggering events of the same type. Therefore, in a real-time active DBMS, the action portion of rules and the acceptable event set must be carefully matched and possibly curtailed. Before discussing further restrictions on the event set, we must analyze the effect of distribution on event composition.

Event composition in its general form depends on the ability to determine the sequence of occurrence of events. This is important not only for operators, such as, sequence, but also for all other operators since the consumption of events directly depends on it. For example, in a distributed system it becomes difficult to determine, whether an event generated at node N1 and detected at node N2, really should be consumed in a chronological consumption policy, or if there is another (older) instance of that event type generated at node N3 that was delayed in the network. The assumption of a 2g precedence clock synchronization model alone does not solve these problems. Particular care must be exercised when specifying a 2g precedence model to keep in mind the underlying assumption of sparse time and a guaranteed upper bound to the precission. If this is not the case, either because dense time is suddenly assumed or because Internet-based distribution is expected, the results will be wrong. Another weakness of applying the 2g precedence model to event composition that is frequently swept under the rug is the implicit treatment of the ambiguities. If two events are not distinguishable because they are not at least 2g apart, they are considered as concurrent, and the event consumption is ambiguous. This ambiguity must either be resolved explicitly through application semantics or it must be made clear to the user that an exception exists. The latter approach must necessarily be taken when developing a generic middleware platform. All the operators that have been defined for the event algebras of centralized active databases must be carefully reexamined and restated with their limitations being made explicit.

Whenever an event is detected, there is an inherent detection delay between the time the event occurred and the time it is detected. The detection delay may depend on system load and may impact the temporal consistency of the data.

This problem was identified in [Branding and Buchmann 1995], where as a first pragmatic approach the complexity of events was drastically curtailed and special high-priority events where proposed for overload situations in real-time active databases. The detection delay is magnified in distributed environments. In [Liebig, Cilia and Buchmann 1999] the detection delay is explicitly taken into account for the case of event detection and composition in distributed active systems. Event composition in a distributed environment under real-time constraints becomes very difficult. Sparse time must be assumed and transmission delays must have an upper bound. This limits the feasible environments to logically distributed systems on single nodes or very tightly controlled specialized LANs. Unless this is the case, only a statistical approach to real-time is possible. Given the temporal fuzziness introduced through the distribution, one would be well advised to avoid event composition in a distributed real-time environment or at least to keep it to a minimum, since the resolution of ambiguities to guarantee correctness is a time-consuming process that real-time applications with tight deadlines may not be able to afford.

Event composition in an open distributed system presents additional problems. In a centralized system every producer of a certain event is known. In an open distributed environment this is not necessarily the case. This means that group communication concepts and mechanisms must be employed to ensure that all potential producers or consumers of an event are encounted for. In the case that event composition is itself distributed over several nodes, the result of event composition is influenced by transmission delays and communication failures. Therefore a correctness criterion for distributed composition of events is needed and a real-time cognizant approach to enforce such criteria must be developed. This problem is even more complicated, when implicit or explicit replication of composite event detection is introduced.

Finally, the crosseffect resulting from rule execution in distributed environments is addressed. In [Ceri and Widom 1992] various paradigms for distributed rule processing are identified and characterized on the basis of whether multisite rules are permitted or not, whether rules may execute before the transaction is done at that site, and whether intersite priorities exist or not. Even in the tightly integrated relational DBMS with a rather restrictive execution model that was analyzed in [Ceri and Widom 1992], distributed rule processing is quite complex and requires locking and coordination among sites to guarantee the same semantics as centralized rule execution. At present, research efforts in distributed active databases are focused on the problems of event handling and composition. A serious discussion of the semantics of rule processing in open distributed systems under timing constraints has not even begun.

4. CONCLUSIONS

An analysis of a set of applications and the base technologies together with the possible cross-effects has shown the futility of attempting to combine the full spectrum of functionality of active databases, real-time systems, and distributed object systems in a single platform. However, some useful subsets, each challenging in its requirements but feasible, can be identified:

- Main-memory databases where distribution can be limited to mirroring for reliability and recoverability with limited active capabilities to provide timing guarantees in real-time environments with tight timing constraints.
- Centralized active databases with high query to update ratios, very high data volumes, and tight but soft deadlines.
- Distributed object platforms with enforcement of timing constraints in tightly controlled distributed environments with little database functionality and limited event sets.
- Time-constrained distributed systems with rather lax but not necessarily soft timing requirements, wide area distribution, and possibly large data volumes.

The biggest challenge will be to move away from application-specific solutions and to provide application-independent platforms for which the underlying assumptions and the resulting semantics are clearly stated and easily understood by the developers of the application systems. To accomplish this we must move beyond a superficial combination of base technologies, and analyze their interactions carefully. As we understand these interactions better we may be able to move the boundaries between the subsets identified above towards the goal of a more generic distributed object platform that offers active and real-time functionality.

5. ACKNOWLEDGEMENTS

The authors wish to thank Mariano Cilia for many interesting discussions.

6. REFERENCES

Arnold, O´Sullivan, Scheifler, Waldo and Wollrath: 1999, *The Jini Specification*, Addison Wesley.

Bestavros, A. and Fay-Wolfe, V. (eds): 1997, *Real-Time Database and Information Systems - Research Advances*, Kluwer Academic.

Bestavros, A., Lin, K.-J. and Son, S. (eds): 1997, *Real-Time Database Systems - Issues and Applications*, Kluwer Academic.

Branding, H. and Buchmann, A.: 1995, On providing soft and hard real-time capabilities in an active dbms, *in* M. Berndtsson and J. Hansson (eds), *Active and Real-Time Database Systems (ARTDB-95)*, Springer, pp. 158–169.

Buchmann, A., Dayal, U., McCarthy, D. and Hsu, M.: 1989, Time-critical database scheduling: A framework for integrating real-time scheduling and concurrency control, *Proceedings Fifth International Conference on Data Engineering*, IEEE, Los Angeles.

Ceri, S. and Widom, J.: 1992, Production rules in parallel and distributed database environments, *Proceedings of the 18th VLDB Conference*, Vancouver, Canada, pp. 339–351.

Chakravarthy, S., Krishnaprasad, V., Anwar, E. and Kim, S.: 1994, Composite Events for Active Databases: Semantics, Contexts and Detection, *Proc. VLDB ´94*, Santiago, Chile, pp. 606–617.

Datta, A.: 1999, Position statement on artdbs, ARTDB-99. Panel Discussion.

Dayal, U., Buchmann, A. and McCarthy, D.: 1988, Rules are objects too: a knowledge model for active, object-oriented database systems, *Proceedings of the 2nd International Workshop on Object-Oriented Database Systems*, LNCS 334, Springer.

Gatziu, S. and Dittrich, K.: 1993, Events in an active object-oriented database system, *Proceedings of Rules in Database Systems*, Edinburgh, pp. 23–39.

Gehani, N., Jagadish, H. and Shmueli, O.: 1992, Event specification in an active object-oriented database, *Proceedings of the International Conference on Management of Data (SIGMOD '92)*.

Harrison, T., O´Ryan, C., Levine, D. and Schmidt, D.: 1998, The design and performance of a real-time corba event service. Submitted to IEEE Journal on Selected Areas in Communications.

Kopetz, H.: 1992, Sparse time versus dense time in distributed real-time systems, *Proceedings of the 12th International Conference on Distributed Computing Systems*, Yakohama, Japan, pp. 460–467.

Kopetz, H.: 1997, *Real-Time Systems - Design Principles for Distributed Embedded Applications*, Kluwer Academic Publishers.

Lamport, L.: 1978, Time, clocks and the ordering of events in a distributed system, *CACM* **21**(7), 558–565.

Liebig, C., Boesling, B. and Buchmann, A.: 1999,
A notification service for next generation
IT systems in air traffic control, *GI Workshop Multicast - Protokolle und Anwendungen*, Braunschweig.

Liebig, C., Cilia, M. and Buchmann, A.:
1999, Event composition in time-dependent
distributed systems, *International Conference on Cooperative Information Systems
COOPIS99*, Edinburgh.

Locke, D.: 1997, Real-time databases: Real-world
requirements, *in* A. Bestavros, K. Lin and
S. Song (eds), *Real-Time Database Systems
- Issues and Applications*, Kluwer Academic
Publishers.

Object Management Group (OMG): 1999, Real-time corba architecture, *Technical Report ptc-99-06-02*, OMG, Famingham, MA .

O´Neil, P., Ramamritham, K. and Pu, C.: 1996, A
two-phase approach to predictably scheduling
real-time transactions, *Performance of Concurrency Control Mechanisms in Centralized
Database Systems*, Prentice-Hall, pp. 494–522.

Paton, N. (ed.): 1998, *Active Rules in Database
Systems*, Springer-Verlag (New York).

Purimetla, B., Sivasankaran, R., Ramamritham,
K. and Stankovic, J.: 1995, Real-time
databases: Issues and applications, *in* S. Son
(ed.), *Advances in Real-Time Systems*, Prentice Hall.

Raatikainen, K.: 1997, Real-time databases in
telecommunications, *in* A. Bestavros, K. Lin
and S. Son (eds), *Real-Time Database Systems - Issues and Applications*, Kluwer Academic Publishers.

Ramamritham, K.: 1993, Real-time databases, *International Journal of Distributed and Parallel Databases* **1**(2).

Schwarz, R. and Mattern, F.: 1994, Detecting
causal relationship in distributed computations: In search of the holy grail, *Distributed
Computing* **7**(3), 149–174.

Stankovic, J., Son, S. and Hansson, J.: 1999, Misconceptions about real-time databases, *IEEE
Computer* **32**(6), 29–36.

Ulusoy, O. and Buchmann, A.: 1998, A real-time
concurrency control protocol for Main Memory database systems, *Information Systems*
23(2), 109–125.

Verissimo, P.: 1993, Real-time communication, *in*
S. Mullender (ed.), *Distributed Systems*, 2
edn, Addison-Wesley.

ASPECTS OF FLIGHT CONTROL SOFTWARE – A SOFTWARE ENGINEERING POINT OF VIEW

Alfred Roßkopf* Theodor Tempelmeier**

* *DaimlerChrysler Aerospace AG, Munich*
** *Fachbereich Informatik, Laboratory of Real-Time Systems,*
Fachhochschule, Rosenheim

Abstract: This contribution reports on some aspects of flight control software. Experience as gained from an actual implementation of control laws in Ada is given. The influence of the chosen programming language, the specification of (parts of) the requirements, principles of the software design, and the concept of tests in various environments are covered. *Copyright © 1999 IFAC*

Keywords: Software engineering for embedded systems, object-oriented techniques, Ada, flight control, safety-critical systems

1. INTRODUCTION

Modern flight control systems use fly-by-wire technology for automatic stabilisation (or stability enhancement) of aircraft. In fly-by-wire systems there is no longer a mechanical link from the pilot's controls to the control surfaces of the aircraft. Instead, only an electrical link exists, with a flight control computer system processing the pilot's commands and the data of the aircraft motion sensors in order to actuate the control surfaces of the aircraft.[1] Such flight control systems are obviously safety-critical, i.e. mishaps in such systems could cause injury or death to humans, and appropriate safety measures have to be taken.

There is literature on flight control systems and on control law design (e.g. Collinson, 1996; Beh and Hofinger, 1994), but relatively few authors deal with the software aspects of actual implementations, for example Frisberg (1998). The present contribution also takes a software engineering point of view and reports on the experience from an actual implementation of flight control laws.

Most of the presentation is based on (simplified) examples, as the whole subject cannot be covered in a single paper.

2. PROGRAMMING LANGUAGE ISSUES

Flight control software is safety-critical. The selection of a suitable programming language or of individual features of the language needs special attention and many criteria must be considered (ISO/IEC, 1998).

Some issues in programming languages may seemingly be simple or trivial at first glance, but eventually exhibit astonishing complexity or surprising effects, at least. In-vogue languages such as C, C++, and Java are contrasted to the programming language Ada in the following. Because flight control systems have to deal with computer arithmetic, simple multiplication operations will serve as examples.

2.1 *Integer Operations*

For integer operations the result of the multiplication seems to be clear. However, C, C++, and

---

[1] It should be pointed out that the automotive industry has already started thinking about drive-by-wire and brake-by-wire systems. So this contribution may very well also apply to automobiles in the near future.

Java actually use modulo arithmetic and compute something like the following function instead of integer multiplication:

$$Int\_Mul(i,j) = ((i*j) \bmod 2^{16})$$
$$-(\textbf{if } ((i*j) \bmod 2^{16}) \geq 2^{15}) \textbf{ then } 2^{16} \textbf{ else } 0)$$
// predefined integer multiplication in
// C, C++, Java for 16 bit operands

In Ada, on the other hand, either the mathematically correct result is computed or an exception is raised.

2.2 *Fixed Point Operations*

When integer operands are associated with an implicit scale factor, the resulting type can be seen as a fixed point type. For instance, a scale factor of 2^{-15} means that 16 bit integer numbers are interpreted as fixed point numbers in the interval [-1.0, 1.0). A multiplication of two such fixed point numbers can be simulated in C, C++, and Java as follows:

$$Fixed\_Mul(i,j) = (\textbf{int } (((\textbf{long}) \, i*j) >> 15)$$
// simulated fixed point multiplication in
// C, C++, Java for 16 bit operands with
// implicit scale factor 2^{-15}

It is true that such fixed point operations can be (and should be) encapsulated in abstract data types, but nevertheless, an implementation effort is necessary to deal with such fixed point numbers. In Ada, mathematically correct operations are predefined and available after a simple appropriate fixed point type declaration such as

type fixed16 **is** delta 2^{-15} **range** -1.0 .. 1.0;
-- Ada fixed point type definition
i, j, k : fixed16;
k := i * j ;
-- fixed point multiplication in Ada

Fixed point types are important in practical situations, if no floating point hardware is available or if input or output values, e.g. from analog to digital converters, require such a format anyway. In the latter case they are also ideal for the specification of the requirements (see next chapter).

2.3 *Floating Point Operations*

It is a nuisance that C, C++, and Java still have different notations for floating point literals depending on the precision of the number. Thus, an innocent looking x * 3.14 will probably evaluate to (float)((double)x * 3.14) because 3.14 is of type double by default. The overhead of the two type conversions and of needlessly carrying out the operation in double precision, may be intolerable in a real-time system. This can easily be avoided by denoting the literal as single precision, i.e. by writing x * 3.14f. Ada's much stronger typing philosophy avoids such nuisances per se.

2.4 *Conclusion*

There are many more "trivialities", which would cause irritations when using inappropriate programming languages. The effort of handling these aspects, especially when interdependencies arise, may be significant. The choice of the programming language Ada renders these aspects almost trivial from a designer's and programmer's point of view. It is thus possible to concentrate on the real problems in the application domain.

3. SOFTWARE REQUIREMENTS SPECIFICATION

Requirements for flight control law software are usually defined using control block diagrams, which essentially describe blocks and data flows between them. Each block represents a transformation of input data to output data. The detailed control law computations are described in an algorithmic language, e.g. in FORTRAN. It is essential that this specification is executable, in order to validate the control law design.

In a small case study the Unified Modeling Language UML (Version 1.0) has been investigated with respect to specifying requirements for typical flight control software. The results are presented here in the form of three examples:

- definition of control surfaces
- control law block diagrams
- definition of external interfaces

3.1 *Definition of Control Surfaces*

The main outputs of a flight control system are commands to the control surface actuators. Therefore, a requirements specification for flight control software would typically include a definition of these control surfaces. A possible UML definition for this is given in figure 1. A control surface "is a" primary or a secondary surface (inheritance relation) and so on. And a delta-canard aircraft "has" one rudder, four flaperons, and so on (aggregation or composition relation).

Figure 2 shows as an alternative a sketch of an aircraft with the primary control surfaces emphasized in white and the secondary ones in dark. It

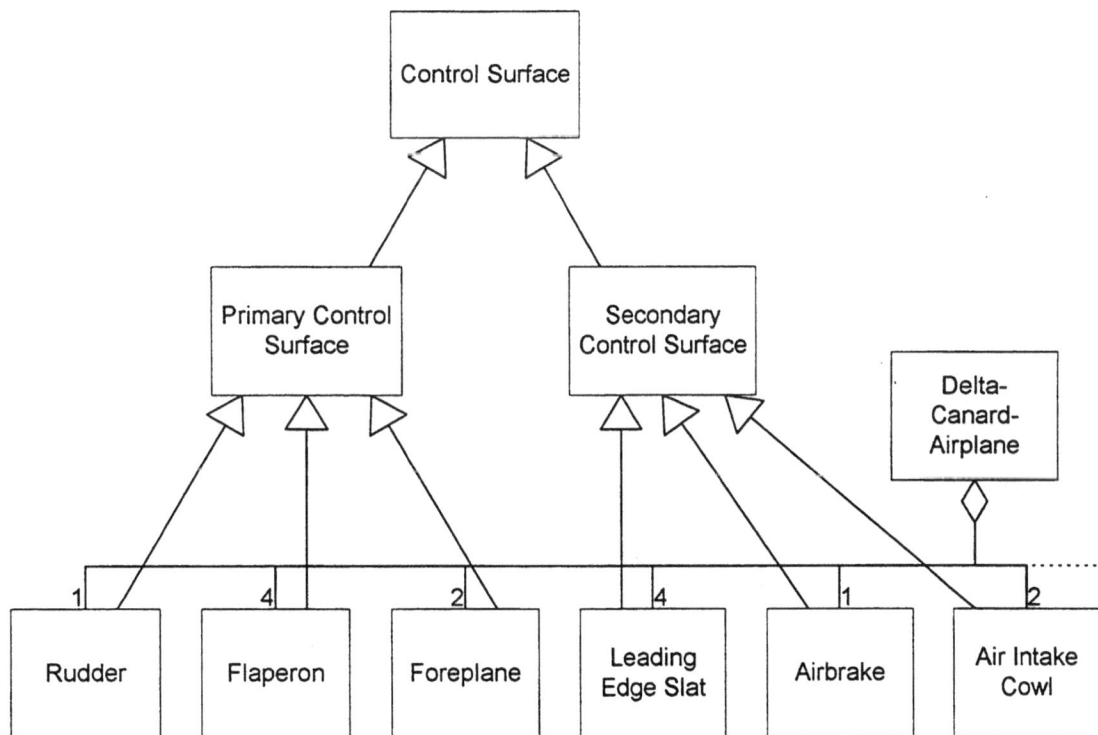

Fig. 1. Control Surfaces of an Aircraft in Delta-Canard-Configuration (UML Diagram)

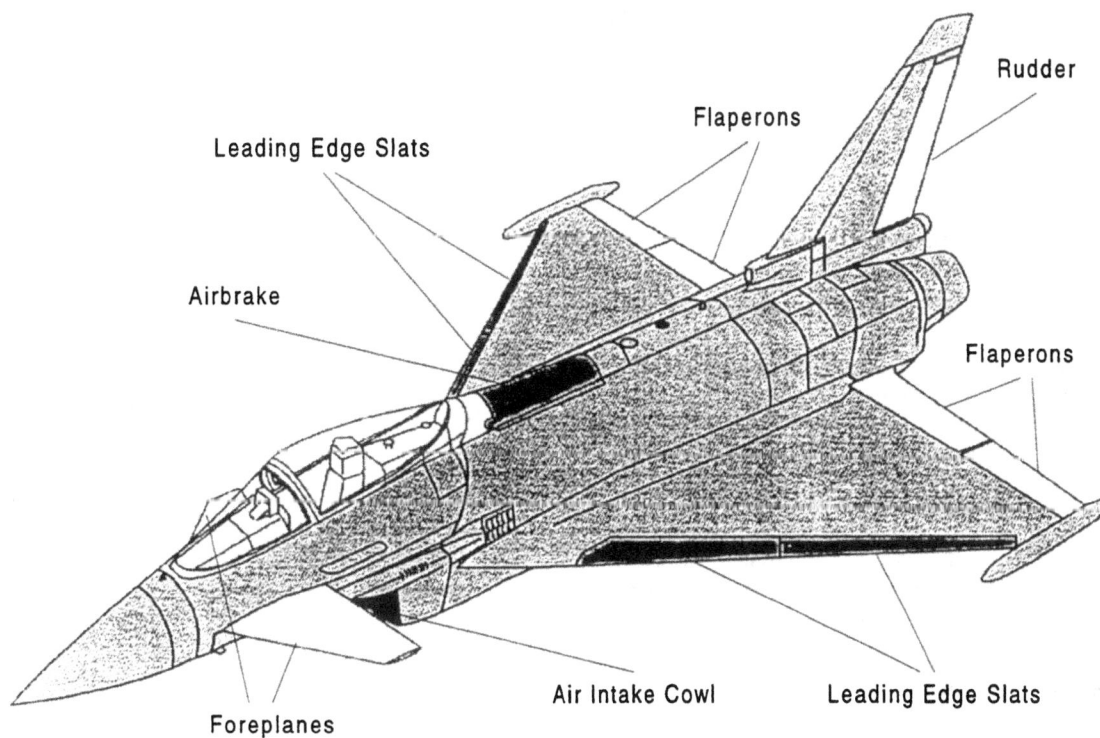

Fig. 2. Control Surfaces of an Aircraft in Delta-Canard-Configuration

can be seen that a simple figure is sufficient to convey at least the same information as the UML diagram. The authors would not consider it very helpful to rephrase such parts of the requirements in UML.

3.2 *Control Law Block Diagrams*

As said above, control block diagrams contain fundamental parts of the requirements for control law software. Figure 3 shows an example (Kaul, 1992). If one would try to rephrase such diagrams in UML, the following difficulties arose.

- Obviously, the control block diagram is not a class diagram. Instead, the blocks constitute multiple instances of classes, e.g. the six "Filtering" blocks. The control block diagram thus must rather be seen as an "object diagram". While it is possible to draw object diagrams in UML, only the association relation can meaningfully be used in this case. However, the association relation is only a line drawn between rectangles with almost no semantic information. UML's object diagrams are thus not a suitable alternative for control block diagrams.
- Class diagrams do not help very much in this case, either. They would reveal surprisingly little information, for instance that a notch filter "is a" filter, etc. The complexity of the control block diagram lies in the functionality, i.e. in the contents of the block diagram elements and in their interdependencies. In contrast, UML seems to be more suitable for applications where the complexity lies in the class relationships, like in database applications, for instance.
- One could try to (mis)use other diagrams of UML, e.g. collaboration diagrams or activity diagrams, but the authors do not see any advantage in doing so, as compared to using well established control block diagrams.

3.3 *Definition of External Interfaces*

Requirements for flight control software must include a definition of the external interface of the software, including the formats of the input and output values. The authors consider Ada a good choice for specifying these formats and advocate its use starting with the requirements definition. The following examples repeat some of these probably well-known features of Ada.

```
type switch_values is (neutral, on, off);
   for switch_values use (neutral => 1,
   on => 2, off => 4);
```

```
Small_180 : constant := 180.0 * 2^{-15};
type T_Fixed_180 is delta C_Small_180
   range −180.0..180.0;
   for T_Fixed_180 'small use C_Small_180 ;
   for T_Fixed_180 'size use 16;
```

Such Ada definitions, firstly, have precise semantics according to the language definition. Secondly, the use of the representation clauses ("for . . . ") allows a precise format definition down to the bit level. For instance, the first definition ties the switch values to their bit representations in say a digital input or output register, the second definition associates fixed point values with the format of an analog digital converter, for instance. Note the use of a delta which is not a power of two in the latter case. This represents a very common situation in practice.

Most importantly, using such definitions in the software requirements specification automatically guarantees consistency between (this part of) the requirements with the final code. The authors feel this method to be superior to rephrasing such requirements in some formal notation[2], and then perform proofs of consistency with the final Ada code. On the other hand, using UML would almost certainly be inferior, because there is no semantics and not even a syntax for describing types in UML.

4. SOFTWARE DESIGN AND IMPLEMENTATION

In a research and technology project of DaimlerChrysler Aerospace, an integrated process for the development of control laws for complex aircraft configurations has been investigated. As part of this project control laws for a typical fighter aircraft have been implemented in Ada. The resulting software is called COLAda (Control Law Software in Ada). In companion publications, various architecture and design aspects of COLAda are reported (Roßkopf, 1999; Roßkopf and Kohlhof, 1999; Kohlhof and Roßkopf, 1999). Here, the focus is on the following three general issues:

- using object-oriented design techniques
- design for multi-processor targets
- using finite state machines

---

[2] The authors are well aware of the capabilities (and the limitations) of formal methods (Tempelmeier, 1997). However, a discussion of these issues is beyond the scope of this paper.

Input Signals Control Law Computation Command Signals to Actuator Loops

Altitude
Mach Number
Stick/Pedals
Pitch Angle
Roll Angle
True Airspeed
Angle of Attack
Pitch Rate
n
Angle of Sideslip
Roll Rate
Yaw Rate

Command Sampling and Filtering
Computation of Demand Signals
g Compensation
Gain Scheduling
Altitude
Mach Number
Filtering
Demand Signal Selection
Error Selection
Gain
Gain
Gain
∫
Trim Scheduling
Signal Distribution
Inertia Coupling Compensation

Leading Edge
Outboard Flaperon
Inboard Flaperon
Foreplane
Rudder
Foreplane
Inboard Flaperon
Outboard Flaperon

Cat. I
Safety Critical
Cat. II
Full Performance

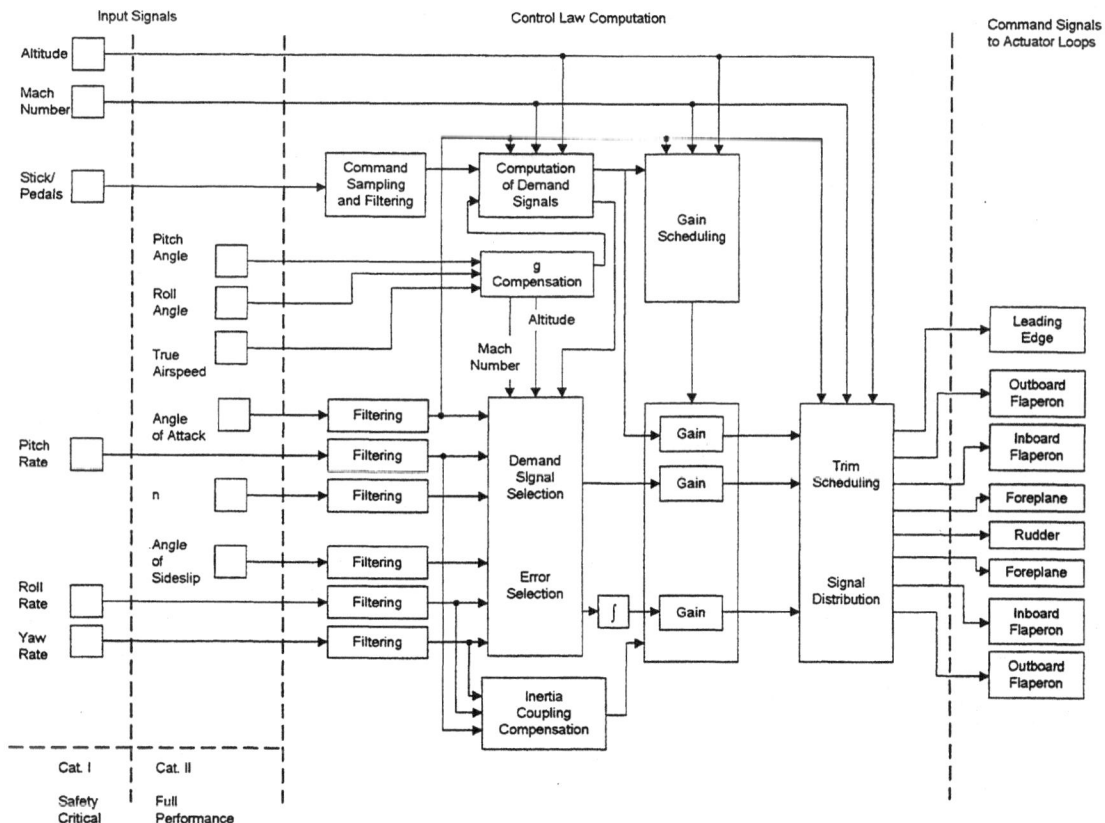

Fig. 3. A block diagram of flight control laws (according to Kaul, 1992)

4.1 Using Object-Oriented Design Techniques

In the design of COLADA object-oriented techniques have been used, where appropriate. Certain control law elements, e.g. filters, have been identified as candidates for objects, and corresponding abstract data types. (e.g. filter types) have been defined. Several objects of these types may be created as required by the control law application – the objects are just instances of abstract data types. The external behaviour of these objects is defined by the operations associated with the abstract data type. Of course all internal data are encapsulated in the objects.

According to a certain terminology, the design might be termed to be object-based, as no type extension ("inheritance") is used. This is for the following reasons:

- Full use of inheritance, in particular polymorphism and dynamic binding (i.e. the use of class-wide types in Ada terminology), may cause certain problems in safety-critical systems (cf. ISO/IEC, 1998).
- There is hardly a need for using inheritance in the context of this project. The cases where variants of certain object types occur (e.g. first order and second order filters) can easily be covered without inheritance.

Though certain aspects of flight control software can be designed and implemented in an object-oriented way, and though the authors very strongly favour such an approach, there must be a warning against hypocrisy with respect to object-orientation: The approach "everything is an object" does not seem very helpful. Over-simplistic approaches like in Coad et al. (1995), where a case study of an object-oriented auto-pilot system is reported on, seem to be impractical for the implementation of real flight control software. And, of course, normal arithmetic operations and static typing are to be used in control law software. This is in contrast to the philosophy of radical object-oriented languages such as SmallTalk and Lisp, for instance, which are considered unsuitable to real-time safety-critical systems.

4.2 Design for Multiprocessor Targets

COLAda has been designed and implemented in a way that the same software implementation can be run on different hardware platforms and under different runtime environments (Roßkopf, 1999), cf. figure 4. Of course, no performance overhead is acceptable in the actual flight control multiprocessor, whereas some overhead is uncritical in the other (single processor) environments.

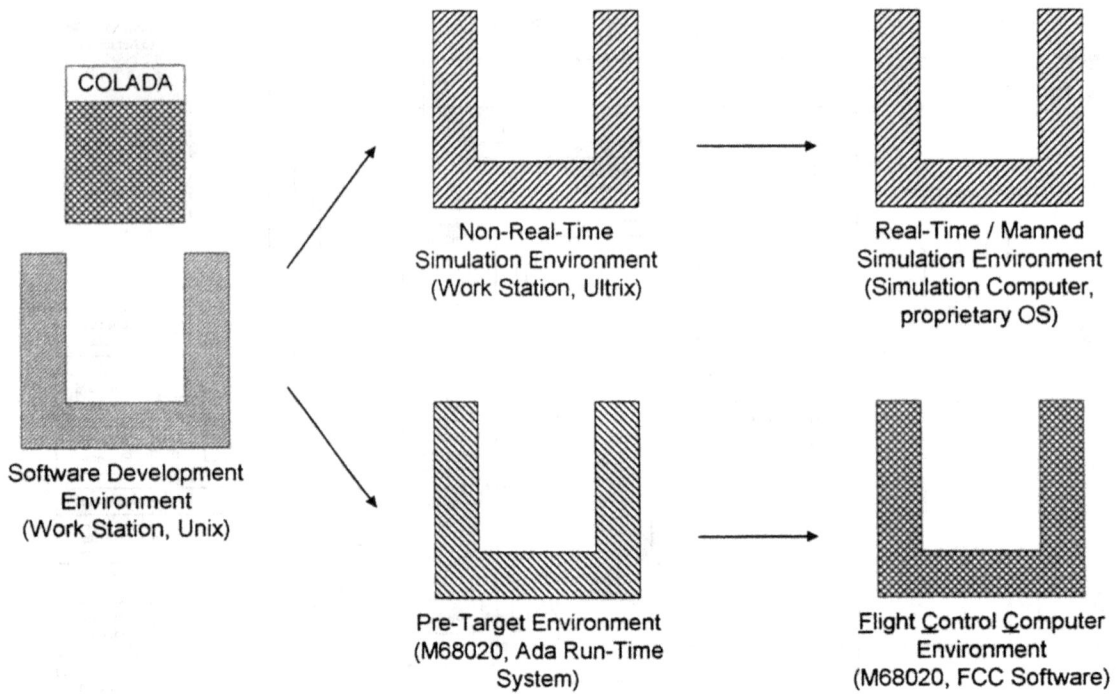

Fig. 4. Runtime Environments of the COLAda Control Law Software (Roßkopf, 1999)

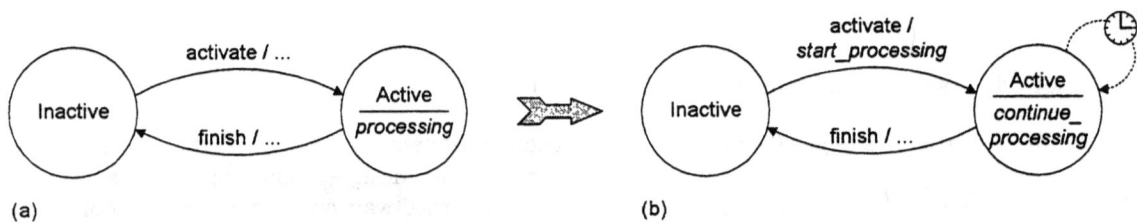

Fig. 5. Usage of Finite State Machines: (a) Moore type, (b) mixed Mealy/Moore finite state machine

An application-specific communication facility has been designed with this efficiency requirement in mind. In essence, it provides procedures to send or receive data to respectively from the other processor via the shared memory of the flight control computer. By the use of procedure inlining, the exchange of data via this interface reduces to a few machine instructions. In environments other than the flight control computer, different implementations of this same interface are used. Instead of shared memory, mailboxes and message passing services are used. These services can easily be implemented with Ada's tasking facilities or protected objects, as known from the standard Ada literature, (e.g. Nielsen and Shumate, 1988; Burns and Wellings, 1998).

4.3 Using Finite State Machines

Some parts of flight control software can be modelled as finite state machines. However, devising state machines that map directly into good software structures is part of the software design process and should carefully evaluate possible design alternatives. Once a suitable state machine has been found, it can easily be translated into (Ada) code in a well-known way (i.e. using nested case and/or if statements).

In the following a general pattern of finite state machine usage in synchronous (i.e. cyclic) control software is outlined. It assumes a state machine with the states inactive and active. Upon the input activate, a transition is made to state active, and the action processing is performed in this state, see fig. 5a. In finite state machine theory this action is the output of the state machine. (Outputs are shown in italics in the diagrams.) Figure 5a shows a Moore type state machine, where the output is associated with the current state. Due to the cyclic nature of control software a mixed Mealy/Moore type state machine might be more appropriate, see fig. 5b. One part of the processing (start_processing) is done (only once) upon entering the active state. Henceforth, another part of the processing (i.e. continue_processing) is executed in every cycle as long as the state is active.

5. CONCLUSION AND OUTLOOK

The development of flight control software demands sound software engineering concepts and methods. Design and implementation of the software must not only satisfy the functional requirements of flight control systems but has also to consider numerous non-functional constraints. These include, for example, the hardware architecture of the flight control computer, specific interfaces to other software components, or the required efficiency and maintainability of the software. Additional constraints are introduced by the requirement of verifiability of the safety-critical software (cf. ISO/IEC, 1998).

Many well-weighed design and implementation decisions have to be made which, in the end, determine the suitability and quality of the software. From the experience of the flight control software implementation reported on in this paper, it seems inconceivable that, in the near future, the target code for a flight control computer could be generated by software engineering tools (like Statemate or MatrixX, for instance). As outlined above, there are numerous design constraints which can hardly be satisfied using these tools. This will probably preclude automatic code generation for such target computers for quite some time.

6. REFERENCES

Beh, H. and G. Hofinger (1994). X-31a control law design. In: *AGARD Conference Proceedings 548. Technologies for Highly Manoeuvrable Aircraft. Flight Mechanics Panel Symposium, Annapolis, Maryland, United States, October 18-21, 1993.* North Atlantic Treaty Organization.

Burns, A. and A. Wellings (1998). *Concurrency in Ada.* 2nd ed.. University Press. Cambridge.

Coad, P., D. North and M. Mayfield (1995). *Object Models. Strategies, Patterns, and Applications.* Yourdon Press. Englewood Cliffs.

Collinson, R.P.G. (1996). *Introduction to Avionics.* Chapman & Hall. London.

Frisberg, B. (1998). Ada in the JAS 39 Gripen flight control system. In: *Reliable Software Technologies – Ada-Europe'98. 1998 Ada-Europe International Conference on Reliable Software Technologies, Uppsala, Sweden, June 8-12, 1998. Proceedings. LNCS 1411.* (L. Asplund, Ed.). Springer-Verlag. Berlin, Heidelberg.

ISO/IEC (1998). *Working Draft 3.8 – Programming Languages – Guide for the Use of the Ada Programming Language in High Integrity Systems. ISO/IEC PDTR 15942, ISO/IEC JTC 1/SC22/WG9.* Internet.

Kaul, H.J. (1992). Flugsteuerungssystem Jäger 90. (Flight control system of the Fighter 90. In German). In: *Jahrbuch 1992 III der Deutschen Gesellschaft für Luft- und Raumfahrttechnik e.V. (DGLR). Deutscher Luft- und Raumfahrtkongreß 1992. DGLR Jahrestagung, Bremen, 29. September - 02. Oktober 1992* (G. Bürgener, Ed.). Deutsche Gesellschaft für Luft- und Raumfahrt e.V. (DGLR). Bonn.

Kohlhof, C. and A. Roßkopf (1999). Flugsteuerungssoftware in Ada – Worst-Case Timing Analysis. (Flight control software in Ada – worst-case timing analysis . In German). In: *Objektorientierung und sichere Software mit Ada. Ada-Deutschland Workshop. 21.-22. April 1999, Karlsruhe.* Forschungszentrum Karlsruhe. Karlsruhe.

Nielsen, K. and K. Shumate (1988). *Designing Large Real-Time Systems with Ada.* Intertext, Multiscience, McGraw-Hill. New York.

Roßkopf, A. (1999). Development of flight control software in Ada – architecture and design issues and approaches. In: *Ada-Europe'99. International Conference on Reliable Software Technologies. June 7-11, 1999. Santander, Spain.* Springer-Verlag. Berlin, Heidelberg.

Roßkopf, A. and C. Kohlhof (1999). Flugsteuerungssoftware in Ada – Konzepte und Methoden für Konstruktion und Verifikation. (Flight control software in Ada – concepts and methods for construction and verification. In German.). In: *Objektorientierung und sichere Software mit Ada. Ada-Deutschland Workshop. 21.-22. April 1999, Karlsruhe.* Forschungszentrum Karlsruhe. Karlsruhe.

Tempelmeier, T. (1997). *Formal Methods - An Informal Assessment. Technischer Report Dasa MT36 SR-1775-a.* Daimler-Benz Aerospace. Ottobrunn.

A TOOL FOR VALIDATING TIMING REQUIREMENTS OF INDUSTRIAL APPLICATIONS BASED ON THE FOUNDATION FIELDBUS PROTOCOL

R. Wild and C. E. Pereira

*Department of Electrical Engineering, Universidade Federal do
Rio Grande do Sul, Porto Alegre, Brazil*

Abstract: This work proposes a tool for evaluating the temporal behavior of the communication in distributed industrial automation systems based on Foundation Fieldbus protocol. It extends existing commercial bus monitoring tools with constructs for specifying timing requirements and with mechanisms to allow the evaluation of the fulfilness of those requirements. It can be used both as a tool for evaluating the temporal behavior of existing plants as well as to provide valuable hints for designers when developing new applications. *Copyright © 1999 IFAC*

Keywords: real-time requirements, timing constraints validation, real-time fieldbus evaluation

1. INTRODUCTION

Timing constraints are often critical in automation and field control real-time systems, in the sense that missing deadlines may result in catastrophes (material or human damage, or even death). Experience has shown that complex systems can hardly be completely formally verified, and it is very difficult to assure that the implementation is error-free, or that an unexpected situation will not arise (unexpected in the sense that it contradicts some assumptions made during system modeling and implementation, therefore endangering the timeliness of the system). This is the main motivation for creating monitoring systems for real-time applications (see (Tsai and Yang, 1995) and (Plattner, 1984)).

Modern distributed industrial automation architectures are characterized by a network of field devices that are usually connected through a communication bus, the so-called fieldbus. Current fieldbus devices are able to perform local processing and to communicate with each other. In this context, the real-time behavior of the adopted control strategies depends not only on the temporal behavior of intra-device processing but also relies heavily on inter-device communication.

This paper presents a contribution to the investigation of some conditions under which communication in the fieldbus meets the timing constraints imposed by control strategies adopted in system design and configuration. It proposes a tool for monitoring, validation and visualization of bus communication. The paper is divided as follows: after a brief discussion on related work given in Section 2, an overview on the Foundation Fieldbus protocol is presented in Section 3. In Section 4 the proposed architecture and tool support to validate timing constraints of Fieldbus is discussed. Developed case studies are presented in Section 5. Finally, Section 6 draws some conclusions and signals directions of future work.

2. RELATED WORK

In a broader sense, monitoring means extracting runtime information, usually not available through 'statical analysis', in order to identify relevant events and their relative ordering or instant of occurrences, so that a validation of the run-time behavior against pre-defined requirements is possible. That means, monitoring encompasses the activities of (i) collecting information about events (ii) processing this information in order to obtain performance figures or to validate requirements (iii) visualizing the obtained information.

Approaches for collecting relevant events from real-time systems can be classified into software-based, hardware-based and hybrid. The first deals with the insertion of additional lines of code to the original real-time software in order to generate and register events (what is called "code instrumentation"). Either the application code (see (Chodrow et al., 1991), (Joyce et al., 1987)), or the operating system code, as proposed by (Miller et al., 1986) and (Tokuda et al., 1988), can be instrumented. Hardware-based approaches makes use of dedicated hardware, which is incorporated to the target architecture, as in (Brantley et al., 1989), (Tsai et al., 1990) and (Liu and Parthasarathy, 1989). Code instrumentation tends to be easier to use and less expensive than dedicated hardware as a monitoring strategy. On the other hand, dedicated hardware monitoring can reduce or eliminate the perturbation in the monitored system behavior due to the overhead caused by the execution of additional code. Hybrid monitoring systems, as (Haban and Wybranietz, 1990), (Mink et al., 1990) and (Dodd and Ravishankar, 1992) aims to attain both benefits of software- and hardware-based approaches.

While several monitoring systems only present the collected events, by performing a post-processing meaningful information can be extracted such as bus occupation rate, jitter evaluation, etc. (see (Tsai et al., 1990)). Just few approaches with similar goals to the present work could be found in the literature, such as (Chodrow et al., 1991), (Raju et al., 1992)).

Textual or graphical presentation of collected data is also a very important feature to help in real-time debugging or validation activities. Systems including interesting visualization modules are presented in (Haban and Wybranietz, 1990), (Joyce et al., 1987), (Tokuda et al., 1988). Interesting work dealing with the topic visualization of real-time systems data are (Tsai et al., 1996), (TimeSys, 1998) and (Girardin and Brodbeck, 1998).

The main goal of this work is to develop a monitoring system for distributed industrial automation systems, which encompasses all steps previously mentioned. Events are collected using a hardware-based approach, in order to minimize the perturbation of the observed behavior. Col-

lected data is post-processed in order to relate messages to significant events in the real-time system. Special care is given on validating specified timing requirements. The proposal also includes graphical presentation of validation results.

3. FIELDBUS MESSAGE SCHEDULING

A key concept in the fieldbus technology is to allow field devices to locally execute data processing and decision making activities, leading to a decentralized control architecture. In a fieldbus with inter-device communication, the control of a process can be made locally, so that the input value measured by a device is passed directly, via the bus, to the acting device. In such an environment, messages are important events to be monitored: the communication is coordinated for the occupation of the bus by each device message, and it becomes important to validate the time behavior of the system, so that each message meets its timing constraints.

Foundation Fieldbus (Foundation, 1998) provides a set of standardized function blocks, which consist of a set of data and algorithms that are intended to provide a given functionality to the application, either standalone or linked to other function blocks. Examples are PID controllers, input and output blocks, transducer blocks, etc. Links among function blocks imply that they have to exchange data. In case of the connected function block reside in different devices, what is quite common in fieldbus applications, messages are exchanged through the bus.

The Foundation Fieldbus defines two types of communication (messages in the bus): periodic and aperiodic. All periodic messages must be scheduled off-line, so that temporal information must be known at configuration time. Nonperiodic message are scheduled for communication during execution time, using the bus resource only during the time intervals where no periodic messages are scheduled. This communication management and medium access control is responsibility of a device that, besides its normal function, is responsible for the bus arbitration. It is called the Link Active Scheduler (LAS) and it has a table with all scheduled messages of each device. According to a scheduling algorithm the LAS determines when a given fieldbus device has to start its communication. Then, at the specific time, it sends a special request message, called Compel Data (CD), to the device that must transmit. When receiving such a message, the target device places its data on the bus and any devices can then read the message (publish-subscriber pattern). During intervals with no periodic messages scheduled, the LAS circulates, among existing devices on the bus, a permission pass message (PT, 'pass token'), allowing devices with pending non-periodic messages to access the bus. Furthermore, the LAS carries out tasks such as live list maintenance (a list with all devices that are currently connected to the network), node activation and device synchronization. Due to its importance, in case of failures of the device running the LAS, automatically another device previously assigned will be responsible for the message scheduling.

The Link Active Scheduler is a key concept in the Foundation Fieldbus technology, being an important step towards achieving a deterministic real-time behavior and overcoming common deficiencies of fieldbus technologies, in relation to timing constraints, high processing load, and communication schedulability. Since the access to the bus is arbitrated by the LAS following the scheduling strategy defined by the scheduler, one can guarantee that periodic messages will meet their deadlines.

Commercially available monitoring tools for Fieldbus are basically message loggers, that associate messages to their time stamps and store them into a file. The major drawback of these tools is that the available features for analyzing the temporal behavior are very poor, since only quite primitive filtering and sorting of stored messages is allowed.

4. VALIDATION OF TEMPORAL BEHAVIOR - PROPOSED APPROACH

4.1 Architecture and General Overview

The tool proposed in this paper can be situated in the project life cycle depicted in Fig. 1. Starting point to the development is the specification of a set of functional and non-functional system requirements. These requirements are usually based upon assumptions about the system and its environment (Jahanian, 1995). Specified timing constraints are usually heavily dependent on the behavior of the system to be automated, its components and environment. Regarding the communication, timing requirements may specify the time that a certain message will spend to be delivered; or the frequency a certain external event will happen. No matter how careful these assumptions are, situations may happen in which they are no longer valid. Very complex systems can hardly be formally verified; and unexpected situations may arise in the environment, out of the reach of control of the system. A runtime monitor can build a system history, made of relevant events and their time instant of occurrence. Then, along with the set of requirements, the temporal behavior of the implemented system can be validated. The system monitoring and validation of runtime

timing constraints aims a twofold role in this case: (i) it can validate if and how imposed constraints are met during runtime, and identify the cases the system has failed to meet these constraints; (ii) it can provide a meaningful graphical display for the information collected and analysed.

Based on the results of the validation process, designers can then adapt the implemented application, by modifying the network topology, by introducing new devices, by re-allocating functional blocks, or even by changing the message scheduling strategy.

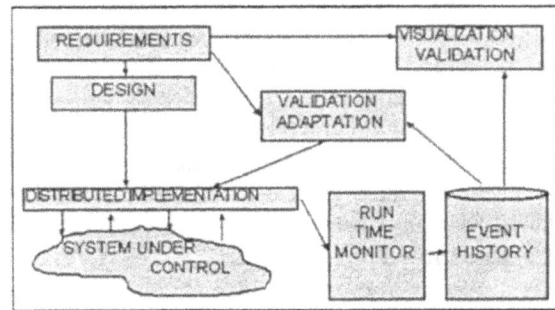

Fig. 1.

The temporal behavior validation proposed in this paper encompasses:

- the formalization of timing requirements imposed by the control strategies and by the desired overall system behavior;
- identification and registration of significant system events and production of an event history;
- a tool to analyze the fulfillment of specified requirements based on the comparison of event history and timing constraints;
- presentation and visualization of collected data and validation results.

4.2 Timing Requirements Specification

In Fieldbus communication, significant events are related to the sending and receiving of messages that are issued by function blocks. Those events have associated timing requirements that are application-dependent. In order to be able to validate given timing requirements, it is necessary that these requirements be formally specified. Notation schemes were surveyed, for criteria such as expressiveness and computational effectiveness of the validation process. A formal description of the validation process is out of the scope of the present work. Another point of interest is the ability of the scheme to state quantitative time relations, not only qualitative. The notation scheme described in (Pereira, 1995), a RTL-based notation (Jahanian and Mok, 1986), has been adopted for the specification. Table 1 summarizes the temporal relations proposed.

Table 1. Proposed temporal relationships

Time Relationship	Description Attribute	Example
Time Sequence	(<ev1> After <ev2>) <duration>	(new-part AFTER old-part) 2s
	(<ev1> BEFORE <ev2>) <duration>	(old-part BEFORE new-part) 2s
Synchronizing	SYNCHRON(<event-list>)<duration>	SYNCHRON(signal1, signal2, signal3) 20ms
Periodic Behaviour	CYCLIC(<ev1>, <cycle>)	CYCLIC(UpdateActuator, 20ms)
Execution Time Limit	DURATION(<element>)　　<interval> where <element> may be state or task	DURATION(IdleState) [20ms, 50ms]

In this scheme, <evn> means event occurrence, and <event-list> means a list of event occurrences. The basic clauses are BEFORE and AFTER, that state the order of occurrence of the events and the minimum amount of time elapsed between them. The SYNCHRON statement states that the maximum time elapsed between occurences of the defined events must be less than some time duration. CYCLIC is for events that must occur at fixed intervals (periodic events), with the period <cycle>. Finally, DURATION states the interval of duration, in the format [minimum, maximum], of an element that can be either a state or a task. This notation is used for expressing timing constraints on the communication events of the Fieldbus system. To establish the correspondence between event and messages exchanged by fieldbus devices, event labels are associated with message types, where the latter are identified by evaluating some fields in the message frame. For instance, by only evaluating the control field of a Foundation fieldbus message, one can identify messages dealing with token passing, such as PT (pass token) and RT (return token). The correspondence between the event label and the message is established by filling out a data structure used by the lexical analyser.

4.3 Event Histories

As previously mentioned, an event is related to the time instant on which a message is sent or received by a function-block running on a given device. Therefore, each message can be associated with a timestamp, which represents the instant of time of its corresponding event. The pairs message-timestamp are stored in the so-called event history. Message timestamping and logging is performed by a commercial system, consisting of a Fieldbus interface card to a PC host and a software package running on the PC for storing the events. Fig. 2 depicts a snapshot, showing a textual visualization of stored messages.

The comprehensive collected events history is then organized into sub-sets of event histories, each one related to a different event type or message.

Fig. 2.

4.4 Timing Requirements Validation

The validation engine has as inputs the specified timing requirements and the stored event history. The output is a graphical display of the temporal behavior, as well as of the validation analysis performed over the input data: requirements that were not satisfied, or which are too tight of the deadlines, or irregular timing can be visualized (see Fig. 3 and Fig. 4).

In the current version of the tool, the validation occurs after a set of data is collected, with a closed event history. The adopted approach is similar to the asynchronous monitoring of constraints as proposed by (Chodrow et al., 1991) and (Jahanian, 1995). The timing validation is independent of the application program, with a separate monitor checking for the satisfaction of the constraints. The validation process is incremental in the sense that it can be performed every time a new event occurs. For performance reasons, the current version only starts the validation after the collection of a given set of events.

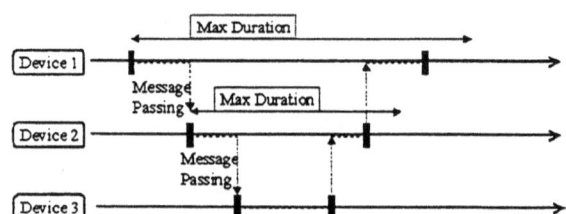

Fig. 3.

The validation generates a list of "event-requirement" entities, which contain references to the event and

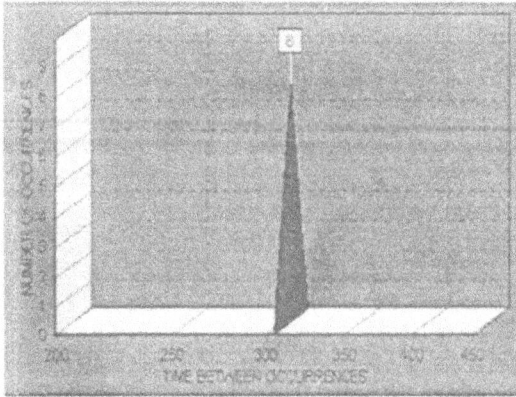

Fig. 4.

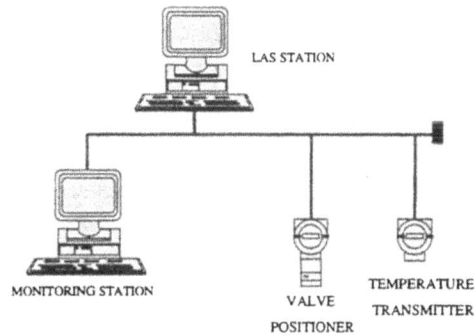

Fig. 5.

to the constraint to be validated, as well some information regarding the end status of the validation (ok, not ok, etc.). The visualization module uses this list for the graphical display of the obtained results. More than one kind of visualization can be run simultaneously over the data, using the same existing list of "event-requirement" entities.

5. CASE STUDIES

For testing purposes, initially a simple architecture with a single control loop and two devices (one temperature transmitter and one valve positioner) was configured (see Fig. 5). For implementing the control loop, three function blocks were used: Analog Input (AI), a PID controller, and an Analog Output (AO). They were connected in the usual way to define a feedback PID control. The AI function block was configured to run in transmitter device, while the PID control and AO function blocks were allocated to the valve positioner. It is important to note that the PID function block could also be allocated to run on the temperature transmitter what could influence the temporal behavior in the communication. Message scheduling was automatically generated using a configuration tool, i.e. the tool automatically defines the cycle of the periodic communication between the AI function block and the PID.

In order to evaluate the jitter in the publication of the measured temperature, it is necessary to establish a relationship between the corresponding event and messages. The temperature available at the AI block output is transmitted to the PID block using the following procedure: (i) the LAS sends a message to the temperature transmitter, the compel data (CD) message; (ii) in response to this message the transmitter publishes the output of the AI block on the bus using the data (DT) message. This periodic message Compel Data (CD) was associated with an event label, "PUBLISH TEMPERATURE", and a constraint "CYCLIC(PUBLISH TEMPERATURE,

0.4s)" was defined, stating that the input temperature should be sent to the PID block at least every 400 ms. The output of the validation tool has shown that in this simple case no jitter occurs. Preliminary results show that, in the tests run, for this simple configuration and time requirements no constraint violation occurred. Further tests for evaluating the impact of additional message traffic (by including additional devices which should communicate with the existing ones) are under development.

A more complex and realistic case study is being developed using a technical plant prototype which has several control loops and a variety of devices. It includes both time continuous process variables including temperature, level, and flow, as well as time discrete ones. In cooperation with industrial partners, additional tests are being carried on using message logs from real applications, in order to allow engineers to better understand what is going on regarding the communication in technical plants consisting of more than fifty devices.

6. CONCLUSIONS

Fieldbus devices and industrial communication protocols, such as the Foundation Fieldbus, have been widely adopted in industrial plants, presenting benefits in terms of reduction of costs in installation, cabling, maintenance, operation, and increasing the technical plant flexibility. In many of automation and control systems requiring this technology, real-time behavior is a fundamental requirement. Unfortunately, existing commercial tools neither allow the specification of timing requirements nor evaluate the obtained behavior. The tool presented in this paper is a proposal to enhance existing commercial bus monitoring tools by extending them with constructs for specifying timing requirements and with mechanisms to allow the evaluation of the fulfilment of those requirements. It can be used both as a tool for evaluating the temporal behavior of existing plants as well as to provide valuable hints for designers when developing new applications. A limitation of the proposed tool is that the validation of timing requirements can only occur when events are observable, i.e. those that appear as messages in the bus. In the current stage, communication between function blocks inside a device can not be validated. An alternative to solve this problem would be to define a special function block for collecting events internally on a device and sending them later to the monitoring tool. In this sense, the event acquisition processing would be considered a hybrid one.

7. ACKNOWLEDGMENTS

This work has been partly supported by CNPq, FAPERGS and FINEP. It has been developed within the context of a cooperative research project on industrial communication protocols, called FINEP-RECOPE, in cooperation with Smar Corp. Int.

8. REFERENCES

Brantley, W. C., K. P. McAuliffe and T. A. Ngo (1989). Rp3 performance monitoring hardware. In: *Instrumentation for Future Parallel Computing Systems* (M. Simmons, R. Koskela and I. Bucher, Eds.). pp. 35–47. ACM Press. New York, USA.

Chodrow, S. E., F. Jahanian and M. Donner (1991). Run-time monitoring of real-time systems. *Proc. Real-Time Systems Symp.* pp. 74–83.

Dodd, P. S. and C. V. Ravishankar (1992). Monitoring and debugging distributed real-time programs. *Software – Practice and Experiencer* **22, No. 10, Oct. 1992**, 863–877.

Foundation, Fieldbus (1998). *Foundation Fieldbus Technical Overview*. Fieldbus Foundation. Available online in www.fieldbus.org.

Girardin, L. and D. Brodbeck (1998). A visual approach for monitoring logs. *Proc. 12th System Administration Conference* pp. 299–308.

Haban, D. and D. Wybranietz (1990). A hybrid monitor for behavior and performance analysis of distributed systems. *IEEE Trans. Software Eng.* **16, No. 2, Feb. 1990**, 197–211.

Jahanian, F. (1995). Run-time monitoring of real-time systems. In: *Advances in Real-Time Systems* (S. H. Son, Ed.). Chap. 18. Prentice Hall. Englewood Cliffs, EUA.

Jahanian, F. and A. K. L. Mok (1986). Safety analysis of timing properties in real-time systems. *IEEE Trans. Software Eng.* **SE-12, No. 9, Sept. 1986**, 890–904.

Joyce, J., G. Lomow, K. Slind and B. Unger (1987). Monitoring distributed systems. *ACM Trans. Computer Systems* **5, No. 2, May 1987**, 121–150.

Liu, A. C. and R. Parthasarathy (1989). Hardware monitoring of a multiprocessor system. *IEEE Micro* **9, No. 5, Oct. 1989**, 44–51.

Miller, B. P., C. Macrander and S. Sechrest (1986). A distributed programs monitor for berkeley unix. *Software–Practice and Experience* **16, No. 2, Feb. 1986**, 183–200.

Mink, A., R. Carpenter, G. Nacht and J. Roberts (1990). Multiprocessor performance-measurement instrumentation. *Computer* **23, No. 9, Sept. 1990**, 63–75.

Pereira, C. E. (1995). Temporal reasoning on object-oriented real-time specification by using constraint propagation techniques. *20th IFAC/IFIP Workshop on Real Time Programming* pp. 147–152.

Plattner, B. (1984). Real-time execution monitoring. *IEEE Trans. Software Eng.* **SE-10, No.6, Nov. 1984**, 756–764.

Raju, S. C. V., R. Rajkumar and F. Jahanian (1992). Timing constraints monitoring in distributed real-time systems. *Proc. Real-Time Systems Symp.* pp. 57–67.

TimeSys (1998). *TimeWiz–User's Manual*. TimeSys Corporation.

Tokuda, H., M. Kotera and C. W. Mercer (1988). A real-time monitor for a distributed real-time operating system. *Proc. ACM Workshop Parallel and Distributed Debugging* pp. 68–77.

Tsai, J. J. and S. J. Yang (1995). *Monitoring and Debugging of Real-Time Systems*. IEEE Computer Society. Los Alamitos, EUA.

Tsai, J. J. P., K. Y. Fang and H. Y. Chen (1990). A noninvasive architecture to monitor real-time distributed systems. *Computer* **23, No. 3, Mar. 1990**, 11–23.

Tsai, J. J. P., Y. Bi, S. J. H. Yang and R. Smith (1996). *Distributed Real-Time Systems: Monitoring, Visualization, Debugging and Analysis*. Wiley-Interscience.

TIMING CONSTRAINTS AND OBJECT-ORIENTED DESIGN

Mark Hermeling, Onno van Roosmalen

Department of Computing Science
Eindhoven University of Technology
P.O. Box 513, 5600 MB Eindhoven, The Netherlands

and

Bran Selic

ObjecTime Ltd.
340 March Road, Kanata, Ontario, Canada K2K 2E4

Abstract: The construction of computer systems that have to comply to certain strict temporal constraints has always been a difficult task. If one considers present design methods and tools, they do not offer systematic techniques to deal with time and timing constraints during the development life cycle. Although timing constraints are identifiable in the requirements phase, there is no systematic way to have them guide design decisions. Usually, satisfaction of timing constraints is verified close to the final stages of software development and during many projects constraints are not met at first by orders of magnitude. This leads to redesign cycles that span a large part of the development trajectory. In this paper we outline a framework to deal with timing constraints and performance guarantees that may remedy such shortcomings. The framework can be incorporated in UML-RT. *Copyright © 1999 IFAC*

Keywords: Object Oriented Design, Real-Time Constraints, Reactive Behavior, UML-RT

1. INTRODUCTION

This paper reports on work in progress that is a logical extension of research conducted at Eindhoven University on timing constraints in object-oriented programming languages and developments taking place at ObjecTime Limited on design languages and tools for real-time systems in the realm of UML-RT.

1.1 Timing constraints in programming languages

A programming language extension has been proposed in (Hooman and van Roosmalen, 1998; Hooman and van Roosmalen, 1997b; Hooman and van Roosmalen, 1997a) that enables the programmer to write platform-independent real-time programs. The major characteristics of this approach are:

(1) Timing constraints can be specified in programs.
(2) Because constraints are specified (not implemented) the resulting program components are strictly compositional, i.e. can be combined on the basis of their interface specification. Compositionality is a requirement for the modular, e.g. object oriented, construction of programs.
(3) The correctness of such programs against the specification, including the timing requirements, can be established independently of a platform that is used for their execution.
(4) The method enables the parameterization of timing aspects of components in a "minimal" way, that is, the programmer is not forced to introduce constraints that are not strictly necessary to satisfy the specification. This

means that only the end-to-end constraints that follow from the specification need to be included in the program and it is not necessary to divide available time to a dead-line over the various program components that are invoked to compute a response. Hence design decisions that are either arbitrary or inspired by platform consideration are unnecessary.

(5) The programming phase is followed by an implementation-generation phase where all the platform dependencies are taken into account. It usually involves scheduling of actions in a way that satisfies the constraints as fixed in the program. The difference with a normal compilation phase of non-real-time programs is that it may end unsuccessfully. Avoiding unnecessary constraints leaves more freedom to the scheduler to find a feasible schedule.

1.2 Design of object oriented real-time systems

ROOM as described by Selic, Gullekson and Ward in (Selic *et al.*, 1994), is a visual modeling language with precise semantics. It enables the designer to model the system structure using structure charts and the behavior of the individual structural components using so called ROOMcharts. The ObjecTime Developer toolset developed by ObjecTime Limited, is the visual modeling tool based on ROOM. It is specifically tailored for specifying, visualizing, documenting and automating the construction of complex, event-driven, real-time systems. ObjecTime Developer can be used to design real-time systems starting at a high abstraction level all the way down to the low level implementation of the system. Although ROOM provides the concept of a timing service that enables a system to react to the passage of time during the execution, it has no facilities to formally specify temporal constraints.

The Unified Modeling Language (UML) (*UML Semantics - Version 1.1*, 1997; *UML Notation Guide - Version 1.1*, 1997) is a general-purpose modeling language for specifying, visualizing, constructing and documenting the artifacts of software systems. UML has a strong set of general purpose modeling language concepts applicable across different domains. The UML is a widely accepted language and it is becoming the de-facto standard for object oriented modeling. UML for Real-Time (UML-RT) (Selic, 1998; Lyons, 1998), co-developed by ObjecTime and Rational Corporation, uses UML to express the original ROOM modeling concepts and their extensions. Models in UML-RT consist of a combination of structural and behavioral descriptions.

1.2.1. Structural Description in UML-RT In UML-RT, concurrent (active) objects are described by capsules. A capsule is an object that has its own thread of execution and that communicates with other capsules through one or more interfaces called ports. A port of one capsule can be connected to a port of another capsule by connectors. A connector is a communication channel that can carry messages between ports. A complex capsule may be decomposed into a structure of collaborating capsules (sub-capsules).

1.2.2. Behavioral Descriptions Behavior of individual capsules is captured using UML's hierarchical state machine formalism that is similar to ROOMcharts. The overall collaborative behavior of the collection of sub-capsules comprising the internal structure of composite capsules can be described in a number of ways in UML using Interaction Diagrams. ROOMcharts are based on statecharts as introduced by (Harel, 1987). Interaction Diagrams come in two forms that are practically equivalent, namely Sequence Diagrams and Collaboration Diagrams. Sequence Diagrams are inspired by Message Sequence Charts (ITU-T, 1996).

1.3 Problem Description

The goal of the work reported on in this paper is to apply the experience on the formulation of timing constraints in programming languages to UML-RT. The project focuses on the construction of an extension of UML-RT that enables designers to include temporal constraints in their design process from the start. The extension should enable a designer to express real-time constraints in a way that helps to guide design decisions and that makes the effect of timing requirements traceable throughout the development process.

Several extensions for dealing with time have been proposed for the mentioned behavioral modeling formalisms. Timed Statecharts (Kesten and Pnueli, 1988; Broy *et al.*, 1997) extend the regular statechart formalism and adds timing expressions to the evaluation of the guards for the transitions. This enables the expression of behavior that explicitly depends on the passage of time. In particular the amount of time that a system is in a particular state can be constrained in such formalisms. However, it is difficult to express constraints that span a number of states, for instance constraints on all the transition that a system makes to generate a response on a particular external stimulus. Message Sequence Charts (MSC's)(ITU-T, 1996) contain a mechanism to set and reset timers on elements of a model. This can be used to express the reaction of a system to the expiration of a timer. Although one can specify behavior depending on the progression of time it does not allow the formulation of constraints like deadlines on reactions. Additionally it is impossible to describe end-to-end timing behavior because MSC's do not allow timers to cross component boundaries.

The aim in the current project is to introduce explicit expressions of timing constraints as they follow from the requirement specification (i.e. end-to-end constraints) without necessitating explicit refinement into constraints on subcomponents. Constraint expressions will reveal the capsules that have an impact on particular constraints. This may help to guide design decisions on the logical decomposition of a system. An important consideration is to minimize the amount of computation that must be performed along the way to a tight deadline. Making decomposition decisions thus requires a quantification of the notion of "amount of computation" that is more precise than computational complexity. Also distribution of components and the resulting communication overhead must be considered in the light of timing constraints. This requires knowledge of available execution resources and their speed.

Two essential ingredients can therefore be distinguished: (1) End-to-end constraints must be an integral part of design models. (2) Tools must support performance estimations to enable direct inspection in how far the requirements are met by the possible implementations of the system components.

To summarize, the aims of this project are to formulate a framework that includes :

- a syntax to express timing constraints on UML-RT model elements,
- a semantics that helps to establish consistency and correctness of timing constraint expressions,
- guide lines for the refinement of models taking these constraints into account, and
- a verification formalism to establish the correctness of the implementation of design models, also with regard to the timing constraints

Although the project targets a practically implementable framework, it should be as general as possible.

2. TIMING REQUIREMENTS

A system that is designed and implemented using UML-RT concepts consists of a number of so called *capsules*. These capsules work together to realize the behavior as described in the requirements specification. In some intermediary design stage a decomposition of a capsule in smaller elements may be given that still need further refine-

ment. Because these elements are not finished capsules, but slots with a specification of functional behavior that has still to be implemented, UML-RT refers to them as *roles*. Roles are realized in a given implementation by capsules. In the rest of this paper these UML-RT terms, role and capsule are used. UML-RT facilitates role-based design. A system is specified using roles and the required behavior is expressed on these roles. In a later stage capsules implement these roles.

From the problem specification a number of so called *reactions* can be derived. A reaction is a sequence of actions that starts with a stimulus usually given by the environment and ends with a response of the system to that stimulus. A stimulus-response pair is called a reaction only if it may be subjected to timing constraint. Although it is possible to refine reaction constraints by carefully dividing them into constraints on sub-(re)actions, it is preferred, in line with the formalisms in (Hooman and van Roosmalen, 1998), to formulate constraints as end-to-end constraints on reactions that are implied directly by the top-level specification. Thus keeping maximal flexibility with respect to the moment at which constraints are refined (if at all). End-to-end constraints will in general affect a number of roles that collaborate to realize the proper reactions of the system. Therefore the Interaction Diagrams of UML which describe communication behavior of collaborating roles, are the most natural place to express them. A proper description of constraints in Interaction Diagrams must satisfy a number of requirements:

(1) A constraint implies a restriction on behavior and a proper expression of a constraint must be unambiguous with regard to the situations in which it applies. This is an issue because Interaction Diagrams usually describe possible behavior, not required behavior.
(2) If a set of collaborating objects must satisfy the constraints of several reactions, it must be possible to show that the combination of constraints is complete, consistent and in accordance with the top-level specification.
(3) It must be simple to check whether an implementation of a role can satisfy the constraints in the collaborations in which it takes part.

As mentioned before, there are two equivalent forms of Interaction Diagrams, Collaboration Diagrams and Sequence Diagrams. This paper focuses on the latter ones.

A reaction is highlighted within the sequence diagram. Examples are given in figure 1 concerning a pump in a mine which must be controlled depending on the water level and methane gas level in a mine: Rhighwater and Rmethane are reactions of role PumpControl on water_high and methane_high messages respec-

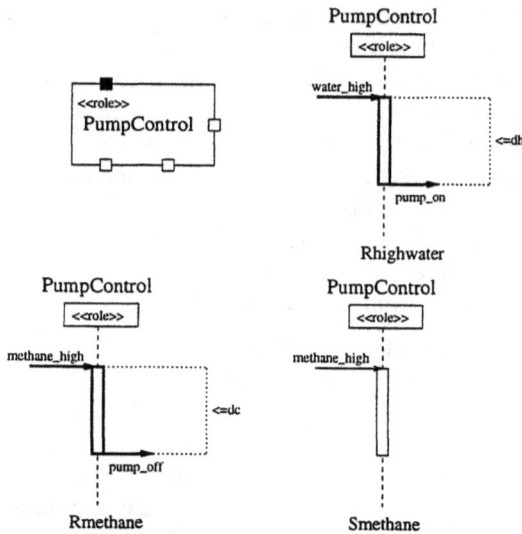

Figure 1. Two reactions and an action that is not a reaction of role PumpControl

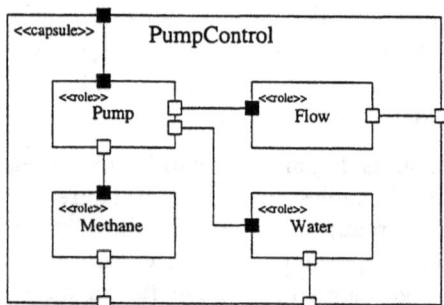

Figure 3. Decomposition of a reaction

noting alternative behavior). Note that a timing constraint only holds when a reaction takes place, i.e. when it is completed as a sequence of sub-reactions as in the example of figure 3. These sub-reactions may have alternatives yielding alternative ways to complete a reaction. The constraints must hold nevertheless. Keeping a sub-reaction highlighted also when it is occurring in isolation or when it is depicted in another scenario than the constrained reaction it belongs to, will remind the designer that it affects some non-functional requirements.

A reaction is a design-time concept, it describes a relationship between one incoming message, a computation and an outgoing message which together may be involved in a timing constraint. Some situations can not be described in this simple way. An example is figure 4 where a computation expects two incoming messages to generate a certain response. Issues such as this one are not addressed here but will be left for a future publication.

tively. A PumpControl capsule filling the pump-control role in figure 1 is decomposed into smaller roles as in figure 2. In such a decomposition the reaction Rmethane is composed of reactions of sub-roles, e.g. Rmethigh and Rpump_off. This is shown in figure 3. Algebraically, such a reaction can be obtained from the parallel composition (indicated by ||) of the reactions of the sub-roles (the notation of (Mauw, 1996) is used here):

Rmethane $= \lambda$ (Rmet_high || Rpump_off)

$= $ Rmet_high . Rpump_off

The operator λ removes terms in which a particular message is received before it is sent, the operator . denotes sequential composition. Note that instead of completing a reaction a system may also decide on the basis of some internal state of a role that a response is not required (e.g. the pump-role has already been notified about the high methane level, or the pump is already off). This means that somewhere in the computation the reaction is not continued. In that case, some sub-role carries out some other action instead of a reaction. An example is the action Smethane in figure 1 (The symbol "S", for "stop", is used just to distinguish the action from the reaction "R"). The behavior of this role may than be described by the expression Rmethane + Smethane (+ de-

3. PERFORMANCE GUARANTEES

In the previous section it was described how timing requirements are expressed as constraints on a design model. However, constraints on collaborations of roles require sufficient performance of the capsules that play a role in the actual system. Reactions of the system are realized by a concatenation of reactions of capsules. For verification purposes the concept of performance guarantees

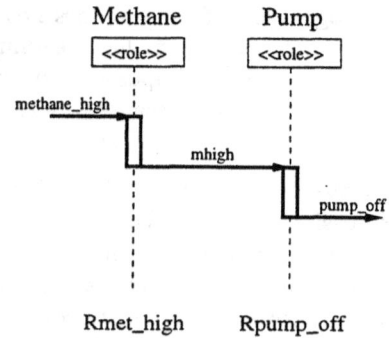

Figure 2. Decomposition of a capsule

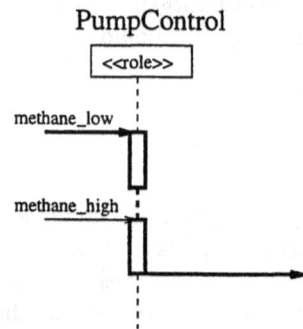

Figure 4. Behavior that can not be captured in a reaction

on reactions of capsules are introduced. It is an estimation of the time needed to execute the reaction, based on the current state of refinement of the capsule in the design. An indication of the timeliness of a model can be obtained by comparing the timing constraints on the reactions of the system with the combined performance guarantees of the involved capsules.

The basic method for computing performance guarantees is as follows. UML-RT has two type of capsules, simple capsules and composed capsules. A simple capsule is a capsule with a state machine that defines its behavior. The performance guarantees of the reactions of a simple capsule can be computed by analyzing this state machine. A composed capsule is a capsule that is composed of one or more smaller capsules. The behavior of the composed capsule is defined by the combined behavior of the smaller capsules. Thus the performance guarantees of the reactions of a composed capsule can be computed by analyzing the smaller capsules that it is composed of.

Three major problems can be identified that make performance estimations difficult, particularly in early phases of a design. (1) Not all roles involved in a reaction are sufficiently refined into capsules. (2) Capsules may be shared by more than one reaction leading to the blocking of progress of one or more reactions. Blocking times must be estimated to obtain a reliable performance guarantee. However, details of the way capsules are shared becomes only apparent when the design has progressed sufficiently. Nevertheless, in any stage of the design, some estimation is better than none, particularly if the designer is aware that certain effects are not taken into account. (3) Parts of the roles in a design may be filled with third party implementations, of which sources are not available for analysis.

There are some partial solutions to each of these problems. For example, the designer can address problem (1) by performing a depth first refinement of crucial reactions of the system. (Note that in the current approach he will be helped by the fact that the reactions and their timing constraints are explicitly present in the design.) An example of a solution to problem (2) is the refined resource analysis in rate monotonic analysis. An example solution to (3) is performance testing of proprietary components.

A characteristic of most such solutions is that they involve constantly evolving techniques, that they depend very much on a particular application area and that the designer is confronted with many details of the implementation of his design models. In other words, from the point of view of the designer there is very little abstraction from the details of the analysis techniques he should

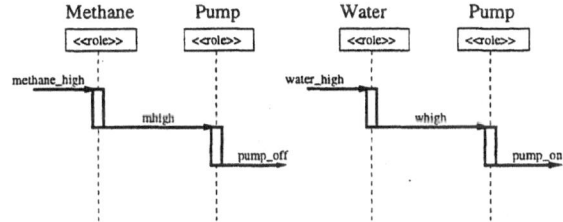

Figure 5. Two interfering reactions

employ as well as the implementation details of his design models. To remedy this, we introduce the concept of a design-time interface by which capsules can provide performance information depending on the role they fill in the design as well as the stage of refinement of the design. The term design-time interface is introduced to stress its use during design. The interface is invoked to obtain information relevant to the designer during the design phase. This concept is not entirely new, it is similar to the reflection API in Java.

For the estimation of the timeliness of a model under design it is often sufficient to compute the performance guarantees for reactions in isolation, i.e. without taking into account effects of resource blocking, and to compare them directly to the timing constraints. The computation of the performance guarantees of a reaction of a simple capsules is straightforward. The computation of the performance guarantee of a reaction of a composed capsule can be done by adding the performance guarantees of the sub-reactions it involves. Using the concept of the design-time interface, it is the responsibility of the design representation of the capsule to compute that performance. In case of a simple capsule this mean its state machine must be analyzed, in case of a composed capsule the design interface pertaining to reactions of capsules it is composed of must be invoked and the results combined.

For the verification of the timeliness of a complete model more is needed than the liberal estimation that is sufficient for a model under design. The system receives stimuli from the environment which start reactions. Hence, the behavior of the environment determines the load on the system and its ability to perform reactions. This behavior must be sufficiently known. If reactions do not interact, i.e. use the same capsules, there is no problem. However, most often the reactions do share at least the platform (which can be also be seen as a capsule). Concurrent requests by one or more reactions on the same capsule will lead to blocking and will necessarily reduce the performance for each of these reactions. Consider the example in figure 5. Both roles Methane and Water use role Pump. Thus, if the environment sends signals to the system that are handled by Methane and Water, the reaction of the system to either of these signals will use role Pump. Using the design interface of capsule Pump, capsules that implement Methane and Water roles can register with the capsule that implements Pump the details of their use of Pump. The design representation of the capsules collaborate in computing a more appropriate performance guarantee.

An advantage of a design-time interface over and above the abstraction from the performance estimation process comes with the use of proprietary software modules. Such modules can often not be easily analyzed on their behavior by third parties that have no access to implementation information. The provision of a design-time interface with the module that is implemented to provide the required performance information can protect the stakes of the supplier of the module while appropriately serving the user.

4. CONCLUSIONS

The most important aspects of the work described in this paper can be summarized as follows.

(1) It is proposed to formulate timing constraints in design models as following from the top-level specification as much as possible.

(2) End-to-end timing constraints are expressed on reactions that span a number of capsules. If timing constraints exist that share involvement of capsules it is not a trivial matter to verify a graphical formulation against the top-level specification. A precise semantics for constrained reactions is required. The formulation of such semantics is part of the current research.

(3) A simple method for estimating the performance that capsules can guarantee is described. It is based on the amount of effort needed to execute a reaction in isolation. Computing such performance guarantees is simple and often adequate in early stages of a design.

(4) It is proposed to offer a design-time interface on capsules that provide the designer with abstraction from the particular mechanisms that are used to compute performance guarantees. Proprietary software components may offer such an interface to enable performance estimations and, at the same time, keep their implementation details hidden. Standardization of such design interface is needed. A design-time interface also helps to make estimations at different levels of accuracy depending on the stage of the design process.

Current activities are aimed at researching remaining open questions and implementing a first version of the framework into the ObjecTime Developer toolset.

5. REFERENCES

Broy, M., R. Grosu and C. Klein (1997). Reconciling Real-Time with Asynchronous Message Passing. *Proceedings of the 4th International Symposium of Formal Methods (FME '97, Graz, Austria), LNCS* **1313**, Springer.

Harel, D. (1987). Statecharts: A visual formalism for complex systems. *Science of Computer Programming* **8**, pp. 231–274.

Hooman, Jozef and Onno van Roosmalen (1997*a*). Platform independent verification of real-time programs. *Joint Workshop on Parallel and Distributed Real-Time Systems (WP-DRTS '97, Geneva, Switzerland, April 1997)*, pp. 1–10.

Hooman, Jozef and Onno van Roosmalen (1997*b*). Timed-event abstraction and timing constraints in distributed real-time programming. *Proceedings of the Third International Workshop on Object-oriented Real-time Dependable Systems (WORDS '97)*, pp. 153–160.

Hooman, Jozef and Onno van Roosmalen (1998). A programming-language extension for distributed real-time systems. *Accepted by Journal of Real-Time Systems (February 1999)*.

ITU-T (1996). Recommendation Z.120 - Message Sequence Chart (MSC). *ITU Telecommunication Standardization Sector (Geneva)*.

Kesten, Y. and A. Pnueli (1988). Timed and Hybrid Statecharts and their Textual Representation. *LNCS* **571**, Springer, pp. 591-620.

Lyons, A. (1998). UML for Real-Time Overview. *http://www.ObjecTime.com/otl/technical*, ObjecTime Limited.

Mauw, Sjouke (1996). The formalization of Message Sequence Charts. *Computer Networks and ISDN Systems* **28(12)** pp. 1643-1657 .

Selic, B., G. Gullekson and P.T. Ward (1994). *Real-Time Object-Oriented Modeling*, John Wiley and Sons.

Selic, B. and J. Rumbaugh (1998). Using UML for Modeling Complex Real-Time Systems. *http://www.ObjecTime.com/otl/technical*, ObjecTime Limited.

UML Semantics - Version 1.1 (1997), The Object Management Group, doc. no. ad/97-08-04.

UML Notation Guide - Version 1.1 (1997), The Object Management Group, doc. no. ad/97-08-05.

An Object-Oriented Approach to Task-Graph Representation

Ami Silberman*, Karthik Sundaram**, Alexander Stoyen***

* Real-Time Computing Laboratory,New Jersey Institute of Technology
silber@homer.njit.edu
**Computer Associates
karthik@homer.njit.edu
Center for Management of Information Technology and Department of Computer Science,
College of Information Science & Technology, University of Nebraska at Omaha
alex_Stoyen@unomaha.edu

Abstract: Task graphs have been used for the specification and analysis of real-time systems for some time. We are currently developing a mobile code management system which uses a task-graph representation for specification, analysis, scheduling, dispatch, and resource allocation decisions. In this paper, we describe our limited hierarchical task-graph model, paying particular attention to its object-oriented representation in the Java language. *Copyright © 1999 IFAC*

Keywords: real-time systems, real-time tasks, software specification, data flow analysis, data-flow

1. INTRODUCTION

Task-graph based representation techniques provide an approach to describing and building complex systems that is suitable for all phases of a software life-style, from specification to analysis of implemented systems. They can be used as the basis for CASE tools, analysis engines, and scheduling tools They are well suited for a distributed environment, since the computation nodes need only be aware of those portions of the task-graph relevant to software modules executing on that node.

ITGS, or the *Integrated Task-Graph System* is an object-oriented cradle-to-grave approach to the specification and description of components, or code modules, for complex systems, in particular for mobile code. A method of an object or a class may be described as a "task", and dependencies between it and other methods may be represented by "edges" between their respective tasks. A collection of such tasks and edges forms a "task graph", which describes part or all of a system. We will include notions of code mobility in ITGS, in order to utilize the powerful task-graph analysis in the context of mobile code management.

Section two of this paper briefly discusses some related work as regards both task graphs in general

and ITGS in particular. Section three discusses the our version of the task-graph model. Section four discusses our implementation representation of task-graphs, Section five briefly discusses the uses to which task-graphs will be put in ITGS, and Section six contains concluding remarks, the current state of implementation of ITGS and our plans for the future.

2. RELATED WORK

A small sampling of works that discuss task graphs (or related large grain data-flow graphs) per se are: (Liu et al. 1993, 1996, Gupta and Spezialetti 1996, Mok and Sutanthayibul 1985). Two early systems that have used task-graphs for specification and analysis are MAX (Rasmussen et al. 1987) and ATAMM (Stoughton and Mielke 1988, Muntz and Lichota 1991) (Many other works that use task graphs for various forms of static analysis.) The task-graph model is related to the macro (large-grain) data-flow model of execution introduced for Mentat (Grimshaw 1988) and adapted to real-time in RTM (Silberman 1997a). The two models have been combined in the ITGS; by allowing the task-graph to be elaborated and altered dynamically, it has subsumed the role of the data-flow graph. Task graphs have been used for some time for system specification and analysis (PERTS, ObjecTime's tools and others), and as an aid in code generation (ObjecTime's tools etc.) Task graphs are capable of a richer general modeling than is capable with modal logic, timed Petri nets etc. For additional information on PERTS, see (Liu et al. 1993, 1996), for its simulation engine DRTSS see (Storch and Liu 1996, Storch 1996) and, for a critique of the commercial version, (Williams 1996). For additional information on ObjecTime's tools, see (ObjectTime 1999), and for a recent review, see (Smith 1997). Note that we make no attempt to be exhaustive here; there are many of papers that use task-graphs for analysis, scheduling, or to specify some limited aspects of a systems behavior. The above works, however, concentrate on the use of task-graphs as the primary means of system specification and as an aid in system development, and hence are of greater immediate relevance to ITGS.

We believe that our approach has the following three merits: First, unlike ROOM/UML(ObjectTime 1999), or CORBA we will provide support for mobile code and process migration, and provide an integrated run-time environment with our specification and synthesis system. Second, use of task-graph extraction algorithms, first presented in (Silberman and Marlowe 1996), we will be able to incorporate legacy code reuse. Third, we are using ITGS to examine issues involved in mobile code management, and believe that our model is both flexible and simple enough to be suitable for such research in an academic setting.

For a more complete discussion of RTM, the immediate predecessor to the ITGS model, see (Silberman and Marlowe 1996, Silberman 1997a, Silberman 1997b). (Silberman and Stoyen 1999) introduces ITGS and discusses its use in mobile-code management, (Silberman et al. 1999) contains a detailed discussion of how ITGS task-graphs can be used to model complex object behavior. Finally, (Stoyen and Petrov 1999) discuss the mobile code management system that ITGS is being designed to support.

3. THE TASK GRAPH MODEL

3.1 *Introduction*

Our task-graph model is essentially a large-grain data flow model, in which the term task refers a unit of work, as opposed to a process model, where task refers to a (potentially) multi-threaded concurrent object. See (Han et al. 1995), (Gillies and Liu 1995), and (Puschner and Schedl 1997), for the use of similar graphs for scheduling analysis.

3.1.1 *Basic Model.* A task-graph is a set of nodes (tasks) and a set of directed edges (dependencies.) A task represents a basic unit of work, but the exact granularity of this "unit" depends upon the problem domain and the use to which the graph will be put. Traditionally, a task has been defined as "the smallest unit of independently schedulable work", but as we will see this definition is inadequate. For now, let us just consider that the granularity of a task lies somewhere between a function and a portion of straight-line code within a function. Tasks may represent either functions belonging to classes or to objects, depending upon context In addition to representing application software, tasks may represent hardware operations, external events, or even system calls. A given task may occur multiple times within a graph. The occurrences, or *task instances*, may represent different invocations or executions of the same piece of code (for example, the same member function of an object), or of functionally equivalent code (for example, the same member function of different objects belonging to the same class.) Tasks are defined in terms of their attributes, such as *temporal properties* (maximum and minimum execution times), how preemptible they are, whether their execution is *optional*, whether they are *periodic* (i.e., iterated with a particular period) etc. (Note that entire portions of the task graph may be optional or periodic. We call such portions an optional or periodic *block* respectively.) A task may also exist in different *versions*, typically the different versions trade-off accuracy for resource usage (including execution time); the choice as to which version to use is made prior to execution of the task. In our system, we allow tasks (as well as edges and resources, discussed below) to have user-defined, domain specific attributes, such as fault rates, security levels etc.

3.1.2 *Dependencies. Edges* represent data, control, and temporal dependencies between tasks. An edge runs between its *source* to its *destination*. Data edges essentially represent message passing, whether explicit or encapsulated as function calls, RPCs etc. Typical edge parameters are: data type, data volume, reliability, and minimum (or maximum) temporal separation. A task has both in-edges, which connect a task's immediate predecessors to the task, and out-edges which connect from a task to its immediate successors. Some in-edges represent dependencies that must be fulfilled before a task may begin execution (initial dependencies) (e.g., parameters passed to a function), likewise some out-edges represent dependencies that are only fulfilled when a task completes (terminal dependencies) (e.g. return values from a function.) Other dependencies may be fulfilled partway through a task's execution, these represent function calls and returns, or messages sent or received during a task's execution. A simple task is one that has at most initial and terminal dependencies.

We will sometimes ignore edges representing dependencies during a task's execution. In particular, we will do so to limit the depth of recursion, or to suppress non-essential detail when representing or analyzing the internal detail of another task. (If we are interested in the details of task *A*, which has a dependency on task *B*, we may ignore edges incident on *B* but not incident on *A*.

A task's *in-type* defines how many of the input edges to a task must have their dependencies satisfied before a task can proceed. Basic in-types are A - meaning all, and O - meaning or. (I.e., one edge must represent a satisfied dependency.) The default is A. A task's *out-type* defines which output edges represent actual dependencies, i.e., which direct successor tasks will receive data, be expected to execute etc. Basic out-types used are A - meaning all, O - meaning or, where the system determines the successor (or successors, an OR-type may be given as *n* out of *m*), and C - meaning conditional, where the successor is dependent upon the results of the task's execution. In practice, a given in- or out- type will be relevant to only a portion of a tasks inputs or outputs. For example, a task might have an in-type of O for its initial input (it may receive data from one of a set of possible predecessors) , and an in-type of A for a subsequent input (an RPC result.) Note that a simple task needs only a single input and a single output interface, and their in-type and out-type respectively are usually capable of being described quite simply. We can use out-types of O to represent optional branches, and out-types of C to produce periodic blocks. Note that we may use "dummy" tasks, here represented as solid dots, to make the graph topology more tractable.

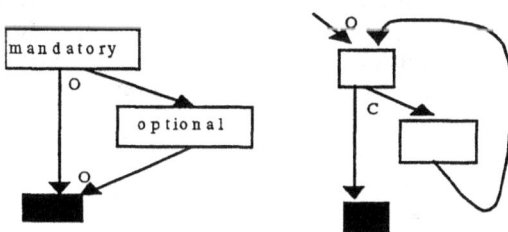

Figure 1: Optional Branches and Period Blocks.

A *resource* represents some system resource, a processor, I/O channel, memory, secondary storage device etc.

3.1.3 *Graphs.*
A given system may have multiple equivalent task-graph representations due to loop unrollings, the duplication of iterated tasks etc. For example, a task representing a commonly invoked function may be duplicated throughout the graph so as to avoid extremely complicated dependency relations. Our approach is to favor simplicity of dependencies at the expense of task duplication. A portion of a task graph is called a task sub-graph, or sub-graph for short. In many systems, an overall task

graph may be difficult or impossible to construct, but portions of the graph, representing subsystems, may be feasible. This is very applicable to mobile code, since the run-time system on a computation node may be given just those portions of the overall task graph that directly concern software modules that are executed on that mode. Mobile code modules will include sub-graphs representing their execution. Finally, a task graph may change dynamically at runtime as tasks migrate, are reconfigured, and have alternative versions selected.

3.2 *Hierarchical Task Graphs*

3.2.1 *Motivation.* The approach outlined above has several serious deficiencies. First, when describing systems based on traditional models of programming (call-and-return semantics), we are faced with a choice of either a granularity at which tasks represent functions and have complicated dependency relationships, or at which they represent portions of functions, in which the dependency relationships are simple, but the containment relationship, (i.e., which task's represent part of the same invocation of a given function) is obscured. Tasks which are part of the same function invocation should be co-located, and data dependencies among them just represent shared state. Secondly, important internal details may be hidden at a large granularity. For example, timing of resource usage, which is necessary for a priority ceiling protocol, or the fact that part of a task is either iterated or optional. Third, there is no natural method for modeling persistent state across multiple heterogeneous tasks (representing methods of the same object, for example.) other than defining a resource type to represent the state. Our solution to this problem is to allow for a limited hierarchical decomposition. Unlimited decomposition and re-composition can result in chimerical tasks containing portions of code from multiple methods and even objects.

3.2.2 *Task Sets and sub-tasks.* A *task set* represents a group of related tasks that share a common state. Typically, it represents either an object or a class, depending upon the role of the task graph in question.

A task may be broken down into *sub-tasks.* Normally, the division is made at points of externally visible behavior. These include preemption points, function calls and returns (or message transmission and reception), changes in resource usage, start or completion of optional or periodic blocks, or of conditional or iterative constructs. Short segments of straight-line code may be ignored, or adjacent behaviors may be "collapsed" in order to avoid proliferation of sub-tasks. See (Silberman and Marlowe 1996) for how this is handled in RTM. A sub-task has two broad categories of dependencies. An *internal dependency* is a dependency on another sub-task within the same task. An *external dependency* is a dependency upon a sub-task belonging to another task. In general, internal dependencies are fixed (i.e., the internal structure of a task is fixed, modulo branching, loops and optional portions.) External dependencies are frequently not fixed, and are context dependent. (For example, a task which represents an encryption function occurs multiple times within a graph but has different external dependencies for each task instance.)

A sub-task is a simple task, with two (possibly null) interfaces. The *input interface* serves as the destination of the edges representing all external dependencies (messages, parameters, temporal dependencies etc.) that must be fulfilled before the sub-task may be executed. The *output interface* serves as the source for the edges representing all external dependencies that the execution of the sub-task fulfills. The interfaces themselves are context-independent, it is the edges that connect them with the interfaces on other tasks that provide the context. Interfaces may also include information as to data-types sent or received, default values (either fixed, or "virtual inputs", i.e., copies of the last data received), and in- and out- types with regard to external dependencies, etc. The in- and out- types of a sub-task are only concerned with internal dependencies; by default a sub-task is always dependent upon its input interface and a sub-task's output interface is always dependent on the sub-task.

Figure 2 depicts asynchronous call-and-return semantics. Boxes labeled **I** represent the interfaces of the calling task. For simplicity, we omit details of the called task, and the tasks predecessor and successor.

4. OBJECT-ORIENTED REPRESENTATION

4.1 *Introduction*

We represent task graphs and their components using an object-oriented methodology in JAVA. JAVA offers an attractive object oriented programming paradigm and above all platform independence, which is of paramount importance for our plans to use ITGS to support mobile code.(Silberman and Stoyen 1999).

Task graphs can be read from a plain ASCII configuration file or the parameters of the tasks can be entered using a graphical tsak-graph editor, which is currently in its final stages of developments. ITGS provides two representations, one as a set of dynamic data objects, the other as flat file representation, and also supplies utility functions to translate between the two representations.

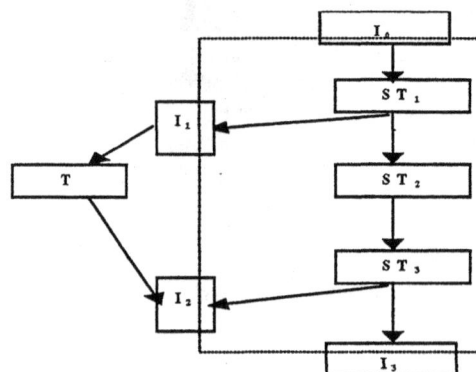

Figure 2: Asynchronous Call-and-Return

4.2 *System Wide Objects.*

4.2.1 *The Complex_System object.* This object forms the abstract representation of the overall complex computer system being specified. It consists of objects representing tasks, edges and resources. *Sub_tasks* are derived (extended in JAVA parlance) from the *task* object. Edges connect one task to another. To provide for a more elegant representation and to isolate the message-passing interface, we require that sub-tasks send and receive external messages only through well-defined interfaces. Each sub_task has at most a single interface for receiving messages from other tasks; likewise it has at most one interface for sending such messages.

4.2.2 *The Resource Object* The *resource* object represents the abstraction of a hardware or a software system resource. Tasks need resources to execute and resources must be allocated to tasks before they begin to execute. Each task will have a reference to each resource it would require. This can be know apriori or at run-time. Important parameters regarding a resource are its acquisition time, de-acquisition time and its context switch time. The acquisition time is the time taken for the resource to move from an idle state to a usable state. De-acquisition time is the time taken by for putting the resource back to an idle state. Another important parameter of resources is preemptibility.. The consumability of a resource indicates, whether a resource is consumed when used or not

4.3 *The Task Object*

The JAVA class *Task* encapsulates the task object, with its variables and methods. The variables include general task information such as *task_id*, *version_number* and *task_name*. Scheduling information includes *start_time*, *ready_time*, *end_time*, *deadline* (absolute or relative). Each of the temporal variables has a *specified* and an *effective* value. (For example, the *effective deadline* is the actual calculated deadline for the current execution of a task.) Task properties such as *pre-emptible*, *optional*, *periodic* are represented as Boolean values. Each task will have a reference to a list of the resources which it will require for execution. This *resource_list* may be stored as a static array if all the resources are known apriori, otherwise resource list can dynamically change its size. Each object in this list will be a reference to the actual resource object. The *vector* object of JAVA is used for this purpose. In a similar manner, an *edge_list* is also maintained on a per task basis. Edges in this list may be determined dynamically.

A *task* object may represent either a method belonging to a class, to a method belonging to an arbitrary member of a specific class, or to a method belonging to a specific member of a class. *Sub_tasks* are derived from *tasks*, and have essentially the same variables. Task objects representing different instances of a task (such as a particular method invoked on different class instantiations) can share the same representation of their contained sub-tasks, since it is only the edges between two tasks that actually differ.

A task may be *pre-emptible* or non-preemptible. A pre-emptible task may be stopped and restarted at a later time. We make further refinements to this model by allowing a task to be preempted only at certain pre-defined points called *preemption points*. A non-preemptible sub-task that is marked as being a preemption point may be preempted just prior, but not during the sub-task's execution.

4.4 *Blocks*

4.4.1 *Optional blocks* Optional blocks are represented as branch with an out-type 'O'. This makes the internal handling of optional blocks simpler since they can now be treated as *or* branches. Each task (and sub-task, by inheritance) will have a Boolean value, which indicates if this (sub-)task is optional or not, as well as if it as an optional block source or a sink. Each optional block will have a unique source and sink. Other optional (sub-)task within the block will be marked as optional but not as a source or a sink. This concept of sinks is fixed into the optional block design. Even if an optional block does not have a single sink, we provide a dummy sub-task to serve as a sink.

4.4.2 *Periodic Blocks* Periodic blocks and their contained tasks are marked in a manner analogous to the above. A periodic block may contain an optional component, hence for a given task or sub-task both flags may be set. The phase of a periodic task denotes the time after which the periodic block is

ready to start to execute. Thus, a phase of n means that, the periodic block's first iteration begins executing n units of time after the task's ready time, and periodically after that. The variation of a task indicates the difference in times between successive iterations. Default values for both variation and phase is zero.

4.5 *Edges and Interfaces.*

Internal edges start and terminate at subtasks belonging to the same task. They are stored as part of the internal representation that is shared by all tasks representing the same method. External edges start and terminate at different tasks. Their source is a particular interface belonging to a sub-task of the originating task, likewise their destination is an interface belonging to a sub-task of the terminating task. Different tasks representing the same method will have different sets of external edges. External edges are represented as a tuple consisting of the edges *source* and *destination* (given as a pair (*interface_number, task_id*), where the task referred to may represent either a class or an object method), the edges *data_volume*, its *minimum* and/or *maximum temporal separation*, and what sort of dependency (data, temporal, and/or control) it represents.

Every sub-task has two *interface objects*, an *input interface* and an *output interface*. These objects allow the sub-task to communicate with other objects and task instances and form a proxies between the sub-task and the outside world. A null input or output interface represents a sub-task with no external inputs or outputs respectively.

4.6 *Task Graphs*

A system as a whole is represented by a collection of task graphs. If the data and control flow between objects is not known at run-time, then a complete graph is impossible. Take, for instance, a system that includes several instances each of a server and a client class, and allows for dynamic instantiation. We will not know a priori which servers are dependent upon which clients, but we are capable of describing the internal structure of the clients' and servers' methods, and the data and control dependencies involved in performing requests.

Since methods invoke other methods, and may even be recursively invoked, we may also limit a task-graph by limiting the amount of recursion depicted. For example, suppose method *A* invokes methods *B* and *C*, method *B* also invokes method *C*, and method *C* may invoke itself recursively. We can cut off the recursion at any level by only considering the initial and terminal dependencies of C and elaborate the graph only as necessary.

5. OTHER TOOLS

The task graph model, as we envisage it, would require a set of tools that would enable easy use of our model. This would be in the form of visual editors and CASE tools for automatic code/template generation and task editing (Silberman and Stoyen 1999, Silberman et al. 1999). We a *task graph editor*, which would enable the user to build task graphs, and a *task graph editor configuration tool* to allow the user to customize the editor for specific domains. A *task graph analyzer* would be used to analyze the task graph for inconsistencies, error and aid in debugging. CASE tools would generate code 'wrappers' and templates, based on the task's internal (sub-tasks) structure and interfaces. These tools would be capable of propagating changes made in the code back to the task-graph description. (For example, adding a function call to a method would result in an additional sub-task and pair of interface objects.) Finally, a *task-graph extraction tool* would be capable of extracting a task-graph description of a system from existing code.

6. CONCLUSION AND FUTURE WORK

The ITGS model provides a rich and comprehensive model for representing a variety of complex, distributed and real time systems. The representation method is robust enough to handle diverse applications and provide support for specification, analysis and run time implementation of complex systems. At the current time, we are completing work on our graphical editor and a prototype of the configuration tool. We are in the initial design stages for our suite of CASE and analysis tools.

REFERENCES

Gillies, D. and Liu, J. W. S.. "Scheduling tasks with AND/OR precedence constraints." *SIAM Journal on Computing*, 24(4):797-810, August 1995

Grimshaw, A., "Mentat: An Object-Oriented Macro Data Flow System," Ph.D. dissertation, University of Illinois, Urbana, Illinois, 1988.

Gupta, R., M. Spezialetti, "A Compact Task Graph Representation for Real-Time Scheduling," *Real-Time Systems*, 11(1): pp. 71-102, July 1996.

Han, C.-C., Lin, K.J., and Liu, J. W. S. "Scheduling jobs with temporal distance constraints." *SIAM Journal on Computing*, 24(5):1104-1121, October 1995.

Liu, J.W.S., C. L. Liu, Z. Deng, T. S. Tia, J. Sun, M. Storch, D. Hull, J. L. Redondo, R. Bettati, and A. Silberman, "PERTS: A Prototyping Environment for Real-Time Systems," *International Journal of Software Engineering and Knowledge Engineering*, pp 161-177, vol. 6, no. 2, 1996

Liu, J.W.S., J. L. Redondo, Z. Deng, T. S. Tia, R. Bettati, A. Silberman, M. Storch, R. Ha, W. K. Shih, "PERTS: A Prototyping Environment for Real-Time Systems," *Proc. 1993 Real-Time Systems Symposium*, 184-188, Dec. 1993

Mok, A.K., S. Sutanthavibul, "Modeling and Scheduling of Dataflow Real-Time Systems," *Proceedings of the 1985 Real-Time Systems Symposium*, pp. 178-187, Dec. 1985.

Muntz, A.H., R. W. Lichota, "A Requirements Specification Method for Adaptive Real-Time Systems," *Proceedings of the 1991 Real-Time Systems Symposium*, pp. 264-273, Dec. 1991.

ObjecTime website: http://www.objectime.on.ca

Puschner, P, and Schedl, A., "Computing Maximum Task Execution Times (A Graph-Based Approach)". *Real-Time Systems, Vol. 13, No. 1*, Kluwer, July 1997

Rasmussen, R.D., N. J. Dimopoulos, C. S. Bolotin, B. F. Lewis, R. M. Manning, "Max: Advanced General Purpose Real-Time Multicomputer for Space Applications," *Proceedings of the 1987 Real-Time Systems Symposium*, pp. 70-78, Dec. 1987.

Silberman, A and Stoyen, A., "Task Graphs for Mobile Code – An Introduction to ITGS", *Proc. Fourth International Workshop on Object-oriented Real-time Dependable Systems*, 1999

Silberman, A, Stoyen A., Sundaram, K, "The Use of Task-Graphs for Modeling Complex System Behavior", *2nd IEEE International Symposium on Object-oriented Real-time distributed Computing*, May, 1999, Saint Malo, France

Silberman, A., "Imprecise Computation in RTM", *Proceedings of the 1997 ACM SIGPLAN Workshop on Languages, Compilers, and Tools for Real-Time Systems*, pp. 82-91, Las Vegas, Nevada, July 1997

Silberman, A., RTM – Design and Implementation, Ph.D. thesis, University of Illinois at Urbana-Champaign, Jan. 1997. Available as Technical Report UIUCDCS-R-97-2040.

Silberman, A., Thomas J. Marlowe, "A Task Graph Model for Design and Implementation of Real-Time Systems," *Second IEEE International Conference on Engineering of Complex Computer Systems*, pp. 432-441, Montreal, Canada, October 1996.

Smith, K.W., "ObjecTime CASE Tool Simplifies Real-Time Software Development," *Dr. Dobb's Journal*, p. 64, Dec. 1997

Storch, M., J. W. S. Liu, "DRTSS: A Simulation Framework for Complex Real-Time Systems," *Proceedings 1996 Real-Time Technology and Applications*, pp. 160-169, 1996

Storch., M.F., A Framework for the Simulation of Complex Real-Time Systems. PhD thesis, University of Illinois at Urbana-Champaign, November 1996. Available as Technical Report UIUCDCS-R-96-1983.

Stoughton, J. W., and R. R. Mielke, "The ATAMM Procedure Model for Concurrent Processing of Large Grained Control and Signal-Processing Algorithms," *Proceedings of the IEEE 1988 National Aerospace and Electronics Conference*, vol. 1-4, 1988.

Stoyen, A. D., P. V. Petrov, "Towards a Mobile Code Management Environment for Complex, Real-Time, Distributed Systems," accepted subject to revision for *J. Real-Time Systems*

Williams, T., "Modeling Tool Turns Up Scheduling Problems in Real-Time Systems," *Computer Design*, v. 35, p. 40, July 1996.

CONCEPTS FROM DEADLINE NON-INTRUSIVE MONITORING

Matthew Harelick* Alexander Stoyen**

* Unipress Software Inc. Edison, NJ
** University of Nebraska at Omaha. Omaha, NE

Abstract: A real-time system must be monitored in a fashion that does not affect its deadlines. Current software monitors are intrusive on the observed process. The intrusive effects of software monitoring can be minimized by the use of idle time that can be identified by static analysis methods. A slightly intrusive software monitor that does not violate real-time deadlines can be effective at observing real-time software without modifying observed behavior. Copyright © 1999 IFAC

Keywords: Instrumentation, Non-intrusive, real-time monitoring

1. INTRODUCTION

Developers and systems users need to monitor a system to identify and repair potential problems. Examples of such problems include programming logic, performance analysis, and environmental interaction. The challenge in monitoring programs or any other natural or artificial process is to insure that the monitor does not affect the behavior of the observed process. This is known as the Heisenberg Principle (Rosenberg 1996). A monitor that does not violate the Heisenberg Principle is known as a non-intrusive monitor. Software engineering methods such as formal verification techniques have been used to insure the correctness of real-time code. Verification performed before run-time is not capable of detecting errors that occur at run-time because it may be impossible to verify certain properties. Timing constraints can be predicted and guaranteed statically, but the behavior of the program may still have logical flaws. A non-instrusive real-time instrumentation system should meet several goals. Task instrumentation can not cause the tasks to miss their deadlines. The monitoring system must allow observation requests while the program is in operation, or at least in a manner that does not require recompilation of the real-time system tasks. The monitor may need to perform rudimentary analysis on collected information before handing it off to a display routine.

1.1 A Solution for Real-Time Systems

An instrumenation system for real-time systems should be composed of hardware and software components. The hardware components are a combination of standard hardware instrumentation devices. Hardware monitoring, through devices such as memory or bus snoops are able to capture information is transmitted over the bus non-intrusively. The information must match a specified mask and usually needs analysis to determine the context of the information received. Software monitoring is able to capture information based on the scope of where the instrumentation is inserted into the program. For instance, software instrumentation inserted into a context switch will enable the capture of process state and registers of a given process. The context of the capture event lets the user know whether a process has started, ended or was interrupted and what its state was at the time. Since hardware probes are now standard components the construction of a non-intrusive software monitor need to be studied.

A software monitor can not behave non-intrusively because it uses the CPU that otherwise would be used by the observed process. Usually a process is observed by a monitor when the users / developers believe there is an error in the system implementation or design. During normal operation, there is no instrumentation used in the system. Instrumentation then will use up processing time that would be used by real-time tasks, changing the behavior of the system. Systems that are instrumented can have different behavior than systems than are not instrumented. A solution is to leave the instrumentation in the system whether under observation or normal operation. Besides the obvious advantage that an observed system has the same behavior as a non-observed system, another advantage is system observation can be initialized or modified during run-time. This philosophy is already in use in (Dodd and Ravishankar 1992) and (Tokuda *et al.* 1988). For real-time systems where too much instrumentation is included, both the observed and normal system will have the same behavior and both will miss their deadlines. Rather than allow instrumentation to be added at arbitrary locations and quantities, context switches and idle code identified by static analysis (Petrov and Stoyen 1997) will be instrumented.

2. SYSTEM MODEL AND ASSUMPTIONS

2.1 *Hardware*

Real-time systems have very strict constraints in order to maintain predictable behavior(Stoyenko and Halang 1991). The hardware is divided into a host and a target system. The Host system performs monitoring and acts as a user interface for the Target system. The Host system allows the user to enter commands that execute on the Target. Real-time behavior is not required for the Host system, however its interaction with the Target can not interfere with the real-time behavior of the Target. The connection between the Host and the Target is through some real-time communication hardware. The Target architecture is a distributed network of processors. A constraint on the processors is that their clocks are synchronized to the same time as described in (Stoyenko and Halang 1991) or (Bhatt *et al.* 1987). Any hardware optimizations that introduce unpredictable timing is absent from the architecture. This includes pipelines, instruction caches and memory caches. Each processor is assumed to have its own local primary memory. Virtual memory has the unpredictable side effect of page faults(Patterson and Hennesy 1994). Therefore each address passed by the processor to memory unit is an actual address in memory. The Target distributed network

is responsible for real-time guaranteed delivery of messages and data. Furthermore, monitoring data can not interfere with the normal flow of data for the system. This can be accomplished through the use of a network backbone dedicated for monitoring. The network and busses are required to be observable through techniques such as a memory bus snoop (Fryer 1973). Packets placed on the distributed network go through a demultiplexer which sends an identical message to the monitoring network to be processed by the memory snooper.

2.2 *Software and Language Issues*

There are three kinds of software to be considered. Host operating system software can be any multi-tasking operating system such as UNIX or VxWorks (Win 1995). The Host operating system is not required to be real-time, but communication between the target and the host must be nonintrusive on the target.

Target Operating System software must be realtime and must be able to facilitate hard real-time behavior. The real-time schedule will initially be static, leaving room for modification by system processes. The primary source for schedule modification will be generated by the central monitor as opportunities arise. The Target Operating System (hereafter abbreviated tOS) needs to be able to insert non-interruptible code dynamically into tasks located on each node.

Target software includes the tasks that codify real-time behavior. Target software is required to be constructed using a high level language with restrictions necessary for the construction of a predictable real-time system. Timing prediction for dynamic memory allocation is infeasible (Kligerman and Stoyenko 1986) and is disallowed. Recursive function calls are disallowed by the compiler because the schedulability analyzer will be unable to predict storage requirements of subprogram at compile time(Kligerman and Stoyenko 1986). Loop bounds must be known at compile time to determine the execution time of the enclosed loop body (Kligerman and Stoyenko 1986). Timing constraints must be expressed explicitly so that schedulability analysis can verify that the real-time tasks will meet the deadlines. There are several examples of languages with these requirements including CRL (Stoyenko *et al.* 1995) and RT-Euclid (Kligerman and Stoyenko 1986). These requirements allow for static verification of timing constraint satisfaction. The form of static verification used is Schedulability Analysis. The analysis uses components of the code to determine worst case running time. The analysis is exponential because conditionals with critical sections in one branch but not the other can have vastly different durations. Clustering Analysis balances the timing of the conditionals by inserting a delay consisting of idle code (no-ops) in the shorter branch (Stoyenko and Marlowe 1992). The inserted delay can be replaced by instrumentation(Petrov and Stoyen 1997).

The following constraints must be used for compiler output. The compilation process must include a front-end schedulability analyzer and provide the information discussed in the previous section. In addition, the compiler must provide us with a symbol table that maps symbols to run-time addresses. The symbol table can be based on standard formats such as the .stabs format used with gcc. Alternatively research compilers such as CRL (Stoyenko *et al.* 1995) can provide symbol tables. Data dependency graphs and control flow graphs generated by the compiler must be available or it must be possible to generate this information accurately from the object code. There will no restrictions on compiler optimizations on the code (other than the clustering algorithm discussed above) as long as these optimizations do not interfere with the deadlines of the processes and there is a known procedure of polynomial efficiency to determine the original form of the source code.

3. DEADLINE NON-INTRUSIVE MONITORING

Task information can be obtained non-intrusively from several sources discussed in the model. Modifications to variables can be determined by using the memory bus snooper. For any task in memory the symbol table will contain the names of the variables and the assigned address, allowing us to determine the change to the variable. The network monitor can provide information on external calls by providing the identifications of the calling process and calling processor, called task and the parameters of the task. Either of these techniques can transmit their information to the host monitor non-intrusively using the dedicated monitoring backplane. While the collection and delivery of this information via passive hardware monitoring is non-intrusive, non-filtered information is useless if incorrect information is snooped at the incorrect time. The filter can be modified using the monitor backplane without interfering with target processors or the target network. Software monitoring can access bundles of information based on contextual information. Software monitoring can for instance grab the parameters to the next subroutine call of the stack frame.

Previous works have included integrated software monitors by using software hooks such as writing information to monitor addresses or changing snoop filters (Haban and Wybranietz 1990). Software monitoring in these cases was intrusive because it delayed the process by a small amount. While these small delays for software monitoring are impossible to avoid whether they are included in the overhead of the context switch or by directly inserting instructions, they can be released at times when their execution does not affect the deadline of the program. Using the delays inserted by clustering analysis, the processor can be used for monitoring during otherwise idle time as described in (Petrov and Stoyen 1997). The location of these delays are recorded by the schedulability analysis performed as part of compilation. The use of delays is intrusive on behavior but not on the deadlines of the real-time program, hence the term Deadline Non-Intrusive Monitoring.

Instrumented Delays can provide important information and perform various tasks. If Delay Instrumentation executes, the branch of a condition being executed is identified immediately. This indicates that a specific critical section was not executed and in the case of linked conditionals, provides with information in advance as to which conditional branches in the very near future will be executed. During a Delay, software monitoring can make a more detail bus pattern selection than a remote monitor. The software monitor can take a snapshot of the state of the machine during the delay. Delay Instrumentation can send the value of variables with the scope of the function being executed. While this information is available to the snooper, the information extracted from the data stream will have to be interpreted while data read by the Delay Instrumentation can be interpreted within the context of the scope of the Instrumentation. Schedulability analysis will be extended by adding one other form of output: a table indicating the location and duration of inserted idle times to balance conditionals. The table will include a unique identifier for the critical section that was balanced. For each idle period an initial timing check will be added that checks the current time against the time that this delay period was expected to begin. It is not possible for the task to arrive at the delay segment late because the tasks were statically verified to meet the worst case timing constraints. If the delay segment is released on time then the instrumentation will processed completely. If there is extra time then the instrumentation will be processed and enough delays to make up for the extra time will be inserted at the end of the instrumentation.

Each context switch will be instrumented by insertion of an instruction at the beginning and the end that writes to a passive monitor register indicating the boundary of the information to or from memory containing process context for the task about to be released or that has just been terminated. The snooper will record the time that the new process is being released. While the two instructions will increase the time of a context switch, this will be included in the overhead calculations during schedulability analysis. During the initial phase of the Delay Instrumentation discussed above the initial release time stored in the bus monitor register will be subtracted from the current time. This value will be compared to the expected duration constant embedded in the Delay Instrumentation as discussed above.

3.1 *Role of the Central Host Monitor*

The Central Host Monitor (hereby abbreviated CHM) will act as interface between the user and passive monitoring devices. The CHM will transmit new or modified event masks to the local passive monitors and perform analysis that is too complex for the local monitors. Event masks specify the kind of memory access the local passive monitors should transmit back to the central host monitor. Any information that does not compare favorably to the event mask will be discarded. The monitor will keep track of the program state by using a copy of the segment trees generated by schedulability analysis. The segment trees will resemble the static schedule. The tree indicates the time that program elements will become available. The control graph will indicate that Process *A* is about to start at a context switch. Once it is started, the CHM transmits to each local monitor the address of the variables in the process that are of interest. The local monitor will then observe memory loads and saves until a variable request matching the address is discovered. The value of the address, the address, a timestamp, the processor number and the operation type (save or load) will be transmitted back to the CHM.

4. EXPERIMENTAL DESIGN

An experiment sufficient to test the capabilities of the system to perform deadline non-intrusive monitoring will have to demonstrate that tasks do not miss their deadlines and that information is retrieved on time and routed correctly to the monitor. The experiment will consist of a simulation of relevant components of a real-time system. The simulation will consist of several levels. At the lowest level will be the real-time system simulation. The next layer will be a simulation of monitoring subsystems including the passive monitor nodes and the dedicated monitoring network. Finally the central host monitor will be a working prototype that will perform updates using a graphical user interface. All internal simulation components will use the same clock object. A simulation of a real-time system will be sufficient to prove that the solution to the deadline non-instrusive monitoring problem. The simulation will be based on the model and assumptions. There are three levels to the simulation, the network level, the processor level, and the operating system level.

4.1 *Real-time system simulation*

Each processor will be simulated at the instruction level. A set of instructions consists of a set of integers representing each instruction or instruction arguments. Associated with each instruction is a measurement in units of the amount of time that each instruction takes. Each instruction is an uninterruptable, atomic instruction. The assumptions specified for the model will be used for the construction of the simulation. For example, any possible temporal inconsistency caused by the existence of a data cache is handled automatically by the operating system. This means that any memory transfer from primary or even virtual memory will take the same amount of time. Processor registers will be represented simply as an array of integers. The network interface will be implemented as an integer array. There may be several network interfaces depending on the processor and network specification. There will be a limited set of hardware interrupts. There will be no explicit simulation of memory. Information stored in a processor's memory will actually be stored in the simulation's memory. The difference in time between primary and secondary memory will not be simulated. All the processors in the simulation will use the same assembly language. Heterogeneous processors will be simulated by mapping a different set of timings for the assembly instructions on each processor. Interrupts will only occur between instructions. If an instruction can not be completed, the processor will wait for the remaining instruction time before executing another. While other systems specify that the arrival of network messages will cause an interrupt, in this case network messages arrive at a periodic rate, and if the processor does not read the information at a given time then the information will be written over by the next message that arrives.

Normally, the interface between a processor and the network is a bus interface. In this case there will be memory areas set aside for both the output and input interface between the processor and the network simulation. These memory areas will not have any form of synchronization or write protection. There are two networks contained in the simulation. The target network is a real-time token ring. A token ring is a reliable real-time network where a packet moves from one node to another in a predictable amount of time. When a packet arrives at a node, its information will be stored in a simulated network buffer and in a monitor in buffer indexed for the target source. Demultiplexer activity is simulated explicitly since its overhead is negligible. The packet will pick up any outgoing messages waiting for delivery and place them in a packet array indexed based on the number of processors available on the distributed system. Since the packet moves at a predictable rate there will be a period of time where the network buffer has not been written. During this time the operating system must transfer this information to a message queue.

A subset of modern operating system functionality will be simulated. The tasks of the Operating System simulation include context switch, execution of the static schedule, allow modification of the schedule from host requests, and manage the network queues. Each node on the distributed system will have its own copy of the operating system and the schedule. Operating System functions to manage temporal inconsistencies such as pipelines or caches will not be simulated.

4.2 Monitoring Hardware Subsystem Simulation

The components to be simulated include the passive monitor node, the monitoring network, and the demultiplexer. The demultiplexer is integrated with the real-time system. The demultiplexer will be simulated by simply copying a message destined for the bus to the passive monitor's input queue and the demultiplexer will be unavailable for a period of time to account for the gate manipulation of the transfer. The passive monitor is more complex than a bus snooper but is not a full coprocessor as described in Gorlick(Gorlick 1991). The passive monitor will have registers that will be written to by either instrumented code running on the task processor, by the operating system simulation, or by results from monitor calculations. The number of registers will be determined by experimentation. The monitor will be able to perform simple operations such as addition, multiplication, and logical operations. This level of complexity is satisfactory to allow the passive monitor to update the event mask from a specification transmitted by the Central Host Monitor and to make time comparisons. The dedicated monitoring network is a two way network between each node connected to the central host monitor. While information traveling towards the central host monitor does not need to be real-time, the information traveling back to the nodes is required to be real-time.

4.3 Programming Language

The language will be sufficient only to write programs to test the monitoring solution. The only form of variables will be integer and the results of all expressions will be integers. Function parameters will be passed by reference and the programmer will not have the capability to allocate memory or directly access the address of any variable or function. Programs will be compiled to the machine language of our simulation and schedulability analysis will be performed on the machine language code segments.

5. CONCLUSION AND FUTURE WORK

Several concepts for Deadline Non-Intrusive monitoring have been demonstrated. Future research will include methods for use of static analysis byproducts such control graphs and data-flow graphs to drive dynamic updates of instrumented delays

6. REFERENCES

Bhatt, Devesh, Adel Ghonami and Ranga Ramanujan (1987). An instrumented testbed for real-time distributed systems development. In: *Real-Time Systems Symposium*. The Computer Society of the IEEE. pp. 241 – 250.

Dodd, Paul S. and Chinya V. Ravishankar (1992). Monitoring and debugging distributed real-time programs. *Software Practice and Experience* 22(10), 863 – 877.

Fryer, Richard E. (1973). The memory bus monitor–a new device for developing real-time systems. In: *AFIPS Conference Proceedings, 1973 National Computer Conference and Exposition*. Vol. 42. AFIPS. AFIPS Press. Montvale, New Jersey. pp. 75–79.

Gorlick, Michael M. (1991). The flight recorder: An architictural aid for system monitoring. In: *ACM / ONR Workshop Parallel and Distributed Debugging*. Association for Computing Machinery. pp. 175 – 183.

Haban, Dieter and Dieter Wybranietz (1990). A hybrid monitor for behavior and performance analysis of distributed systems. *IEEE Transactions of Software Engineering* 16(2), 197 – 211.

Kligerman, Eugene and Alexander D. Stoyenko (1986). Real-time euclid: A language for reliable real-time systems. *Transactions on Software Engineering* SE-12(9), 941 – 949.

Patterson, David A. and John L. Hennesy (1994). *Computer Organization and Design The Hardware / Software Interface*. Chap. 7. Morgan Kaufmann.

Petrov, Plamen and Alex Stoyen (1997). Compiler support for non-intrusive monitoring and debugging of real-time systems in the crl environment. In: *18th IEEE Real-Time Systems Symposium*. IEEE.

Rosenberg, Jonathan B. (1996). *How Debuggers Work: Algorithms, Data Structures, and Architecture*. John Wiley And Sons.

Stoyenko, A. D. and T. J. Marlowe (1992). Polynomial-time transformations and schedulability analysis with restricted resource contention. *Real-Time Systems*.

Stoyenko, A. D., T. J. Marlowe and M. Younis (1995). A language for complex real-time systems. *Computer Journal*.

Stoyenko, Alexander and Wolfgang Halang (1991). *Constructing Predictable Real-Time Systems*. Kluwer International Series in Engineering and Computer Science. Kluwer Academic Publishers. Norwell Massachusets.

Tokuda, Hideyuki, Makoto Kotera and Clifford W. Mercer (1988). A real-time monitor for a distributed real-time operating system. In: *Proceedings of the ACM Workshop in Parallel and Distributed Debugging*. Association for Computing Machinery. pp. 68 – 77.

Win (1995). *VxWorks Programmers Guide*. version 5.3 ed.

BUILDING RE-USABLE COMPONENTS USING FORMAL SPECIFICATIONS FOR COMPLEX EVOLVING SYSTEMS

G. Tsai* and W. A. Halang†

*Fairleigh Dickinson University, Computer Science, Teaneck, NJ 07666, U.S.A.,
tsai@alpha.fdu.edu
†FernUniversität Hagen, Faculty of Electrical Engineering, 58084 Hagen, Germany,
wolfgang.halang@fernuni-hagen.de

Abstract: A complex evolving (evolutionary) system is one which must adapt to changes in the environments during development and after deployment. A key to success in building an evolving system lies in the construction of the manager for the evolving system, i.e., the repository and its search engine. An approach to identify software components with semantic and syntactic similarity is presented as a key step in constructing such a repository. We then define levels of semantic matches for use in retrieval and re-use of software components; these can then be used to construct a hierarchy of components in the repository. Finally, initial steps toward an integrated set of formal methods tools for developing and maintaining evolving system managers is suggested. *Copyright © 1999 IFAC*

Keywords: Real-time systems, temporal logic, formal methods, software re-use, analogical reasoning.

1. INTRODUCTION

A complex system is a system with a high level of complexity, typically in number of entities, in heterogeneity of components, concepts, or procedures, and/or in organisation or performance criteria. A complex evolving (evolutionary) system is a complex system which must adapt to changes in the environment, both during development and after deployment. Environmental changes include, but are not limited to, changes in requirements (perhaps resulting from changes in the application domain or client base), changes in available platform components or availability of services, or even perhaps changes in personnel or in applicable law.

Small or simple systems can typically be re-designed to cope with changes, often incorporating re-use and incremental modification by hand. In contrast, complex and long-lived systems would greatly benefit from a semi-automatic process for evolution; such tools can also support re-use with inheritance of component clusters (e.g., a client-server model, or an information database). One would like a tool and a repository which could ensure that a system was easily evolvable and extensible, remained maintainable and error-free, and that changes were compatible with retained components.

However, any but the most elementary tools will require additional annotations on components (either attached to the component itself, or within the repository), a reasonable repository with one or more search engines (likely using pattern matching, expert knowledge, and heuristic search techniques), and sophisticated analysis techniques. Annotations may arise from compilation (function and class signatures), analyses (such as control-flow, data-flow, or schedulability analysis), be provided by specification, or perhaps even originate from profiling, testing, and other validation techniques; in some cases, as for typestate (Strom, 1991), a combination of techniques may be required.

Given an existing application, a needed change, and a repository of annotated components, the search engines will propose a number of possible component matches based on semantics and syntactic constructs. At this point, analysers will be called to determine if any of the proposed components actually provide a suitable match. We expect that, for most classes of applications and many classes of changes, it will be possible to create sound repositories and analysis engines, which can guarantee matches, but may suffer from false negatives, i.e., fail to recognise a correct match, or complete engines, which will recognise all matches, but may suffer from false positives; however, obtaining both soundness and completeness is impossible, since the problem is clearly undecidable. (By use of both types of annotations and analyses, one can obtain somewhat more, but hardly perfect, certainty.) Formal methods are clearly one of the tools which will form a part of any reasonable analysis engine. In this paper, we discuss how formal methods can be used to manage signature matching in a repository.

Constructing a repository and its search engines will borrow heavily from tools proposed for software libraries. The techniques proposed for constructing software libraries include natural language (Maarek et al., 1991; Helm and Maarek, 1991) and formal specifications (Cheng and Jeng, 1992; Jeng and Cheng, 1992; London, 1989; Nishida et al., 1991; Rich and Waters, 1983; Weide et al., 1991; Wing, 1990). The latter differ in the formal languages used and in the types of re-usable components allowed. Formal specifications, if necessary extended with additional algebraic domains, allow precise annotation of components with their specification and observable behaviour, making classification of re-usable components and retrieval easier; therefore, we use formal specifications to represent software components in the repository.[1]

The main contribution of this work is (1) providing an approach to a repository of components for complex evolving systems using formal specifications, (2) describing how ITL, a formal language with an interval representation of time, can be used to specify components, (3) classifying and retrieving components according to syntactic and semantic similarity, and (4) proposing a suite of tools, largely based on formal methods, to predict the adaptability of an existing system when a change to its requirements occurs. We target the domain of complex evolving systems, so *components with timing properties and constraints* are allowed in the repository.

To facilitate retrieval, re-use, and substitution, repositories are constructed with two layers, each inducing a partition of the components. The upper layer is composed of components with semantic similarity; this partition is refined in the lower layer to components with syntactic similarity. Definitions of exact match and relaxed match of component specifications are given for building the relationships among the components. Incorporation of the layered structure into a prototyping environment facilitates verification of mutability in evolving systems following incremental changes.

Beyond issues of component matching and proofs of correctness or mutability, an evolving system manager and the repository itself may need additional analysis and transformation tools, to specialise components, to synthesise the observable behaviour of a changed system from those of the old system and of the new components, and to construct new annotations. In the balance of this paper, we indicate where such tools may be invoked, but not the tools themselves, nor precisely how they are integrated.

----

[1] Clearly, the precision in specification of observable behaviour is at best that of the analyses used.

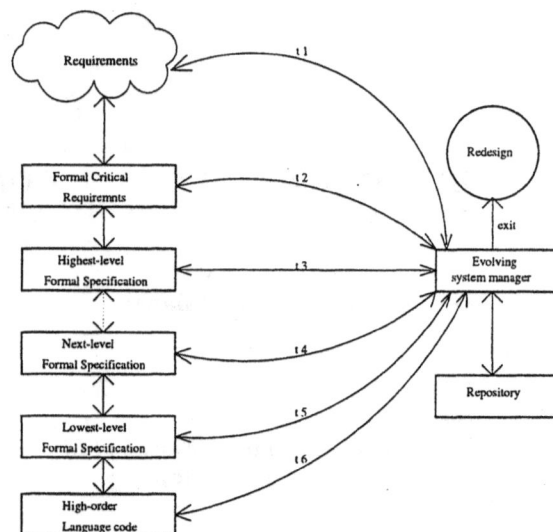

Figure 1: Formal Development Process

This paper is organised as follows. Section 2 contains an overview of a formal development process for complex evolving systems. In Section 3 we describe the discipline of specifying an evolving manager. Section 4 covers the construction of an upper-layer hierarchy using semantic similarity. Section 5 describes the integration of formal methods tools for developing and maintaining an evolving system. Section 6 concludes the paper and discusses future work.

2. A HIERARCHICAL FORMAL DEVELOPMENT PROCESS

For large complex systems, detailed specifications are hard to understand and error-prone; it is also often difficult to show that a high-level language code is consistent with the requirements. We can, however, ease this problem to a significant extent by using a hierarchical development process, decomposing a system into several levels. We then represent each level using a formal notation, and verify the design at each level, and if necessary check the interfaces between levels. We may also use multiple abstractions and abstraction granularities in representing and proving system properties at a given system level.

In Figure 1, the top entity is the set of (informal) requirements provided through discussions/negotiation with the customer. We then express the preliminary design in a formal notation, analyse the design, and state and prove formal properties of the preliminary design. As a more detailed design is developed, we formally verify consistency with the abstract preliminary design, which we assume has been shown to satisfy the more abstract desired properties, such as safety or security. This process is called design verification or specification verification.

The evolving system manager is for the evaluation of success in adaptability of changes during development and maintenance. It consists of a search engine, together with an analyser, specialiser and transformer. After an update/change, it determines whether a system can evolve and continue operation successfully after a update/change. For example, let transition t1 denote a change to the requirements. The following steps are required to verify the change.[2]

1. Perform an initial analysis, comparing the change in requirements against system specifications, and resulting in a (perhaps partial) specification of the needed new components. In many cases, this analysis is trivial, and the needed components are completely described by t1 itself; for simplicity, we suppose this situation in the following discussion.

2. Invoke the search engine, which performs search on the repository/component library for a semantic match with transition t1, which describes the change to the requirements. If reusable components with appropriate semantics are found, use them to establish the feasibility of incremental change.

3. The output of step 2 is fed to the analyser of the search engine. The analyser then decides where the components found by the search engine are suitable for verifying incremental change. The specification of t1, including the signatures, typestate, pre-condition and post-condition, will be evaluated against the components found in step 2.

4. Trigger the specialiser and the transformation engines to modify/specialise the component and update annotations, as required and possible. Otherwise, use current information to decide whether to continue operation or re-design. (It should, in addition, be possible to extend the semantic matcher to suggest components which might be suitable after minimum re-design.)

3. SPECIFYING SOFTWARE COMPONENTS

First-order predicate logic (FOPL) is commonly used to specify programs (Hoare, 1969; Dijkstra, 1976; Wing, 1990; Cheng, 1994). *Many-sorted* logic adds the equivalent of types to first-order logic, and timing accommodated by adding a formalism, such as Interval Temporal Logic (ITL), to a FOPL L. A *sort hierarchy* begins with primitive data types, such as integer, float and double, and is constructed recursively with type constructors such as array, set, and record structures. (In addition, there are arrow (function) sorts, but these are of less significance in the procedural or object-oriented languages used for real-time systems than in functional languages such as ML, where sorts in any case represent something more like meta-types.) A sort hierarchy thus mirrors the

relationships among the corresponding implementation language types. Due to lack of space, the details of order-sorted Interval Temporal Logic (OSITL) are presented in (Tsai, 1998).

We use order-sorted specifications to describe software components. The relationship between two components is based on the sort information, together with the pre-conditions and post-conditions of the components. This relationship of components can be considered as the re-usability of one component against the other.

The specification of a software component consists of the specification of an abstract data type and a set of methods that operate on the data type. Each method is denoted by a *signature*, a *pre-condition* and a *post-condition* (and perhaps additional annotations, e.g., timing properties or resource requirements). The signature describes the types of input and output parameters. The pre-condition (post-condition) describes the state of the variables of the method before (after) the execution of the method. The pre-condition and post-condition are represented using ITL formulae which may contain timing properties and constraints.

method method_name $((Var : DomaninSort)^*$
$\rightarrow Var : RangeSort)$
  requires pre-condition
  ensures post-condition

The pre-condition and post-condition are specified based on order-sorted ITL. The **requires** clause describes the restrictions on input parameters and static variables when the method is invoked (e.g., typestate annotations (Strom and Yellin, 1993)). The **ensures** clause puts constraints on the behaviour of the method. These two clauses refer to the state when the method is called, and the state after it terminates. Figure 2 gives the component specification of Stack (Cheng and Jeng, 1994), where the methods **create** and **destroy** are the constructor and destructor of this component, respectively; the method **push** puts an element to a stack; the method **pop** deletes an element from a stack; the method **topElement** retrieves the top element of a stack. This specification could easily be extended to include predicate methods such as **isEmpty** and **isFull**.

The software components are organised according to semantic and syntactic similarity. Different levels of exact and relaxed match are given to place the most general components at the top of a hierarchy and the more restrictive components at the bottom, which forms the upper layer of the hierarchy. The lower layer is built according to syntactic similarity. Hierarchical clustering algorithms are used to group similar components together. The construction of a lower-layer hierarchy was presented in (Tsai, 1998).

4. UPPER-LAYER HIERARCHY

The objective of constructing an evolving manager is to facilitate browsing, retrieving and searching of

---

[2]Since some dynamic semantic properties cannot necessarily be represented or proven from static annotations on components, depending on the nature of requirements and annotations, some human intervention may be required after a match is found by the system manager. We do not pursue this issue further in this paper.

```
component = Stack
  method create: () → stack : Stack
    requires true
    ensures(stack') = NULL_Stack

  method destroy: (stack:Stack) →
    requires true
    ensures trashed(stack)

  method push: (stack:Stack, newElement:Object)
      → stack' : Stack
    requires ¬ full(stack)
    ensures top(stack', newElement) ∧ size(stack')
      = size(stack) + 1

  method pop: (stack:Stack, topElement:Object)
      → stack' : Stack
    requires ¬ empty(stack)
    ensures top(stack, topElement) ∧ size(stack')
      = size(stack) - 1

  method topElement: (stack:Stack)
      → topElement:Object
    requires ¬ empty(stack)
    ensures top(stack, topElement)
```

Figure 2: Component specification for **Stack**

software components which satisfy a given query and adapt the components found to fit its environmental constraints. Sometimes we need to decide whether one component matches another. Or we may want to replace one component with another one, and must decide whether a component in the library can be adapted to fit the needs of a given system. Therefore, we need to have different levels of semantic matches.

In this paper, we assume that software components are procedures/functions, plus user-specified or system-defined data structures. Informally, two components match if their signatures match, and their specifications match according to some semantic matching definition. The following defines various matching criteria (Zaremski and Wing, 1996). There are basically two dimensions to the match — how pre-conditions and post-conditions are compared, and when two predicates are considered equivalent.

4.1. Pre/Post Matches

For a function specification, F, we define its pre- and post-conditions as F_{pre} and F_{post}. F_{pred} is defined as $F_{pre} \Rightarrow F_{post}$, which means that if the pre-condition F_{pre} holds when the function is invoked, then F_{post} will hold after the execution of the function.

Definition 1. (Generic Pre/Post Match)
Two functions, S and Q, are said to satisfy a generic pre/post match, iff $match_{pre/post}(S, Q) = (Q_{pre}\ R_1\ S_{pre}) \wedge (\hat{S}\ R_2\ Q_{post})$, where S_{pre} and Q_{pre} are the pre-conditions of S and Q, S_{post} and Q_{post} are the post-conditions of S and Q, R_1 and R_2 denote some relations, and $\hat{S}$ is either S_{post} or $S_{pre} \wedge S_{post}$, as defined below.

The relations R_1 and R_2 are either ⇔ or ⇒. Figure 3 summarises five types of generic matches. $\hat{S}$ can be either S_{post} or $S_{pre} \wedge S_{post}$, to allow for the ca-

ses where we need to include information about the pre-condition in the post-condition. These five types of matches are listed from the strongest to the weakest. Relaxing the match allows for comparisons of less related components. For instance, two components with plug-in post match are not completely equivalent but have the same behaviour for some subset of inputs. This is useful when we need a relaxed match.

Exact Pre/Post Match. Two components are interchangeable if they are equivalent, or their pre-conditions are equivalent and their post-conditions are equivalent. In this case, we can replace one component by the other without changing the observable behaviour.

Plug-in Match. Under plug-in match, R_1 and R_2 are relaxed from ⇔ to ⇒. Given a query Q, we search for a component S with weaker pre-condition and stronger post-condition. In other words, Q_{pre} leads to S_{pre}, and S_{post} will lead to Q_{post} after the execution of S. We say that S is equivalent to Q, since we can obtain the same observable behaviour by plugging in S for Q. However, it is not symmetric, since we may not obtain S by plugging in Q.

Plug-in Post Match. This match considers only the post-conditions or effects of two functions, i.e., $match_{plug-in-post}(S, Q) = (S_{post} \Rightarrow Q_{post})$ with R_1 dropped and R_2 instantiated to ⇒.

Guarded Plug-in Match. In this case, the post-condition relation holds only for some values allowed in the pre-condition. S_{pre} is used as a guard to limit the conditions to prove $S_{post} \Rightarrow Q_{post}$.

Guarded Post Match. We drop the pre-condition term, and check $(S_{pre} \wedge S_{post}) \Rightarrow Q_{post}$.

4.2. Predicate Matches

Definition 2. (Generic Predicate Match) *Two predicate specifications, S and Q, are said to satisfy a generic predicate match, iff $match_{pred}(S, Q) = (S_{pred}\ R\ Q_{pred})$, where S_{pred} denotes $S_{pre} \rightarrow S_{post}$ and Q_{pred} denotes $Q_{pre} \rightarrow Q_{post}$.*

The relation R is equivalence (⇔) for exact match, implication (⇒) for generalised match, or reverse implication (⇐) for specialised match.

Definition 3. (Exact Predicate Match) *Two predicate specifications, S and Q, are said to satisfy an exact predicate match, iff $match_{E-pred}(S, Q) = (S_{pred} \Leftrightarrow Q_{pred})$.*

Definition 4. (Generalised Match) *Two predicate specifications, S and Q, are said to satisfy a generalised match, iff $match_{gen-pred}(S, Q) = (S_{pred} \Rightarrow Q_{pred})$.*

Definition 5. (Specialised Match) *Two predicate specifications, S and Q, are said to satisfy a specialised match, iff $match_{gen-pred}(S, Q) = (S_{pred} \Leftarrow Q_{pred})$.*

Exact Pre/Post

↓

Plug-in

↙ ↘

Guarded Plug in   Plug-in Post

↘ ↙

Guarded Post

↓

True

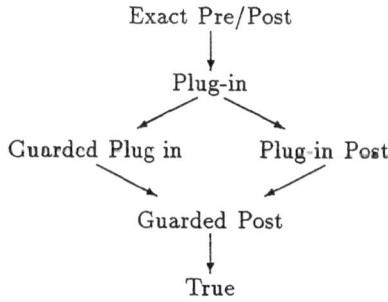

Figure 3: Function specification matches

Note that exact pre/post match is stronger than exact predicate match, i.e., $match_{E-pre/post} \Rightarrow match_{E-Pred}$. Each pre/post and exact predicate matches are equivalent when $S_{pre} = Q_{pre} = true$. Also, generalised match is weaker than plug-in match (i.e., $match_{plug-in} \Rightarrow match_{gen-pred}$). Figure 3 summarises the relationships of various specification matches (Zaremski and Wing, 1996).

4.3. Module Matching

Function matching addresses the problem of matching individual functions. There are some situations that require comparisons of collections of functions, i.e., comparisons of modules. Modules, such as Ada packages or C++ classes, are often used to group a set of related functions together. Assuming that two modules, $module_S$ and $module_Q$, satisfy signature matching (Zaremski and Wing, 1995), the algorithm in Figure 4 can be used for matching module specifications. The one in Figure 5 describes the construction of an upper-layer hierarchy of component specifications. Furthermore, according to function specification matches of Figure 3 we classify related/similar components into clusters, which is shown in Figure 6.

5. PROTOTYPING ENVIRONMENT

The goal of our work is to develop a prototyping environment for the analysis of evolutionary systems. To support evolution and maintenance, the following are required. (1) Formal representation of software components to facilitate search, re-use and retrieval; identify/classify re-usable formal specifications for the verification of changes. (2) Formalisms for representing changes from an existing to a proposed system. (3) Satisfiability checker to check the feasibility of changes, and to detect various types of inconsistencies, e.g., an inconsistency between different stages of development, as a complex system evolves.

The evolutionary system engine contains component library, ITL analyser, search engine, and analyser/specialiser/transformer. The ITL analyser is a verification tool, a satisfiability checker, which evaluates the satisfiability of ITL formulae. In this situation, it evaluates a change represented using ITL formulae against system constraints likewise represented as such formulae. The component library consists of two layers, one with semantic similarity, and the other with syntactic similarity.

Algorithm: More_General_Module
Input: Two modules $module_S = \{S_1, S_2, \ldots, S_m\}$
    and $module_Q = \{Q_1, Q_2, \ldots, Q_n\}$.
Output: The generality relationship between
    $module_S$ and $module_Q$.
```
begin
  I = 0
  while module_Q ≠ ∅
    select any Q_i ∈ module_Q
    module_Q → module_Q \ Q_i
      if there exists j, such that
      match_gen-pred(S_j, Q_i),
      then I = I + 1
  endwhile
  if I = n, return ("module_Q more general")
```

Figure 4: The generality relationship between two modules

Algorithm: Pairwise_Comparison
Input: A set of components
    $SET = \{C_1, C_2, \ldots, C_n\}$
Output: A hierarchy of components based on the
    generality relationship
```
begin
  while SET ≠ ∅
    select some C_i ∈ SET
    SET → SET \ C_i
    set → SET
    while set ≠ ∅
      select C_j ∈ set
      set → set \ C_j
      if match_gen-pred(C_j, C_i), make C_i
        a parent of C_j
      if match_gen-pred(C_i, C_j), make C_j
        a parent of C_i
    endwhile
  endwhile
end
```

Figure 5: Building an upper-layer hierarchy

Algorithm: Clustering components based on
    semantics
Input: A set of components
Output: One or more clusters
```
begin
  select each component k from the set
  the related/similar components of k are
  case 1. components a, b, c: if ∃ a, b, c, such
    that match_plug-in(a, k)
    and either match_guardedplug-in(b, k),
      or match_plug-in-post(c, k).
  case 2. components a, b, c: if ∃ a, b, c, such
    that match_E-pred(a, k),
    and either match_spel-pred(b, k)
      or match_gen-pred(c, k).
end
```

Figure 6: Clustering components with similar semantics

Regardless of the actual source of a change (client-requested changes, resource availability, regulations, standards, bugs), the system becomes aware of it when the user enters the change, which may be to the requirements, platform, system objective, or a combination thereof. The search engine will search in the component library for related/re-usable components for incremental verification of the change. Next, the analyser will perform an analysis to see if the type of change can be handled incrementally, and how it affects the system (via some initial analysis), and if so, whether new components can be found and the modified system be incrementally verified — invoking the ITL analyser for this purpose. The specialiser or transformer may be triggered in cases of imprecise match or when minimum re-design is required.

The search engine includes the matcher, evaluator and search algorithms. The matcher examines the component library for predicates and inferences with similar semantics. Note that we do not address type/class in the matcher, although this is arguably still another dimension of a match. The nature of the evaluator, which makes choices when multiple matches are found, and of the search algorithms proposing candidates, are likewise beyond this paper's scope.

6. CONCLUSION

In this paper, we presented an approach to develop and maintain the component library for an evolving system. Software components are classified according to semantic and syntactic similarity. Different levels of semantic matches are defined for retrieval and re-use of software components to construct a hierarchy of components. Components with similar syntactic constructs are organised into another hierarchy. These two hierarchies are incorporated into a prototyping environment to aid in the development of complex evolving systems.

This work allows an evolving system manager to search for related/similar components to make and verify incremental changes. In the future, we shall examine the techniques used in the analyser/specialiser/transformer, for the purposes of (1) extraction of timing properties for integrated systems from those of old and new components, and of new components or interfaces from system properties, (2) specialisation and other modification of components or interfaces which "nearly" meet a search pattern, (3) use of ITL in verifying properties of integrated systems which are not directly representable in ITL, and (4) integration of replication and similar systems-level modifications to support additional incrementality.

7. REFERENCES

Cheng, B.H. (1994). "Applying formal methods in automated software development," *Journal of Computer and Software Engineering*, Vol. 2, No. 2, pp. 137–164.

Cheng, B.H., and Jeng, J.-J. (1992). "Formal methods applied to reuse," in *Proceedings of the Fifth Workshop in Software Reuse*.

Cheng, B.H., and Jeng, J.-J. (1994). "Reusing analogous components," Tech. Rep. MSU-CPS-94-28, Computer Science Depeatment, Michigan State University.

Dijkstra, E.W. (1976). *A Dicipline of Programming*. Englewood Cliffs, NJ: Prentice Hall.

Helm, R., and Maarek, Y. (1991). "Integrating informational retrieval and domain specific approaches for browsing and retrieval in object-oriented class libraries," in *Proceedings of OOPSLA'91*, pp. 47–61.

Hoare, C.A.R. (1969). "An axiomatic basis for computer programming," *Communications of the ACM*, Vol. 12, No. 10, pp. 576–580.

Jeng, J.-J., and Cheng, B.H. (1992). "Using automated reasoning to determine software reuse," *International Journal of Software Engineering and Knowledge Engineering*, Vol. 2, No. 12, pp. 523–546.

London, R. (1989). "Specifying reusable components using Z: Realistic sets and dictionaries," *ACM SIGSOFT Software Engineering Notes*, Vol. 14, No. 5, pp. 120–127.

Maarek, Y., Berry, D., and Kaiser, G. (1991). "An information retrieval approach for automatic constructing software libraries," *IEEE Transactions on Software Engineering*, Vol. 17, No. 8, pp. 800–813.

Nishida, F., Takamatsu, S., Fujita, Y., and Tani, T. (1991). "Semi-automatic program construction from specification using library modules," *IEEE Transactions on Software Engineering*, Vol. 17, No. 9, pp. 853–870.

Rich, C., and Waters, R. (1983). "Formalize reusable software components," in *Proc. Workshop on Reusability in Programming*, (Newport, RI).

Strom, R. (1991). *Hermes: a Language for Distributed Processes*. ACM Press – Addison Wesley.

Strom, R., and Yellin, D. (1993). "Extending typestate checking using conditional liveness analysis," *IEEE Transactions on Software Engineering*, Vol. 19, No. 5, pp. 478–485.

Tsai, G. (1998). "Building Reusable Components Using Formal Specifications for Complex Evolving Systems," Tech. Rep. CSIS-98-1, Computer Science and Information Systems, Fairleigh Dickenson University.

Weide, B., Ogden, W., and Zweben, S. (1991). "Reusable software components," *Advances in Computers*, Vol. 33, pp. 1–65.

Wing, J.M. (1990). "A specifier's introduction to formal methods," *IEEE Computer*, Vol. 23, No. 9, pp. 8–24.

Zaremski, A.M., and Wing, J.M. (1996). "Specification matching of software components," *submitted to ACM TOSEM*.

Zaremski, A.M., and Wing, J.M. (1995). "Signature matching: A tool for using software libraries," *ACM TOSEM*, Vol. ?, No. 4, pp. 8–24.

Building Safety-Critical Real-Time Systems with Synchronous Software Components

Michael Gunzert

Institute for Industrial Automation and Software Engineering (IAS)
University of Stuttgart, Pfaffenwaldring 47, D-70550 Stuttgart, Germany
gunzert@ias.uni-stuttgart.de

Abstract: In this paper a new method for the development of distributed safety-critical real-time systems is presented. The method is based on the synchronous approach for designing reactive systems and a time-triggered communication architecture. Synchronous software components consisting of a reactive and a transformational part are used to specify the behavior of the system. The reactive part of a synchronous component is specified in the synchronous language ESTEREL. In the design model, hardware and software components are composed graphically on a high level of abstraction. From the graphical design specification executable code can be generated automatically. Due to the synchronous execution model, the code is deterministic and can also be simulated and verified. *Copyright © 1999 IFAC*

Keywords: safety-critical systems, synchronous reactive systems, ESTEREL, time-triggered architectures, component-based development

1. INTRODUCTION

Safety-critical real-time systems must be correct and deterministic as they must always react the same way to the same inputs. The design of the application software for such systems is a complex task. The reason is that software is becoming dominant in embedded system design as fast hardware is becoming cheaper. Earlier, custom hardware might have been required because of performance requirements, but now cheap, fast general purpose microcontrollers are well suited for many of these jobs. However, fast cheap hardware allows larger, more powerful systems to be built and leads to greater system complexity. The growing complexity influences the design process of the systems since it makes it much harder to build systems correctly.

Synchronous languages like ESTEREL (Berry and Gonthier, 1992) provide mathematically defined semantics based on the *synchrony hypothesis*. On this basis it is possible to generate deterministic code and apply formal optimization and verification techniques.

Unfortunately, it is not always practical to build synchronous systems. In particular, distributed systems with long communication times are difficult to make synchronous. However, the synchronous approach can be used advantageous in combination with a time-triggered communication architecture.

Another point is, that synchronous languages like ESTEREL need a host language like C for the interfaces and data-handling parts of the application software. Component-based techniques can be used to make the features of synchronous languages applicable for non-specialists. Libraries of verified components can also be provided. In our approach we use a high-level description of both software and hardware components. A system can be built by instanciating and connecting software and hardware components graphically.

2. THE SYNCHRONOUS APPROACH

Embedded real-time systems are *reactive systems* (Harel and Pnueli, 1985) which are computer systems interacting continuously with their environment. Synchronous languages have been developed to simplify the programming of reactive systems. They are based on the *synchrony hypothesis* (Berry and Gonthier, 1992) which makes the following abstractions:

- The computer is infinitely fast.

- Each reaction is instantaneous and atomic, dividing time into a sequence of discrete instants. Different reactions cannot interfere with one another.

- A system's reaction to an input appears at the same instant as the input.

A real system can behave synchronously if it is fast enough. It must always finish its computations before more events arrive from the environment. This requires knowing both the minimum inter-event time as well as the worst-case execution time. The synchrony hypothesis is a generalization of the synchronous model used for digital cirquits where each reaction must be finished in one clock cycle. The synchronous model of time simplifies the design of correct systems. Temporal details are hidden during specification and so the behavior of the system is also simplified. Non-deterministic behavior caused by the interference of parallel actions cannot occur. Deterministic systems are one order of magnitude easier to specify, analyze and test as non-deterministic ones.

The main languages based upon the synchrony hypothesis are ESTEREL, LUSTRE (Halbwachs et al, 1992), SIGNAL, STATECHARTS, SML, SAGA and ARGOS (Halbwachs, 1993). ESTEREL is an imperative language while LUSTRE and SIGNAL are declarative and STATECHARTS (Harel, 1987) and ARGOS are graphical languages.

2.1 *The synchronous language ESTEREL*

ESTEREL is a synchronous language based on a mathematically precise defined semantics developed specifically for deterministic reactive systems. The programming model of ESTEREL is the specification of modules which are assumed to be executed in parallel. The modules communicate one with another and with their environment through signals. The signals are broadcasted and can be received from each module. A signal has a state and can optionally have a value of an arbitrary type in each instant of time. Sending and receiving signals is performed instantaneously according to the synchrony hypothesis. ESTEREL only allows the specification of deterministic behavior. The input signals for each reaction step determine uniquely the output signals (and their values) to be sent in this instant as well as the input output behavior of the program. An ESTEREL program can be compiled to a deterministic finite state automaton. In fact, the translation into automata justifies the synchrony hypothesis. If reactions are not instantaneous they are as fast as they can be as they include only actions that must be done at run-time. Process-handling and synchronization are done at compile-time, therefore produce no actions. ESTEREL only supports some basic data types (boolean, integer, float, double and string) and operations. For other data types and their operations,

external implementations must be provided in a host language.

2.2 *Synchronous Software Components*

On the basis of the synchrony hypothesis it is possible to define components which can easily composed to larger systems. Since the components communicate through signals being sent by broadcast, the components are not required to make any assumptions about each other. They are working independently like software ICs.

A synchronous software component consists of a reactive part and a transformational part. The reactive part is specified in a synchronous language. The transformational part is optional and consists of data-type specifications and several data-handling functions written in the host language (Fig. 1). We use ESTEREL for the reactive part and C for the transformational part of the components.

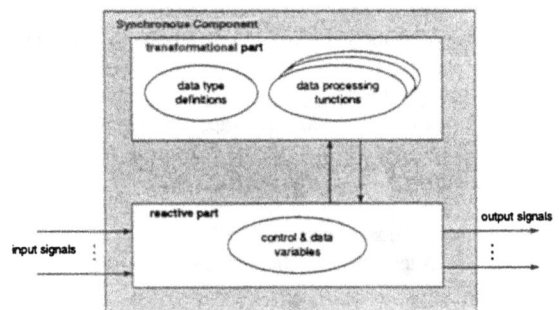

Fig. 1. A Synchronous Software Component

The interface of a component consists of input and output signals. Synchronous components can be hierarchically composed to larger modules.

3. DISTRIBUTED SYSTEMS

Many reactive systems have to be distributed on several computing locations. The reasons in most cases are distributed location of sensors and actuators and fault tolerance requirements rather than performance improvements.

3.1 *Event-Triggered vs. Time-Triggered Architectures*

There exist two different paradigms for the design of distributed real-time systems: *event-triggered architectures* and *time-triggered architectures*. In an event-triggered system, all activities – activation of tasks, sending of messages etc. are triggered by events. The advantage of an event-triggered architecture is flexibility, the disadvantage, however, is non-determinism. For safety-critical systems, this is not acceptable. In a time-triggered system all activities are driven by the progression of a global time base. All tasks and communication actions are periodic. In a distributed system all clocks have to be synchronized with a known precision. Time-triggered systems rely on stronger assumptions about the regularity of the processes to control in their environment and are therefore less flexible but far easier to analyse and test.

The combination of the synchronous approach with a time-triggered communication architecture allows for distributed systems to be built completely synchronous and deterministic as communication times are strictly bounded.

3.2 *The Time-Triggered Protocol (TTP)*

The Time-Triggered Protocol (Kopetz and Grünsteidl, 1994) is an integrated communication protocol for time-triggered architectures. It has been designed to fulfill the specific requirements of safety-critical real-time applications and therefore provides services needed to build distributed fault-tolerant systems:

- deterministic message transmission

- clock synchronization service (global time-base)

- integrated network management

- error detection with short latency

- distributed redundancy management

A node in a TTP real-time system consists of a host processor and a communication processor (TTP-controller). This entity is called a *Fail Silent Unit (FSU)*. The interface between host and communication subsystem is a dual-ported RAM called *Communication Network Interface (CNI)*. Two Fail Silent Units can be composed to a *Fault-Tolerant Unit (FTU)*. The TTP-protocol has been designed to tolerate any single physical fault in any one of its constituent parts (nodes, bus). The communication network topology is a broadcast bus where bus access is controlled by a static TDMA[1] scheme.

4. COMPONENT-BASED DEVELOPMENT: THE VIPER METHOD

The name ViPER stands for **Vi**sual **P**rogramming **E**nvironment for **E**mbedded **R**eal-Time Systems. ViPER is both a method and a tool environment for visual component-based development of embedded real-time systems.

The concept consists of three models:

- the *design model*,
- the *generation model* and

---

[1] Time Division Multiple Access

- the *execution model.*

These three models are shown in Fig. 2 and are described in the following paragraphs.

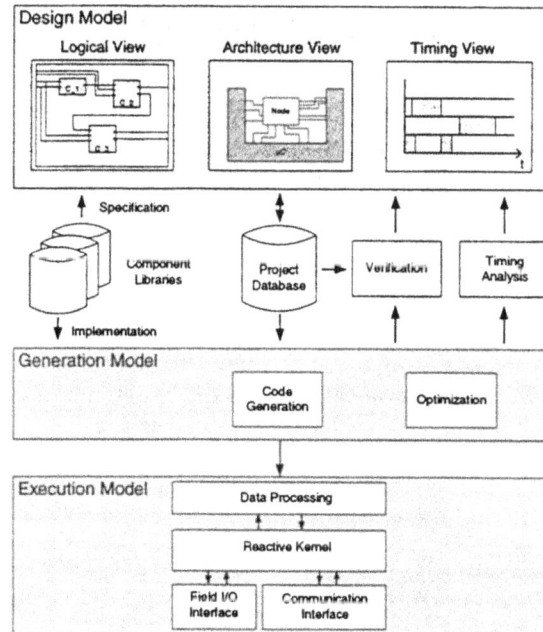

Fig. 2: ViPER Concept

4.1 *Design Model*

In the *design model*, application software can be constructed from components provided in several libraries. At this level of abstraction only high-level descriptions of components are used. Components are specified with their interfaces and parameters. For this purpose a specification language has been developed covering the specification of components with their interfaces and parameters as well as larger modules and configurations. The implementation of the components is not visible to the user. For the development of embedded systems the hardware platform must also be considered. For each platform a hardware library has to be provided. This library includes the specification of the hardware components which can be used from the software. A good example is a microcontroller which typically provides several hardware components as timers, ports or puls-wide modulation units. This allows the interconnection of software and hardware components on an abstract level and platform-specific code-generation as well.

The design model itself has three different views – the *logical view*, the *architecture view* and the *timing view*. In the *logical view*, the logical structure of the system is specified. The components are connected and data and control flow are defined. For this purpose block diagrams are used which can be hierarchically composed. In the *architecture view* software components can be connected with platform-specific hardware components. This includes connections with hardware ports to which sensors or actuators are connected as well as communication connections between the nodes of a distributed system. The *timing view* specifies the temporal behavior of the system. In a time-triggered system time is divided in time-slots in which certain actions are performed. The global behavior of the system is defined first in a global message schedule by a separate tool. This schedule is cyclic and defines in which time-slot a certain node is sending. After that, a specific schedule for each node is specified in the timing view, that describes the actions on the individual node and meets the global requirements. The timing view displays the results of the timing analysis. This is the sequential execution order of I/O- and data processing functions and the calculated worst-case execution time for a single reaction step. The graphical specifications made in the design model are stored and main-tained in the project database.

4.2 *Generation Model*

The *generation model* describes the process of generating executable code from the high-level graphical specifications. In order to meet the requirements of a specific application, the code produced must have minimal overhead. This can be achieved using the synchronous model, generative techniques and code optimization. Moreover, it must also be possible to generate code for simulation and verification purposes. In our model we generate only the glue code needed to connnect the several components of the application. The generation process consists of several stages where ESTEREL code, C-

code and makefiles are generated. The intermediate code is a finite state machine (FSM) driving action tables that call the data processing functions. On this intermediate format several existing verification tools (e.g. XEVE, VIS) can be applied to verify safety properties (Bouali, 1997). These tools perform a reachability analysis over the reachable state space of the FSM. Optimization of the FSM is also possible using tools for sequential circuit optimization (Sentovic et al, 1992). For a specific hardware platform, a platform-specific C-compiler and a development environment has to be provided to compile, link and download the generated code to the targets.

4.3 *Execution Model*

The *execution model* is an execution framework for a single node in a time-triggered system. It provides a generalized view on a distributed time-triggered application. For a concrete application, the component-based design is mapped into this framework by the generation model automatically where the overall structure needs not to be changed. This can be reached by clearly separating platform-specific and application-specific parts from platform- or application-independent parts. Figure 3 shows the execution model for a single node in a distributed TTP-system.

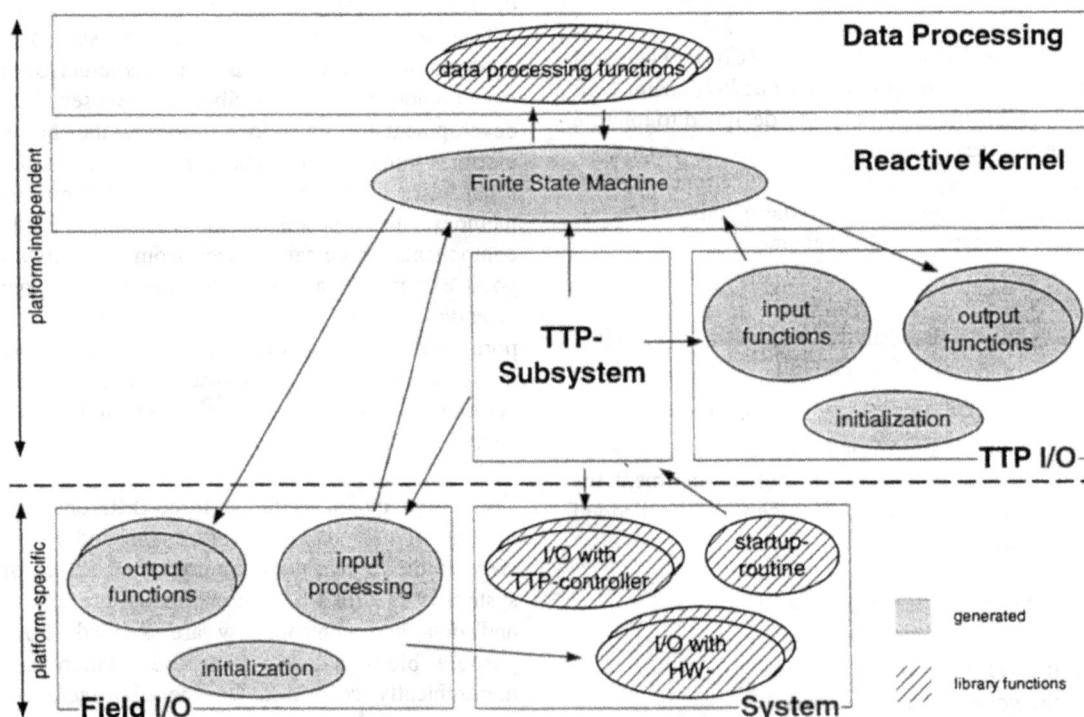

Fig. 3: Execution Model

The reactive kernel is a deterministic finite state automaton generated from the reactive parts of the synchronous software components and additional automatically generated ESTEREL modules needed to connect the components.

The TTP-Subsystem controls the application of the node and acts as a kind of operating system. It processes the interrupts from the communication system, reads sensor values and messages and sets the input signals for the reactive kernel. After that, it calls the kernel for the next reaction step. In normal operation, there is only the time interrupt occurring at predefined points in time. An application programming interface is provided to integrate the communication system into the application.

The data processing layer consists of library functions, constants and user-defined data types and functions used by the synchronous application components. There is no access from the data-processing layer to the hardware or the TTP-Subsystem. All data-processing actions are controlled by the reactive kernel. Data and control flow are separated in the execution model.

The TTP I/O processes input and output signals of the reactive kernel to be sent or received as TTP messages. The Field I/O processes the input and output signals from and to local hardware devices and sensors and actuators. The system layer includes platform-specific code modules from the hardware library. These are a platform-specific boot-up routine, I/O driver functions for accessing the hardware devices and platform-specific macros for accessing the communication network interface (CNI) of the TTP-controller.

5. CONCLUSION AND FURTHER WORK

The synchronous model of time and the use of a synchronous language like ESTEREL has significant advantages in software development for safety-critical embedded systems over traditional techniques used in industry. The main advantage lies in the rigorous analysis performed by the ESTEREL-compiler. This analysis is only possible on the basis of the mathematically precise defined semantics (Jagadeesan et al, 1995b). The combination of the synchronous approach with a time-triggered communication system allows for distributed systems to be built completely synchronous and deterministic. In our point of view, this is a necessary condition of every safety-critical system.

The use of component-based techniques can simplify the design process of complex software and libraries of standard components can be provided. Abstract specifications of software and hardware components can be used to construct systems on a very high-level and allow for small and efficient platform- and application-specific code to be generated.

Safety properties which stipulate that „something bad never happens" are sufficient to describe most of the important properties of real-time systems (Manna and Pnueli, 1992). In this class of properties, responses are required within a specified and bounded interval of time can be automatically verified by model-checking tools. This is not possible on programs written in a language without formally defined semantics such as C. Verification allows design errors to be found earlier and to reduce the effort for testing and debugging significantly (Jagadeesan et al, 1995a). At the implementation level the synchrony hypothesis must be justified by a worst-case execution time analysis.

The ViPER method with its tool environment is currently under development at our institute. A first prototype implementation is now ready to work with small models. A first version of the code-generator has been implemented and tested with a distributed TTP-system on two different hardware platforms. But much work needs still to be done. Further work concentrates on building interfaces to commercial tools like Matlab/Simulink and to a new fault-tolerant operating system. We are evaluating our concepts by means of a steer-by-wire model at our institute. And we have applied some of the concepts to an industrial case study – a brake by wire system for a car (Gunzert and Nägele, 1999) and (Ringler et al, 1998). The results were promising but more research is necessary to improve the method.

6. REFERENCES

Berry, G. and Gonthier, G. (1992). The ESTEREL Synchronous Programming Language: Design, Semantics, Implementation. *Science of Computer Programming*, 19(2): 87-152.

Bouali, A. (1997). XEVE: an Esterel Verification Environment, Technical Report, INRIA, 1997.

Caspi, P., Girault, A. and Pilaud, D. (1994). *Seventh International Conference on Parallel and Distributed Computing Systems* (PDCS'94) ISCA, Las Vegas, USA.

Clarke, E.M, Emerson, E.A, Sistla, A.P. (1983). Automatic Verification of finite state Concurrent Systems Using Temporal Logic Specifications: A practical Approach, Department of Computer Science Report, Carnegie-Mellon University.

Gunzert, M. and Nägele, A (1999). Component-Based Development and Verification of Safety Critical Software for a Brake-by-Wire System with Synchronous Software Components, To Appear in Proc. of *Int. Symposium on Parallel and Distributed Systems Engineering PDSE 99*, Los Angeles, CA, USA.

Halbwachs, N. (1993). *Synchronous Programming of Reactive Systems*, Kluwer Academic Publishers.

Harel, D. and Pnueli, A. (1985). On the Development of Reactive Systems: Logic and Models of

Concurrent Systems, Proc. NATO Advanced Study Institute on Logics and Models for Verification and Specification of Concurrent Systems, *NATO ASI Series F*, vol. 13, Springer-Verlag, pp. 477-498.

Harel, D.(1987). Statecharts - A visual Approach to Complex Systems, *Science of Computer Programming*, Vol. 8-3, pp. 231-275.

Halbwachs, N, Lagnier, F and Ratel, C. (1992). Programming and verifying real-time systems by means of the synchronous data-flow language LUSTRE, *Transactions on Software Engineering*, 18(9):785-793.

Jagadeesan, L.J., Puchol, C. and Von Olnhausen, J.E. (1995a). A formal approach to reactive systems software: A telecommunications application in ESTEREL, *In: Proc. Workshop on Industrial-strength Formal Specification Techniques*.

Jagadeesan, L.J., Puchol, C. and Von Olnhausen, J.E. (1995b). Safety Property Verification of ESTEREL Programs and Applications to Telecommunications Software, *In: Proc. of the Seventh Conference on Computer-Aided Verification*.

Kopetz, H and Grünsteidl, G. (1994). TTP - A Protocol for Fault-Tolerant Real-Time Systems, *IEEE Computer* Vol 27, No. 1.

Manna, Z. and Pnueli, A. (1992). *The Temporal Logic of Reactive and Concurrent Systems*, Specification, Springer-Verlag.

Manna, Z. and Pnueli, A. (1995). *Temporal Verification of Reactive Systems - Safety*, Springer-Verlag.

Ringler, Th., Steiner, J., Belschner, R. and Hedenetz, B. (1998): Increasing System Safety for by-wire Applications in Vehicles by using a Time Triggered Architecture, *In: Proc. Safecomp*, Heidelberg, Germany.

Sentovich, E.M. et al (1992). *SIS: A system for sequential cirquit synthesis*. Technical Report, U.C. Berkeley.

A Completely Integrated Approach to Developing, Implementing, Evaluating Distributed Active Database Management and its OS Support

Horst F. Wedde
University of Dortmund, Germany

Kwei-Jay Lin
UC Irvine, USA

Aloysius K. Mok
University of Texas, Austin, TX, USA

Krithivasan Ramamritham
IIT Bombay, India & UMass, Amherst, Mass., USA

Abstract

This paper constitutes a work-in-progress report on the first, mostly conceptual phase of a major international effort in building and evaluating a distributed testbed for database application systems in safety-critical real-time environments. Given that safety/ reliability requirements and real-time constraints are in conflict there cannot be a closed form solution or design, and all system functions have to dynamically adapt to the unpredictable environmental situation in safety-critical real-time systems. Over the past years groups headed by the authors had pursued novel concepts and approaches on a novel tailored LINUX kernel, safety-critical application operating system support (MELODY project) and Active Database levels (concurrency control under data replication). In the research project a complete integration of these achievements from the basic OS kernel through the application levels is the major theme. While quite a number of research and design problems stemming from the partial insights or incomplete functionality on the various levels posed serious challenges the integration itself requires, and gives rise to, extensions, modifications, or refinements of the functions involved for ensuring system survivability. The paper will describe the technical details of these implications from the integration for the whole project as well as the stepwise design and evaluation of the different functions and models.- To our knowledge this is the first fully integrative attempts in the area of distributed safety-critical real-time systems.
Copyright © 1999 IFAC

Keywords: distributed operating systems, real-time systems, safety-critical systems, concurrency control, active databases, similarity

Address for correspondence: *Prof. Dr. Horst F. Wedde, Informatik III, University of Dortmund, 44221 Dortmund / Germany, wedde@cs.uni-dortmund.de*

1. TASKS IN SAFETY-CRITICAL SYSTEMS.

In *safety-critical systems,* (such as nuclear power plants, distributed cooperation of autonomous robots in Outer Space or on the plant floor, automated aircraft landing systems for bad weather conditions, etc.) tasks not only have to meet deadlines, but most are critical in the sense that the system would not survive in case of a task-specific number of deadline failures of subsequent task instances. In such a critical stage, an instance is said to have become ***essentially critical***. However, beyond this special real-time responsiveness (here successful handling essentially critical task instances which was termed ***survivability*** (Wedde, Lind(1997)), safety-critical systems must also satisfy rigid ***dependability*** requirements. A high amount of ***adaptability*** of system functions is demanded to meet these ***conflicting requirements***, since the unpredictable environment does not allow for a static trade-off between these classes of restrictions.

Organization of the Paper. The theme of the paper is a multi-level integration of research work in real-time operating system kernels, operating system support for safety-critical applications, and real-time adaptable transaction processing. In order to substantiate, and further elaborate and expand on, the concepts addressed a novel application scenario will be described in section 2. Section 3 gives a brief survey over the novel MELODY operating systems functions. At the same time, it is an introduction into our novel design and analysis methodology termed Incremental Experimentation which is central for the whole project. Section 4 is a short intimation of our transaction model and its refinement through criticality and sensitivity. In section 5 we sketch a similarity concept studied previously. We indicate its use for refining both MELODY functions (File Server and File Assigner) and the concurrency control algorithms described in section 4, also pointing to the potential benefits for system survivability in the integrated model.

2. AUTOMATED LANDING SYSTEMS

Corrective actions in automated aircraft landing systems (ALS) contribute to adjustments for approach and landing of an aircraft. The environments of these systems are categorized regarding the ability of the pilots to assume control of the aircraft. Category 1 conditions are those in which the line of sight is at least 600 meters and results in a decision height (maximum height at which a decision to abort landing could be made) of 200 meters. Category 2 conditions reduce the line of sight to 400 meters which results in a decision height of only 100 meters. In category 3 conditions the line of sight is further reduced to 200 meters which effectively reduces the decision height to 0. Typically, all aircraft are grounded (even military aircraft) in category 3 weather conditions in order to avoid the loss of lives.

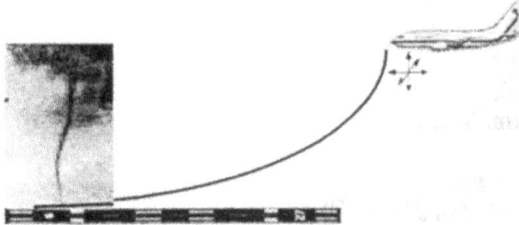

Fig.1: Bad Weather Landing

Automated landing systems could help to allow flights to land even during category 3 weather conditions such as depicted in fig.1 . Corrective actions to bring the aircraft back on course, or keep it there, are computed by tasks which use information periodically gathered by sensors at the aircraft periphery and filtered/ structured by distributed preprocessors. The response times of the mechanical control systems are considerably long (up to more than 4 sec). During this time the environmental situation may have changed drastically which may render a correction obsolete, or even dangerous, for the safety of the plane. Therefore such actions have to be designed to cause considerably *small-scale changes* which, in the negative case mentioned, are unlikely to have disastrous consequences. The computing tasks would be performed whenever necessary (typically aperiodically).

A key idea is that in a series of small-scale corrections occurring with a high frequency, each of them may outbalance the effect of the previous ones while contributing to gradually forcing the aircraft back on course. Also the information on the environment would be more frequently taken into account resulting in more up-to-date reactions. For this purpose the deadlines of the corresponding task instances are set very tight. When a task instance is about to miss its deadline it will be aborted since we assume subsequent task instances may make up for the previous ones to a certain extent. However, as the aircraft may meanwhile get considerably off its course it becomes more and more critical to successfully complete a correction in due time. These actions, while then potentially more and more rigid, might eventually cause the plane to be destabilized. Therefore, for every corrective activity, the tasks are to be designed such that

(2) there is a preset number of tolerable consecutive deadline failures of task instances. Beyond this the next deadline is *hard* or *essentially critical* (see section 1).

(3) whenever during this series of instances a deadline has been met – the associated corrective action is assumed to have the outbalancing effect mentioned – it has the same frame of tolerance as the previous successful instance.

We have formalized this into a concept of *(relative) task criticality* (Wedde, Lind(1997)). The more critical it is for a corrective action to be performed in due time the more tolerable it is for the survivability of the system to trade up-to-date information – which may not be available in time in our type of distributed systems – for information that is *nearly up-to-date* while available at the computer site where it is needed (by a reading task). So, beyond a preset number of consecutive deadline failures the use of nearly up-to-date information is considered satisfactory. This has been formalized into a concept of *(relative) task sensitivity* (Wedde, Lind(1997)).

3. THE ADAPTABLE SAFETY-CRITICAL OPERATING SYSTEM MELODY

Task criticality and sensitivity are crucial operating system features in the MELODY project to guarantee survivability of the application. We will explain MELODY's major functional novelties in this section.

As both the adaptability and the conflicting constraints of real-time responsiveness and dependability were to be respected in MELODY (see section 1) we started the project development with a rather simple model for studying novel adaptive file system functionality *(phase 1)* in order to cope with this enormous complexity. In order to provide for a variable file replication a novel function, the *File Assigner (FA)*, was created. Not only was the number and location of mutually consistent file copies (termed *public copies*) to be arranged according to the needs of local tasks. There was also a relaxed distributed protocol introduced by which certain copies would be refreshed after each update of a file *(i.e. of its public copies)* was completed. The latter copies were called *private copies*. They represent nearly up-to-date information (see section 2) as long as there were only gradual changes to the file. According to the varying weight of the real-time and dependability constraints the local File Assigners executed a distributed consensus protocol to manage the number, location, and quality of file copies, at the same time reflecting that read tasks (in phase 1!) would use public or private copy information interchangeably, depending on local availability only. (Write tasks write to the public copies only, by definition.)

Since there is no overhead for maintaining consistency among private copies a model which provides for private copy replication only (one public copy only) *(Private Copy Model)*, could be expected to have clear performance advantages (regarding the number of deadline failures) over a model which allows for public copies only *(Public Copy Model)* as *long as there* is a high dominance of read tasks (which entails a very low number of public copy updates, hence very infrequent private copy refreshments. (Public copy maintenance

requires a considerably higher overhead.) Conversely, under a high write dominance the frequent refreshments of the private copies (sending a fresh copy to the private copy site) would probably jam the network while the overhead for mutual consistency could be assumed to be approximately linear with the number of public copies and competing tasks.

This was investigated in comparative simulation experiments. Both models were, of course, compared with MELODY but also with two benchmark models: the *Base Model* (no FA) and the *Ideal Case Model* (FA costs set to zero). Some results are depicted in fig. 2 and 3. They clearly showed that the expectations mentioned for the Private and Public Copy models were valid. Even more importantly, MELODY outperformed both of them throughout all simulation experiments. Its higher overhead (compared to the simpler models) was more than outweighed by its higher flexibility.

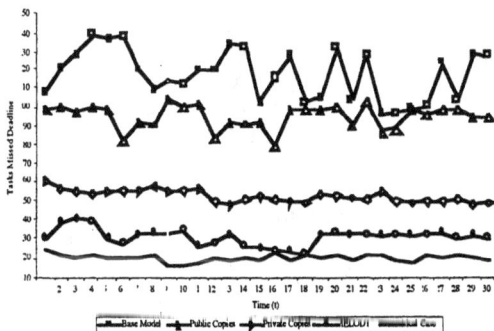

Figure 2: Read Dominance Task Profile

With these results in mind the MELODY task model was refined by including criticality and sensitivity (see section 2). The FA functions had e.g. to be modified in that sensitive read task instances would *always* require public copy information. This constituted *phases 2 & 3* of MELODY. Technical details will otherwise have to be omitted here, due to page limitations. They can be found in (Wedde, Lind(1997)).

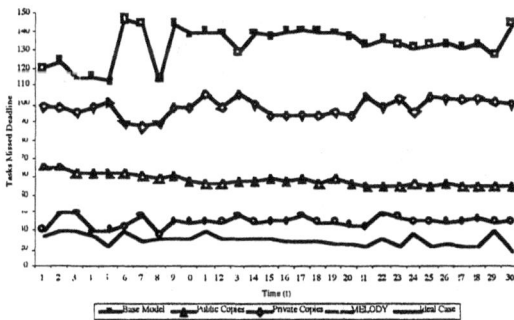

Figure 3: Write Dominance Task Profile

In order to connect the insights from phase 1 to the refined model, extensive simulation was done such that the task profiles chosen for the new MEL-ODY stage were solely consisting of non-critical and non-sensitive *(robust)* tasks. Our expectation was that the previous experimental results (see fig. 2,3) could (and should) be mimicked in the new MELODY model. This was the case throughout (Wedde, Lind(1997)). While in a second round criticality and sensitivity values were widely varied it turned out that with increased sensitivity the roles of the Public and Private Copy models were surprisingly flipped (fig. 4) thus revealing a strong influence of sensitivity on the deadline failure rate performance. Criticality, in turn, has a remarkable impact on the system *survivability* (ability of meeting all essentially critical or hard deadlines). Again, we refer to (Wedde, Lind(1997)) for technical details.

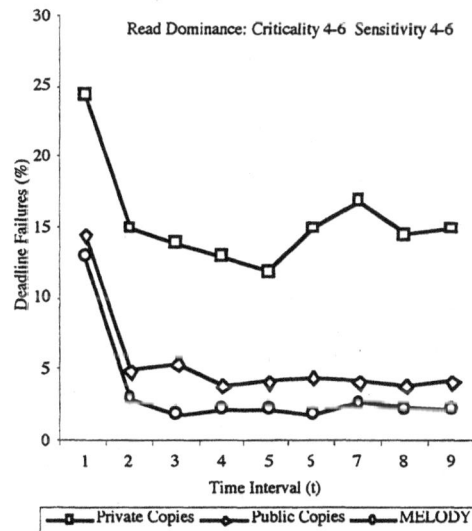

Figure 4: Read Dominance and Sensitive Tasks

These three MELODY development phases discussed represent the major features of our heuristic and experiment-driven design methodology called *Incremental Experimentation.* It leads to an iterative procedure of refinement and extension which follows a scheme of using experimental insights in one phase, leading to concepts and expectations for system refinement or extension in the subsequent phase, evaluating the previous insights within the refined/ extended model context, deriving new insights by utilizing the evaluation just mentioned etc. An example for the efficiency of our method is that the results of the impact of criticality and sensitivity on the system performance described for phases 2 & 3, could hardly have been achieved without doing phase 1 before building the refined model. So far, we have successfully competed completed six phases in the MELODY project.

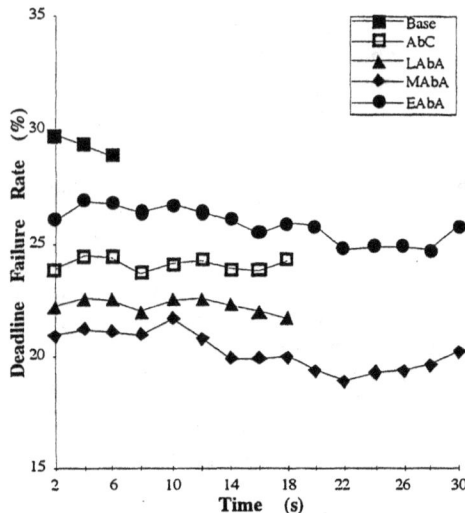

Figure 5: Medium Competition Task Profile

As phases 2 and 3 were being completed, novel distributed resource scheduling algorithms (Priority Insertion, Delayed Insertion and On-the-Fly protocols) had been defined and validated (through extensive comparative simulation experiments) in *phase 4* (Daniels(1992)). After a complete comparative evaluation the Delayed Insertion protocol was chosen to implement the distributed resource scheduling in MELODY.

Different from traditional operating system design the distributed task execution on distributed resources in MELODY led us to *reverse the order of task and resource scheduling* (**phase 5**). To make up for the inaccurate information the Task Schedulers were left with the *Run-Time Monitors (RTM)* were introduced which, at each site, would abort tasks as early as possible, based on precise information from the local FS. At the same time competing task instances at remote sites would benefit from resources being locked as late as possible in the task execution course at a given site. Technical details are in (Wedde et al.(1998)).

The local RTM was also put in charge of a *dynamic integration of Task Scheduling and File Server activities.* These measures proved to be very effective for MELODY. Details on the very extensive simulation studies can be found in (Wedde et al.(1998)).

In a second effort we investigated and refined the Task Monitoring function of RTM considerably. The idea behind was again to trade earliest possible task abortion (carrying a measurable decision inaccuracy) for an abortion based on accurate information but occurring possibly much later in the task life phases. The key for this approach is that wrong abortion decisions could be tolerated *to a certain degree* (see section 2). So, beyond the previous RTM function to abort before computation (AbC model) we provided for three more models in which abortion decisions would even be made during the initial part of the resource acquisition phase: *the Latest Abort before Acquisition (LAbA), Medium Abort before Acquisition (MAbA),* and *Earliest Abort before Acquisition (EAbA).*

Since at this time all MELODY servers had been fully and explicitly integrated (Wedde et al.(1996),

Wedde et al.(1998), Wedde et al.(1999b)) simulation did not appear any longer as an adequate evaluation method since we had no longer a sufficient understanding of the impact coming from pre-specified parameters like task execution, communication, resource acquisition times etc. This left us without a decent basis of judgment about the quality of our highly complex service and integration mechanisms. Therefore the TM evaluation, as well as the previous evaluations, was done through *distributed experiments,* in *phase 6.* (An in-depth technical discussion can be found in (Wedde et al.(1999b)) based on our own method of internal time synchronization (Wedde et al.(1999a)).

While the new protocols outperformed the AbC model throughout the MAbA model performed best looking over the whole range of task profiles, both under the deadline failure rate and survivability performance criteria, and was therefore adapted for MELODY (see figure 5). As a unique result, although the EAbA model was worse than both the AbC and LabA models regarding the deadline failure rate, it clearly survived its two competitors for both the medium and high competition profiles. While this appears counterintuitive at first glance it proves in fact that deadline failure performance and survivability are independent real-time measures.

4. DISTRIBUTED ACCESS TO REPLICATED DATA

Recently novel concurrency algorithms have been constructed for replicated data access under firm deadlines, and extensively evaluated under simulation (Xiong et al. (1999) The idea was to define a new Optimistic Distributed Concurrency Control protocol (OCC) and a new Distributed Optimistic Two-Phase Locking algorithm (O2PL). They operate on transactions modeled as depicted in fig.6. Assuming that preemption is used on the master level only a number of data conflict resolution mechanisms (Priority Blocking, Priority Abort, Priority Inheritance) and combinations are used to avoid, or minimize, priority inversion.

If we identify cohorts as tasks the MELODY task model includes the cohort model in fig.6 in which updaters at different sites take care of the replicas. (MELODY tasks executions are not limited to replication of updates but could be considered as small-scale transactions.) No provision is taken for adaptability in safety-critical environments. In the proposed international project, we will introduce criticality and sensitivity on the cohort level assuming that – like deadlines – these parameters may have different values (or otherwise all cohorts have the same value), thus creating adaptability to the unpredictable environmental requirements on the transaction level. The new protocols will be evaluated in *distributed experiments.* They will be compared to each other, and with their non-adaptable yet simpler versions, and with any other adaptable protocols (so far they will be available during this advanced project phase).

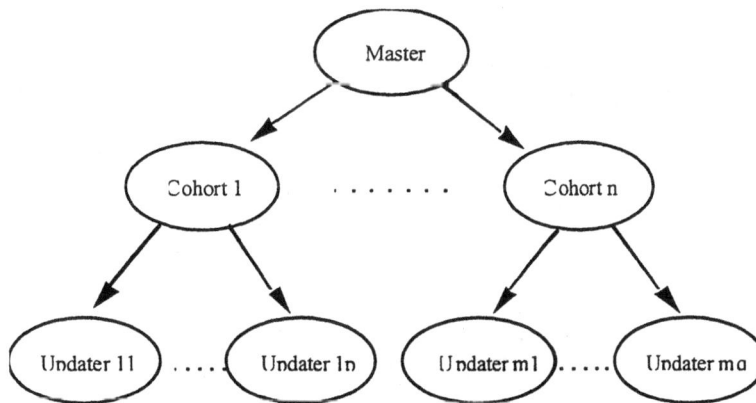

Figure 6: Transaction Model for Replicated Data References

5. DATA AND TASK SIMILARITY

In a distributed safety-critical system such as depicted in fig.1 there are sensors at the periphery of the application which input a continuous stream of data into preprocessors. Here periodic tasks filter, digitize, and structure these data. To each data object O originating from this process – which is a template for its instances – a *similarity bound sm(O)* will be assigned. (More technical details can be found in (Kuo et al.(1997)).)

(0)   Note that this bound could be dynamically reset for different task instances (as e.g. the aircraft in fig.1 approaches a more critical phase of landing). Typically the bound would be lowered as the environment gets more unpredictable.

(1)   Every task t_j will also be assigned a *similarity bound sm (t_j, O)* which is related to the effect of t_j using O for an execution instance. (In a first phase we will assume that all sm $(t_j.O)$ are equal for the objects which t_j needs. An easy example is a cohort updating replicas of one object.) $sm(t_j, O)$ will be sent to the site from which O originates. We will assume that

$$sm(O) \le min\{sm(t_j) \mid t_j \text{ needs O}\}.$$

(2)   If for a data object O generated at a preprocessor P_i, or another processor, a subsequent instance is no longer similar to the previous one (as specified by sm(O)) – say it differs by $a(O) > sm(O)$ - then the object manager at P_i will compare $a(O)$ to $sm(t_j)$ where t_j needs O. If $a(O) \le sm(t_j)$ then tj would not be invoked. If $a(O) > sm(t_j)$ then a message will be sent to the site holding t_j to invoke the next task instance.

(3)   The condition under (1) makes sure that task instances will be invoked whenever necessary, i.e. whenever $sm(t_j)$ is exceeded, since the alerting procedure at

the site generating O would be invoked in time. At the same time this sets up, in an explicit model *for safety-critical systems and their environment,* the principle of *typically aperiodic tasks* as it underlays MELODY.

Note: Sensitivity and similarity are closely related concepts in the MELODY context: Both refer to the difference between earlier and later information. However, they are still independent parameters: First, sensitivity is no measure for invoking a task instance. Also, as the situation for the computer system becomes more and more unpredictable (and subsequent task instances would fail more and more frequently) the relative sensitivity will be more and more relaxed (see section 2) while similarity bounds get tighter instead (as was remarked in (0)). Another difference is that sensitivity directly relates to task management while similarity reflects the data management.

Similarity will be defined for tasks in MELODY as well as for cohorts in the transaction model introduced in fig.6. It will be utilized in the project to further modify and improve the adaptability on different levels:

OS level (MELODY): Upon an update of a file f, i.e. of the public copies of f (see section 3), if the change to f is within preset similarity bounds then the update operation which refreshes the private copies would not be executed at this time. Also, when updated data objects are similar to the previous instance tasks which operate on these objects would not be invoked.

Database level: The concurrency algorithms mentioned in section 4 will be assigned similarity values for every cohort, or for the master (if the cohort values are set to be all equal).

All actions combined, the following project phases emerge:

- Creating an adaptable real-time LINUX version to handle the shell-level MELODY functions;
- Extending MELODY functions FA, FS by including similarity; realize similarity of files, tasks, according to an appropriate application study;
- Extending the transaction model by introducing criticality and sensitivity of cohorts or of the master;
- Extending the transaction model by introducing similarity of cohorts or of the master;
- Evaluating the different modified concurrency control protocols, at the same time comparing them with the less adaptable versions, in distributed experiments which are incrementally performed, according to the principles of Incremental Experimentation (see section 3)
- Investigating existing application studies, in particular in automated flight control.

This constitutes a mid-range research effort (5-6 years) involving four institutions from Europe, India, and the US. From initial studies we derived the hope that the more sophisticated models of our completely integrated computer system will significantly outperform the earlier system models, in particular regarding the most crucial measure: *survivability*.

6. REFERENCES

D.C. Daniels, "The Design and Analysis of Protocols for Distributed Resource Scheduling under Real-Time Constraints"; *Ph.D. Dissertation*; Wayne State University, June 1992.

T. Kuo, S. Ho, A. Mok, "Similarity-Based Load Adjustment for Real-Time Data-Intensive Applications"; Proc. of the 18th IEEE Real-Time Systems Symposium; San Francisco, December 1997

M. Xiong, K. Ramamritham, J. Haritsa, J.A. Stankovic, "Regulating Concurrent Access to Replicated Data in Distributed Real-Time Systems"; Proceedings of the IEEE Workshop on Dependable and Real-Time E-Commerce Systems (DARE'98),Denver, Colorado, June 2, 1999

H.F. Wedde, J.A. Lind, "Building Large, Complex, Distributed Safety-Critical"; *Real-Time Systems* **Vol. 13 No. 3** (1997)

H.F. Wedde, C.Stange, J.A. Lind; "Integration of Adaptive File Assignment into Distributed Safety-Critical Systems"; *WRTP '96, 21st IFAC/IFIP Workshop on REAL TIME PROGRAMMING*; Gramado, RS, Brazil, November 1996

H.F. Wedde, J.A. Lind, "Integration of Task Scheduling and File Services in the Safety-Critical System MELODY"; Proceedings of the EUROMICRO '98 Workshop on Real-Time Systems, Berlin, Germany, 1998

H.F. Wedde, J.A. Lind,., G. Segbert, "Achieving Internal Synchronization Accuracy of 30 μs Under Message Delays Varying More Than 3 msec"; Proc.of *WRTP '99, 24th IFAC/IFIP Workshop on REAL TIME PROGRAMMING*, June 1999;

H.F. Wedde, J.A. Lind,., G. Segbert, "Distributed Real-time Task Monitoring In The Safety-Critical System MELODY"; Proceedings of the EUROMICRO '99 Workshop on Real-Time Systems, to be published 1999

TOWARDS A GENERAL REAL-TIME DATABASE SIMULATOR SOFTWARE LIBRARY

Juha Taina* Sang H. Son**

*Department of Computer Science,
University of Helsinki, FINLAND*

**Department of Computer Science,
University of Virginia, USA*

Abstract: A real-time software simulator library scope is presented. It consist of hardware, software, transaction, and event generator components that can be concatenated to form a core real-time database simulator. The goal of the library is to allow simulator designers to use a well tested and documented software to implement basic database simulator functionality. All additional and research-specific functionality can be added on top of the library components. *Copyright © 1999 IFAC*

Keywords: Real-Time Databases, Simulation, Software Reuse

1. INTRODUCTION

Software reuse is one of the key principles of software engineering community. Generally 80% of new software can be reused from existing software components. Only 20% of code is new. The same principles can be applied to real-time database simulators (and of course to real-time database management systems.) A proper simulator library package can simplify the process of generating new real-time simulators.

With an object-oriented approach it should be possible to develop a set of real-time database simulator classes that can be used to generate different types of simulators; starting from simple main-memory database simulators and ending to complex real-time distributed parallel processing database architecture simulators. While the extremes do not seem to have much in common, they still share common components such as concurrency controller, scheduler, and transaction manager. All those components, and others as well, can be modeled with object-oriented classes.

In this paper we give a general scope of such a library. We concentrate on the necessary components and introduce the first steps of our development effort. We trust that this work will benefit the whole real-time database research community.

2. RELATED WORK

Although most real-time research papers include simulators, few describe their simulator software. Simulation parameters are well described. A classical paper in this field is (Abbott and Garcia-Molina, 1988). Some recent papers that have good simulator environment descriptions include (Ulusoy and Buchmann, 1998), (Lam et al., 1995), (Gupta et al., 1997), and (Chen and Gruenwald, 1996). A comprehensive description of a real-time database management system implementation is in (Aranha et al., 1996).

Recent simulator studies in real-time systems include (Audsley et al., 1994) and (Kim, 1997). The former is a hard real-time system simulator. The latter describes a new object model to support real-time systems and simulators.

Audsley et al. describe their hard real-time system simulator as a tool to evaluate scheduling and resource management algorithms. It is based on a set of CASE-tools along with a window-based user interface and a language to describe simulated environments. The simulator interprets a special language that is used to describe the simulated environment. Their approach is good and useful to variety of simulation problems. However, their approach is for hard real-time systems while our library is for real-time databases.

Kim describes a new real-time object structure that has a concrete syntax structure and execution semantics to specify temporal behavior of modeled subjects (Kim, 1997). Although the new object type can be used in real-time simulators, it is first intended for simplifying design and implementation of real-time systems. Nevertheless their approach can be used to implement a real-time database simulator, although a model alone does not give enough support for new real-time database simulators.

RADEX, under development at the University of Massachussetts, is one of few well documented RDB simulators (Sivasankaran *et al.*, 1998). The RADEX software consists of a complete active and temporal real-time database simulator software.

While RADEX project is similar to ours, they are more concentrated on a full simulation software. RADEX implements a set of simulated database managers and their relationships. Our plan is to design a simulator library that supports simulated hardware, managers, event generators, and transactions. Hence, our plan is to give enough support that new simulator softwares can be implemented with less effort. This allows designers to concentrate on actual research instead of mechanic simulator coding.

3. LIBRARY COMPONENTS

A real-time database simulator software can be divided into four types of components: hardware components that model the underlying hardware architecture, software components that model database managers, transaction components that model transactions, and event components that generate events to the real-time database simulator. Object inheritance allows to add new components to the library. For instance, a new scheduler can inherit the functionality of the general scheduler component. Totally new software components inherit functionality from the software component. The architecture must be flexible enough to allow a simulator designer to easily expand library functionality.

3.1 *Hardware components*

At least the following basic hardware components must be modeled: CPU, Disk, Memory, Bus and Network, and Node.

- *CPU.* CPUs are used to model transaction execution.
- *Disk.* Disks model all disk traffic. They include a disk manager that serves disk requests.
- *Memory.* Memories model all main memory hardware. They are the main source of CPU-processed data. It is possible to simulate non-volatile memory.
- *Bus and Network.* A Bus connects hardware elements. Basically it is a data transporter between different components but it can also be used to simulate bus behavior. A special case of a Bus is Network. A Network is similar to a Bus but it connects distributed database nodes.
- *Node.* A Node functions as a compound element to hardware elements. It combines the elements together to form a larger entity. Nodes can include nodes when a parallel architecture is simulated.

3.2 *Software components*

Software components offer core functionality for the database simulator. They include database managers that model different tasks in a database management system. At least the following components are needed: Schedulers, Concurrency controllers, Cache managers, Transaction managers, Query optimizers, Distribution managers, Replication managers, Directory managers, Active operation managers, Watchdogs, Recovery managers, Log managers, and Data model managers.

- *Scheduler.* The Scheduler component offers services to schedule, commit, and abort transactions. The scheduled transactions consist of low-level operations: reads, writes, and command executions. Reads and writes affect specific data items, command executions simulate operation execution by using simulated CPU instructions.
- *Concurrency controller.* The Concurrency controller component offers services to allow transactions to execute in parallel. Its functionality is equal to the modeled concurrency control method.
- *Cache manager.* The Cache manager components manage both disk and main-memory data.
- *Transaction manager.* The Transaction manager component accepts transaction requests and processes them accordingly.

- *Query optimizer.* Although not mandatory, this component is nevertheless useful. It offers services to calculate query optimization processes. No actual query optimization is made here, only simulated behavior is included.
- *Distribution manager.* This component offers services for both distributed and parallel transaction management. It processes low level operations. Higher level distributed operations are left to suitable data model managers.
- *Replication manager.* This component offers services that manage replicated elements, both in a parallel and in a distributed architecture. Again, it processes low level operations.
- *Directory manager.* This component offers services that are related to directory modeling. It does not implement actual directory management operations. Instead, it allows other components to use directories and have their transaction management measures tuned accordingly.
- *Active operation manager.* Active operation manager component offers services that are related to active databases. Various triggering operations are forwarded to it for further processing.
- *Recovery manager.* Recovery managers work together with Watchdogs. The Recovery manager simulates database behavior once a recovery process has started. A Watchdog simulates processing which checks that other controllers are alive.
- *Watchdog.* Watchdog component offers services that support database fault-tolerance and recovery simulation. Together with a Recovery manager it simulates database recovery processes.
- *Log manager.* Log manager component offers services to simulate database log reading and writing, either in a main-memory or disk-based environment (or both.)
- *Logical data model manager.* This component offers services for high-level logical data model simulation. Some simulators may want to model high level operations, such as method calls in object-oriented databases, in which case a Logical data model manager translates them to low level operations.
- *Security manager.* Security manager component offers services to simulate database security.

All the listed components are measurable. The actual measured items depend on the component. For instance, a Transaction manager component can keep track of committed and aborted transactions. A Concurrency controller component can keep track of transaction conflicts.

3.3 *Transaction components*

A transaction in a simulated database behaves mostly the same way as in a real database. It is generated from a request, its operations are executed, it can wait in several wait queues, and it is finally committed or aborted.

Two types of transactions are needed: real-time transactions and regular transactions.

- *Real-time transactions.* Naturally a real-time database simulator must have real-time transactions with absolute or relative deadlines. Soft, firm, critical, and hard transactions are all supported. This class also supports transactions that read or modify temporal data.
- *Regular transactions.* Some simulations may need transactions that do not have deadlines.

A transaction has one or more executed operations. The operations can be modeled at a low level as reads and writes, or at a higher level as logical model operations. The high-level operations need the Logical data model manager to translate them to low level operations.

3.4 *Event generator components*

The Event generator components offer services to feed events to the simulator. The Event generators include Transaction event generators, Control event generators, and Failure event generators. A simulator may have several Event generators, each of which generates different types of events.

The Event generators are connected to a Node; usually to the high-level node that encapsulates the full distributed architecture, but also to a subnode in the architecture. A high-level generator simulates global requests. A low-level generators simulates requests that are specific to a single node.

At least the following types of Event generators are needed:

- *Predefined transaction generators.* A predefined transaction has known operations to execute. The parameters of operations can be randomized. For instance, it is possible to use a transaction core and change the accessed database items in read and write operations.
- *Random transaction generators.* The transaction request is generated randomly. No predefined transaction core exists. However, rules can be used to define semi-random

transaction generators. For instance, a rule can specify that usually two or three read operations are needed for one execution and one write operation. Other rule can specify that 80% of read operations occur to 20% of data items.

- *Online transaction generators.* An Online transaction generator generates a transaction request where the generated transaction will receive new operations during its execution. This simulates regular online processing where a user executes database operations from a console.
- *Control event generators.* A Control event generator generates such events as when to start measuring a specific attribute and when to finish simulator execution.
- *Failure event generators.* A Failure event generator generates highly irregular events that usually imply a system failure. Also hardware and software components can include attributes to simulate failures without an external event.

4. DESIGN ISSUES

Starting from the previous definitions, we are developing our real-time database simulator library COLISART (COmponent LIbrary for Simulating Active and Real-Time databases). The library is based on object inheritance that is extended from the defined hierarchy in Section 3. The whole class hierarchy is in Figure 1.

Most of the library hierarchy is based on the definitions in Section 3. The base class of the library is the Component. It offers basic functionality that is common to all classes. Under it are the basic component types: Hardware, Software, Transaction, and Event generator. Every subclass has its own subclasses that further describe library functionality.

Communication between Components occur via streams. Each Component subclass has the following stream types: input streams, output streams, measurement streams, and control streams. An event is generated every time data is sent to a Component stream.

An input stream receives related information to a Component. The Component processes this information and forwards the result to zero or more output streams. The type of both input and output streams depends on the Component. A Component can have different types of input and output streams. For instance, a Concurrency controller component accepts both transactions and transaction operations. Hence, it has both transaction-based and operation-based streams.

All Components have acknowledgment streams. They are used as command acknowledgments (accepted, rejected, delayed, etc.)

The output streams can be connected to other Components' input streams. This results Component pipelines the same way as in a Unix environment.

A measurement stream is used for measuring parameters of the Component. The stream accepts read requests from an external element (where the stream is connected) and sends back measurement values of the Component. The responses from the Component describe the current status of itself. They can be used to draw statistical conclusions. The measurement stream can be automated to send measurements at specific intervals without a request.

A control stream is connected to an external software that is used to control the behavior of the simulator. It can be used to modify the attributes or internal structures of the component.

Hardware components simulate data traffic in the simulated database (for instance, disks (or disk controllers) offer services to move data between themselves and main memory). Software components simulate data processing in the database. Transaction components simulate transaction behavior, and Event generators feed new events to the simulator.

Perhaps the most interesting area in the library are the Software components. The Software class is further divided into three subclasses: Control-based, Transaction-based, and Data item-based Software components. These classes describe what kind of data each manager processes.

The generated and accepted events in the Software components depend on the class hierarchy. This restricts the necessary events to minimum since we do not have to define a new event type between all combinations of managers. Inheritance is used to define composite event classes that support special types of Software components. The composite event classes follow the library hierarchy: control events, transaction events, and data-item events.

Control-based components accept only events that control the whole database functionality. Such events include hardware and software failure events and database security events. The Components do not directly see transactions or transaction operations.

Transaction-based components accept events that concern whole transactions. For instance, a Transaction manager receives events to create new transactions.

Component

Transaction Hardware Software **) Event Generator

Real-time Regular

Transaction Control Failure

Predefined Random Online

Node CPU Store Transport

Network node Bus node

Memory Disk Bus Network

**) Software

Control-based Transaction-based Data-item -based

Recovery Security

Transaction manager Query optimizer Active operations manager Concurrency controller Cache manager Distr. manager Repl. manager Direct mana

Scheduler

Watch-dog Recovery manager Security manager

Logical data model manager

Log manager

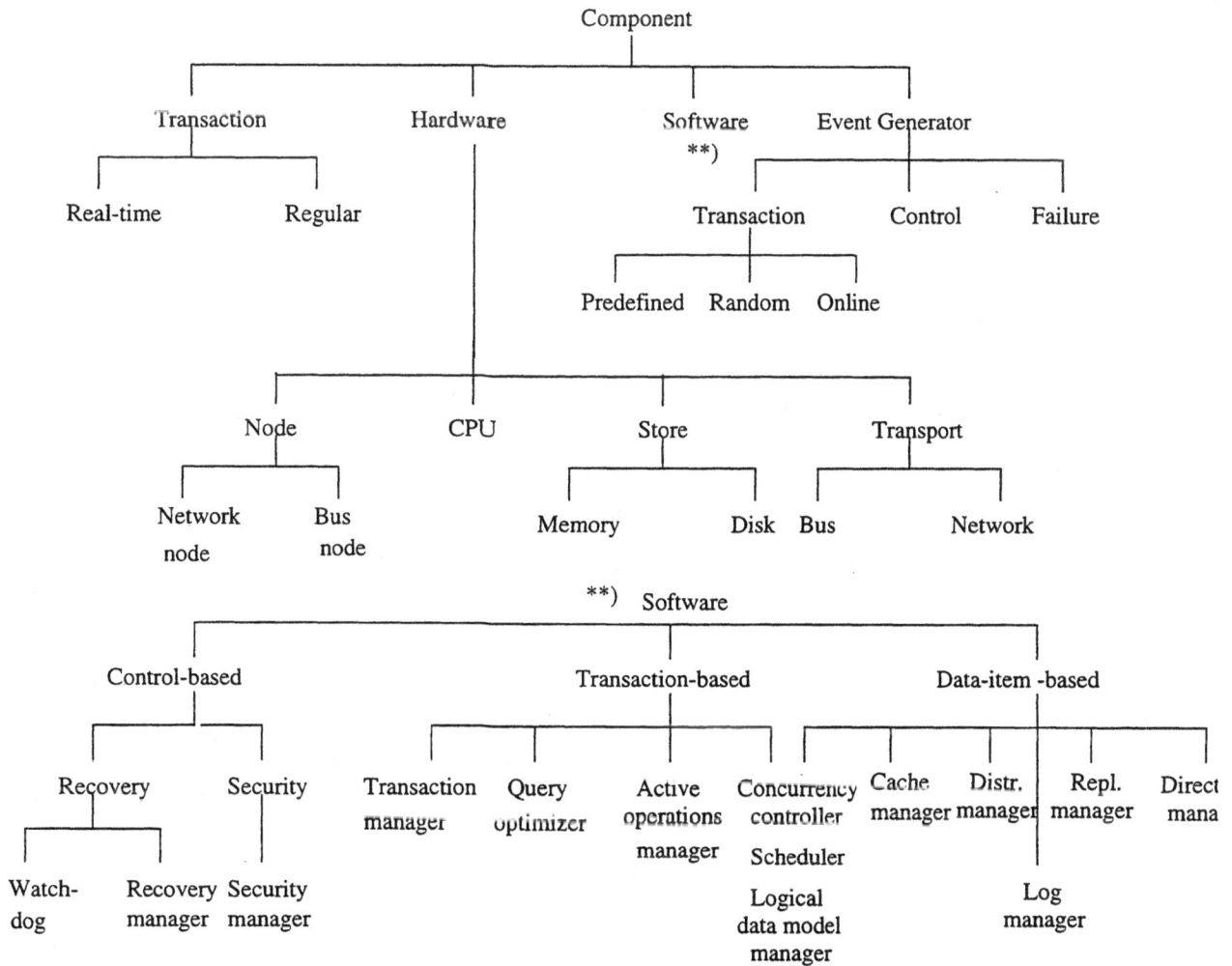

Fig. 1. COLISART Object Inheritance

Data item-based components accept events that concern a single data item in a transaction or database. Most Components are of this type since most operations in a database occur at a data item level.

Most Transaction-based components also need data item functionality. Hence, multiple inheritance is needed for those classes. The Concurrency controller, the Scheduler, and the Logical data model manager need both transaction and data item processing functionality.

The inheritance tree can be changed especially between Transaction-based and Data item-based components. In some simulators some Transaction-based components, such as the Transaction manager and the Active operations manager, may need Data item-based component functionality. Moreover some components, such as the Active operations manager, may need Control based component functionality.

5. LIBRARY EXAMPLE

The simplest possible real-time database simulator simulates a main-memory database management system. As an example we show the library components and stream connections that are needed to support this simulator.

Our simulated system hardware consists of one CPU and main memory. Moreover we need a bus to connect them and a node to encapsulate the components.

Our simulated system software consists of a Transaction manager, a Scheduler, a Cache manager, and a Concurrency controller. The whole architecture is in Figure 2.

The Transaction manager accepts new transaction requests and informs the simulator about committed and aborted transactions. The Scheduler decides which transaction gets the CPU. It sends transaction and operation information to the Concurrency controller which in turn forwards requests to the Cache manager.

On the route back the Cache manager acknowledges the Concurrency controller about the operation status. The Concurrency controller decides if the scheduled operation and transaction can be accepted. It forwards this decision back to the Scheduler.

The input streams of software components accept either transactions, transaction operations, or ac-

Hardware components:

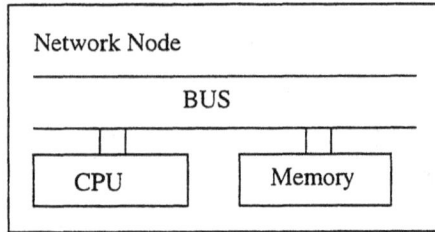

Network Node

BUS

CPU   Memory

Software Components:

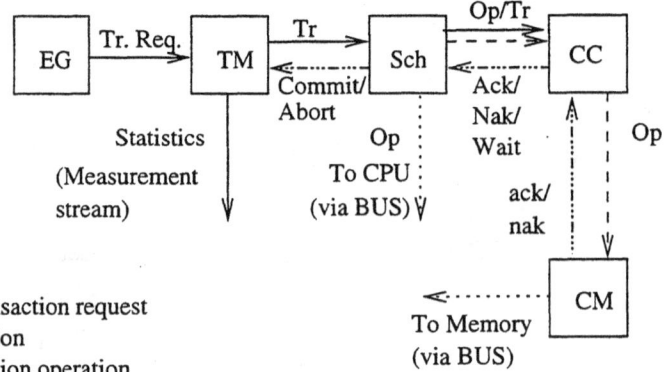

EG   Tr. Req.   TM   Tr   Sch   Op/Tr   CC

Commit/
Abort

Ack/
Nak/
Wait

Statistics

(Measurement
stream)

Op
To CPU
(via BUS)

ack/
nak

Op

CM

To Memory
(via BUS)

EG = Event generator
TM = Transaction manager
Sch = Scheduler
CC = Concurrency controller

Tr Req = Transaction request
Tr = Transaction
Op = Transaction operation
CM = Cache manager

Transaction stream   Operation stream

Hardware stream   Accknowledgement stream

All these are input/output streams

Fig. 2. Main-Memory simulator example

knowledgment information. These three types of streams are sufficient in the example.

6. SUMMARY AND FUTURE WORK

In this article we have described the first steps in defining a general real-time database simulation library. Most of the description is for the elements of the library. This is a natural approach that allows different types of implementations. The implementation we have described is by no means the only possible one.

The next step in COLISART project is to design and develop the first prototype version of the library. At the same time we will write tools to simplify the development of the next prototypes. Both requirements and design will probably change during this iteration. Still, without a proper prototype a project like this can easily fall into its own complexity.

7. ACKNOWLEDGMENT

This work was supported in part by project RO-DAIN in University of Helsinki, Department of Computer Science, and Ministry of Information and Communication of Korea.

8. REFERENCES

Abbott, Robert and Hector Garcia-Molina (1988). Scheduling real-time transactions. *ACM SIGMOD Record* **17**(1), 71–81.

Aranha, Rohan F. M., Venkatesh Ganti, Srinivasa Narayanan, C. R. Muthukrishnan and S. T. S. Prasad (1996). Implementation of a real-time database system. *Information Systems* **21**(1), 55–74.

Audsley, N. C., A. Burns, M. F. Richardson and A. J. Wellings (1994). Stress: a simulator for hard real-time systems. *Software – Practice and Experience* **24**(6), 543–564.

Chen, Yu-Wei and Le Gruenwald (1996). Effects of deadline propagation on scheduling nested transactions in distributed real-time database systems. *Information Systems* **21**(1), 103–124.

Gupta, Ramesh, Jayant Haritsa and Krithi Ramamritham (1997). Revisiting commit processing in distributed database systems. In: *Proceedings of the ACM SIGMOD'97 Conference on Management of Data*. pp. 486–497. ACM.

Kim, K. H. (1997). Objet structures for real-time systems and simulators. *IEEE Computer* **30**(8), 62–70.

Lam, Kwok-Wa, Kam-Yiu Lam and Sheung-Lun Hung (1995). An efficient real-time optimistic concurrency control protocol. In: *Proceedings of the First International Conference on Active and Real-Time Database Systems (ARTDB-95)*. pp. 209–225. ACM.

Sivasankaran, Rajendran, Bhaskar Purimetla, Jack Stankovic, Krithi Ramamritham and Don Towsley (1998). Design of radex - real-time database experimental system. Technical report. University of Massachusetts. Amherst.

Ulusoy, Özgür and Alejandro Buchmann (1998). A real-time concurrency control protocol for main-memory database systems. *Information Systems* **23**(2), 109–125.

HOW TO CONSTRUCT PREDICTABLE RULE SETS

Esa Falkenroth * Anders Törne *

* Real-Time Systems Laboratory,
Department of Computer and Information Science,
Linköping University, Sweden
{esafa, andto} @ida.liu.se

Abstract: This paper addresses the problem of unanticipated rule interactions. It characterizes problems with active rules in dynamic manufacturing environments, and formally defines a number of interaction properties such as cascades, rule cycles, and race conditions. The inversion of these properties defines criteria for interaction-free rule sets that form the basis for a compiler that automatically generates active rules. A feasibility study was conducted where an active database system was embedded into a manufacturing control system. A compiler translate high-level manufacturing operations into queries and active rules for the embedded active DBMS. Since the generated rule sets conform to the inverse criteria, they cannot lead to rule cycles, non-termination, or non-determinism. Copyright © 1999 IFAC

Keywords: Active and real-time databases, rule-termination analysis, automatic rule generation, software analysis and design methods for predictable behaviour

1. INTRODUCTION

The present work [1] has its origin in an investigation of how active database technology can support data management in control systems. Real-time system research has traditionally focused on developing mechanisms to support predictable execution of periodic control tasks with minor data dependencies. These mechanisms are not well suited for complex control systems that involve large volumes of data. Active database systems have been proposed an alternative solution to the data management problems of time-critical applications. Particularly control applications require the sort of reactive behaviour that is provided by active database systems. The HiPAC project (Dayal et al., 1988) pioneered this area and introduced the coupling modes. During the 90s, concrete systems such as REACH (Buchmann et al., 1992) and DeeDS (Andler et al., 1996)

followed. Within the EERTS project (Extendable Embedded Real-Time Systems), we investigated how active database systems can be embedded into layered manufacturing control systems. The work focused on the use of active rules for specification of high-level manufacturing operations.

After an initial study in 1994, we concluded that applications based on active rules are difficult to develop. Even a very small number of dependencies between rules will make it difficult to predict the behaviour of the rule set for arbitrary transactions and database states (Widom, 1994). Since then, the research community has given this problem considerable attention. A majority of the efforts has focused on refined execution models (Zhou and Hsu, 1990), schemes for exhaustive compile-time analysis (Baralis et al., 1998), and methods for detecting non-termination at runtime (Widom and Ceri, 1996). This paper suggests an alternative approach which allows rule sets with less entangled behaviour to be constructed. Rather than analyzing termination and other

---

[1] The EERTS project was financed by the Swedish National Board for Industrial and Technical Development.

properties of arbitrary hand-coded rule sets, the idea is to construct rule sets according to sound principles that guarantee that the final rule set will be confluent and terminate. The foundation for the construction principles is a formal definition of the interaction properties in a generic rule model (GRM). Based on an inversion of these properties sufficient and necessary criteria for interaction-free rule sets can be obtained. With very few exceptions, these criteria are complex and do not lend themselves to analysis. The main problem is that necessary criteria are quantified over database states, rule orderings, and update transactions which makes any verification of a rule sets intractable. However, stronger versions of the criteria can be analyzable. The aim of this work has been to use these strengthened inverse properties to define constraints for rule sets that can be syntactically verified. The results have been applied in the EERTS project where a compiler automatically transforms high-level programs into rule sets that conform to the constraints, and therefore are guaranteed termination and other desirable properties.

2. MOTIVATION FOR COMPILER APPROACH

In an early phase of the EERTS project, a number of problems with active rules were discovered concerning semantics, rule-base organization, and development methodology. Many of the problems were caused by the desire to encode too much functionality using active rules. Another cause was the highly dynamic nature of manufacturing applications. This section lists the encountered problems.

Sequences of operations: Rule-based systems are suitable for unpredictable environments or when the operational behaviour of a system is not predetermined. Manufacturing environments, on the other hand, are at least partially determined and the operational behaviour is a fixed set of sequences with synchronization points. To use rules for manufacturing operations, the concept of sequence must somehow be reintroduced. To our knowledge, there are three ways of handling sequences using active rules. Simple sequences without internal synchronization points can be coded as a procedure in the rule action. More complex sequences can be handled either by introducing state variables or by defining application events. When there are several possibly parallel sequences, neither of these approaches provides a natural way to specify sequences since they introduce numerous unintuitive state variables and composite event specifications.

Interactions and liveness: Active rules may have problems with unanticipated and unwanted interactions. Common interactions include situations where the action of one rule undoes the effect of another, or when the actions lead to firing of additional rules. If the propagated firing form cycles, the firings may not terminate. Such rule interactions are known to be difficult to analyze.

External interactions: Not all functions are suitable to encode with rules. For example, computations, input-output, and graphical user interfaces, are best programmed with traditional programming techniques. A common approach is to combine external applications with active rules. The active rules are allowed to send signals (call-out events) to the external program and the external program is allowed to update the database and to raise application events. The problem is that the updates may cause rule firings which in turn may perform call-out to the external application, *i.e.*, an additional possibility of unwanted interactions has arisen. An analysis of this interaction has to combine rule analysis as well as program analysis.

Reactivity timeliness: A computer based control system must be able to sense and react to changes in the environment. Based on sensor data, the system must make decisions and respond within a bounded time so that the response can safely operate the system in each monitor-respond cycle. For typical control applications, the response deadlines are in the millisecond range up to several minutes. Missed deadlines may result in serious damage to humans, material, or the system itself. Since the consequences can be catastrophic it is vital that active rules involved in the monitor-respond cycle also fulfill the timeliness requirements.

Efficiency: The performance of many rule processors implementations is still too low. Even if it is possible to encode large portions of the application semantics using active rules, this might not be wise from a performance point of view.

Dynamic rule sets: Control applications need to group activities into modes that are changed during the execution of different tasks. Each mode is associated with a set of conditions or situations that should be monitored. When a task enters a mode, the corresponding rules should be activated and when the task exits the mode these rules should be deactivated. Since most active DBMSs implementations do not support mode changes, this functionality has to be implemented by introducing additional state variables in the rule conditions. Often a rule will refer to several different state variables, which make the rules complex and unintuitive.

Non-atomic transactions: Because users rarely understand the effects of spawned information, implementers of transaction systems have been reluctant to provide notification across transaction boundaries. However, some rule implementations provide transactional leaks that violate the atomicity property and make rollback incomplete. Although this is necessary for maintaining audit trails that include aborted transactions, it poses intricate development problems for safety-critical system.

Rule-base organization: Normally, software organization involves an introduction of abstraction and a structuring of entities that belong together into logical group. However, active rules are by nature a flat organization, and do not usually support abstractions or logical groups. Notable exceptions exist (Stonebraker *et al.*, 1989)(Widom and Ceri, 1996)(Sköld *et al.*, 1995). For large and complex rule sets, this implies management problems.

Development methodologies: We lack a systematic approach to rule design. The available development methodologies and supporting tools are limited. Few systems provide tools that analyze rule interactions. Even fewer systems can guide the rule developer in choosing which part of the functionality should be encoded as active rules and which parts should stay in the application.

Maintainability: The maintainability involves intricate details. Changes to the rule set may require modifications to the external program and vice versa. It is generally non-trivial to determine how any changes will affect the total system.

Portability: When active DBMSs become more widespread, the issue of portability will be important. Given the heated discussions that preceded the ADMBS Manifesto (Dittrich *et al.*, 1995), it is unlikely that the rule processing semantics will be standardized. Instead, the rule developers must protect their investment by designing rule sets that are easy to port. The design of rule sets should not rely on peculiarities of any rule processor semantics or implementation. Furthermore, efforts to make the rule sets independent of rule processing semantics will make the rule set easier to port between ADMBSs.

Verifiable operation: A verifiable operation is essential for manufacturing applications. Unfortunately, verification of complex rule sets is at best tedious, at worst undecidable. It is hard to predict what rule firings an update will result in; especially when the rule conditions are allowed to be parameterized in combination with a high degree of expressiveness. In addition, most active DBMS rule semantics provide opportunities to write non-deterministic rule sets, where the final database state depends on the ordering in which the rules are processed. Testing is another unexplored area. Without good testing methods and the possibility of automatic instrumentation (test hooks) the coverage of the test-cases becomes limited.

All the problems taken together indicate that active rules are not suitable for direct specification of activities in highly dynamic environments such as manufacturing control. We believe that rules are a too low-level mechanism that is too far from the application domain (manufacturing), to be used as a direct specification language. Instead, we have adopted a compiler approach. A compiler automatically transforms specifications of manufacturing operations into queries and active rules.

3. GENERIC RULE MODEL

To understand the rule interactions we need a formal vehicle to analyze rule properties without involving the peculiarities of various rule system implementations. This section introduces the Generic Rule Model (GRM) which was developed for this purpose. It covers most of the major features of active DBMSs in a single coherent framework.

In GRM, an active database is represented by a 13-tuple $< \mathcal{Q}, q_0, \Sigma, \delta, \mathcal{E}, \mathcal{C}, \mathcal{A}, \Re, \xi, \psi, \Lambda, \lambda, \Gamma >$ where $\mathcal{Q}$ is the set of possible database states, q_0 is the initial database state, Σ is the set of allowed database operations from external applications or users. δ maps an old database state and a string of updates and actions to a new database state according to $\delta : (\Sigma \times \mathcal{A})^* \times \mathcal{Q} \rightarrow \mathcal{Q}$, where $\mathcal{X}^*$ denote a finite sequence of elements from the set $\mathcal{X}$. The model also includes events, conditions, and actions. Here, $\mathcal{E}$ is the set of allowed event specifications, $\mathcal{C}$ is the set of allowed conditions (normally boolean expressions) over the database state, and $\mathcal{A}$ is the set of allowed actions, including updates and call-out to applications. $\Re$ represent the active rules which are encoded as triplets $\langle e, c, a \rangle$, where $e \in \mathcal{E}, c \in \mathcal{C}$ and $a \in \mathcal{A}$.

The GRM does not assume any specific rule language. Two predicates ξ, ψ and the δ operator abstracts on the action execution and the evaluation of conditions and event specifications. The predicate $\xi(c, q)$ holds iff the database state q satisfies the condition c, and the predicate $\psi(e, \sigma)$ holds iff the transaction σ generate events that satisfy the event specification e. Finally, Λ is the set of legal queries over the database state $\mathcal{Q}$,

Γ is the set of possible query results, and $\lambda : \mathcal{Q} \times \Lambda \to \Gamma$, represent the mapping from queries and database states to query results (query processing). For simplicity we assume that all database states are reachable through an appropriate sequence of operations. Another assumption is that rule actions are executed in a serial or serializable manner.

To understand the effect of rule interactions the previous properties must be formally defined. Based on the definitions, the relations between properties can be determined. The taxonomy resulting from the definition and analysis of the rule interactions facilitate the construction of well-behaved rule sets. Due to space limitations the formal definitions are only given for rule cascades and race conditions for actions. Here, $XX_{ECA}^{DD}(\Re)$ is shorthand for that interaction type XX applies for mode DD (deferred-deferred) for a ECA rule set $\Re$.

DEFINITION 1. Action interference (AI) occurs iff two rule actions can write the same set of data. More precisely, this interference occurs between two rules when the state of the database depends on the order in which their actions are processed. Consider a pair of rules $r_1 = \langle e_1, c_1, a_1 \rangle$ and $r_2 = \langle e_2, c_2, a_2 \rangle$. Formally r_1 and r_2 interfer iff $\exists q \left[\delta(a_2, \delta(a_1, q)) \neq \delta(a_1, \delta(a_2, q)) \right]$.

DEFINITION 2. Event interference (EI) occurs iff a rule can signal an event that is monitored by another rule.

DEFINITION 3. Condition interference (CI) occurs iff the action of one rule can affect data that is read by the condition of another rule.

DEFINITION 4. Rule interference (RI) is a collective name for action interference, event interference, and condition interference.

DEFINITION 5. Action race (AR) occurs when two or more simultaneously fired rules write the same set of data. AI between two rules becomes an action race (AR) if the rules may fire simultaneously. The resulting database state will be unpredictable. Formally, a rule set can lead to AR iff

transaction s (Line 4). Finally, the race occurs iff different orderings of the actions result in different database states (Line 5).

DEFINITION 6. Condition race (CR) occurs when the an action race concerns data read by conditions of other rules.

DEFINITION 7. Event race (ER) occurs iff two or more simultaneously fired rules generate events monitored by some affected rule and the ordering in which the event occur result in the affected rule either being fired or not.

DEFINITION 8. Condition enabling (CE) occurs iff a rule action can change the database state in such a way that a previously unsatisfied condition of some other rule becomes satisfied.

Condition enabling is not classified as a cascade since the affected rule does not fire directly, but when the event occurs later, this rule will have been partially responsible for the firing.

DEFINITION 9. Condition disabling (CD) occurs iff a rule action can change the database state in such a way that a previously satisfied condition of some other rule becomes unsatisfied.

Condition disabling inhibits a firing of a rule that otherwise would have been fired given the appropriate sequence of events. In a sense, CD is the opposite of cascades. Note that CD cannot occur in causally dependent decoupled (CDD) mode.

DEFINITION 10. Event enabling (EE) occurs iff an action of one rule satisfies the event specification of some rule. This involves signalling previously missing events that are monitored by other rules.

DEFINITION 11. Event disabling (ED) can occur in deferred rule processing system with composite event specification languages. It occurs iff a rule action emits an event that makes a previously satisfied event specification false. Consider an event specification $a \wedge \neg b$. A single a event would satisfy the expression, but iff the b event occurs later, the event specification would become false.

DEFINITION 12. Rule cascade (RC) occurs iff the action of one rule causes the firing of another that would not have fired unless the first rule action was executed. Formally RC occurs in a deferred-deferred rule processing systems iff

$$AR_{ECA}^{DD}(\Re) \equiv \exists r_1 r_2 q \sigma \left[r_1, r_2 \in \Re \wedge q \in \mathcal{Q} \right.$$
$$\wedge r_1 = \langle e_1, c_1, a_1 \rangle \wedge r_2 = \langle e_2, c_2, a_2 \rangle \wedge r_1 \neq r_2$$
$$\wedge \sigma \in (\Sigma \cup \mathcal{A})^* \wedge \xi(c_1, q) \wedge \xi(c_2, q) \wedge \psi(e_1, \sigma)$$
$$\wedge \psi(e_2, \sigma) \wedge \delta(a_2, \delta(a_1, q)) \neq \delta(a_1, \delta(a_2, q)) \right]$$

For race conditions, it is sufficient to check groups of rules that can have their conditions satisfied for the same database state. Any race condition will have at least two rules firing simultaneously (Line 3). Another premise is that event specifications of the two rules should be satisfied by some

$$RC_{ECA}^{DD}(\Re) \equiv \exists \sigma r_1 \ldots r_M r_{M+1} \ldots r_N q_0 \ldots q_M \left[\right.$$
$$N = |\Re| \wedge r_1 \ldots r_N \in \Re \wedge q_0 \ldots q_M \in \mathcal{Q} \wedge$$
$$\wedge \sigma \in \Sigma \wedge \forall w (1 \leq w \leq N \Rightarrow r_w = \langle e_w, c_w, a_w \rangle) \wedge$$
$$\wedge \forall xy (1 \leq x \leq N \wedge 1 \leq y \leq N \wedge x \neq y \Rightarrow r_x \neq r_y)$$
$$\wedge \forall u (1 \leq u \leq M \Rightarrow \xi(c_u, q_0) \wedge \psi(e_u, \sigma)) \wedge$$
$$\wedge \forall v (M + 1 \leq v \leq N \Rightarrow (\neg \xi(c_v, q_0) \vee \psi(e_v, \sigma)))$$
$$\bigwedge_{s=1}^{M} [q_s = \delta(a_s, q_{s-1})] \wedge \forall gh (1 \leq g \leq M \wedge$$
$$\wedge 1 \leq h \leq N \Rightarrow \xi(c_h, q_g) \wedge \psi(e_h, a_g))]$$

The formula is quantified over update transactions (σ), fired rules ($r_1 \ldots r_M$), non-fired rules ($r_{M+1} \ldots r_N$), the initial state q_0, and the subsequent intermediate states ($q_1 \ldots q_M$) resulting from the applications of rule actions ($a_1 \ldots a_M$) of all fired rules. First, preliminaries are defined. All rules belong to the rule set $\Re$, all rules in $\Re$ should be accounted for ($r_1 \ldots r_N$), and the update transaction (σ) should belong to the set of legal transactions. The third line identifies the events, conditions, and actions of each rule in the rule set. The next part assures that each rule in the quantification is unique (Line 4), and the conditions ($c_1 \ldots c_M$) and event specifications ($e_1 \ldots e_M$) for each fired rule ($r_1 \ldots r_M$) is satisfied at commit time (Line 5). Line 6 states that all non-fired rules ($r_{M+1} \ldots r_N$) must either have their conditions ($c_{M+1} \ldots c_N$) unsatisfied, or the required events ($e_{M+1} \ldots e_N$) should not have been generated during the firing transaction (σ). In line 7, the intermediate database states are computed by applying each action ($a_1 \ldots a_M$) to the database state as produced by the initial transaction (q_0). The final part of the formula dictates that a rule cascade occurs iff there is some action ($a_1 \ldots a_M$) that simultaneously satisfies the condition and event-specification of any rule in the rule set ($r_1 \ldots r_N$) including itself. This defines the additional firings.

Note that the RC^{DD} formula does not handle situations where all rules fire. Technically, this is merely a special cases and can be obtained by omitting $r_{M+1} \ldots r_N$ and adding M=N to the formula.

DEFINITION 13. Rule cascade with limited depth. RC(LD), is when the depth of the firings (the length of the chain of indirect cascades) is limited to N. Rule sets with depth limited to 1 are denoted $RC(1D)$. $RC(0D) \Rightarrow \neg RC$.

DEFINITION 14. Rule enabling (RE) $RE \equiv EE \vee CE$.

DEFINITION 15. Self disabling (SD) describe rule sets where every rule-action makes its own rule-condition false.

DEFINITION 16. Cascade cycle (CC) occurs iff a rule is fired due to its own previous actions. A cascade cycle can span over one or more rules.

DEFINITION 17. Non-termination (NT) occurs iff rules repeatedly fire themselves and other rules in a pattern that never terminates.

Taxonomy of rule interactions

Figure 1 below shows how the interaction properties are related to each other. An arrow from X to Y should be interpreted as X is a special case of

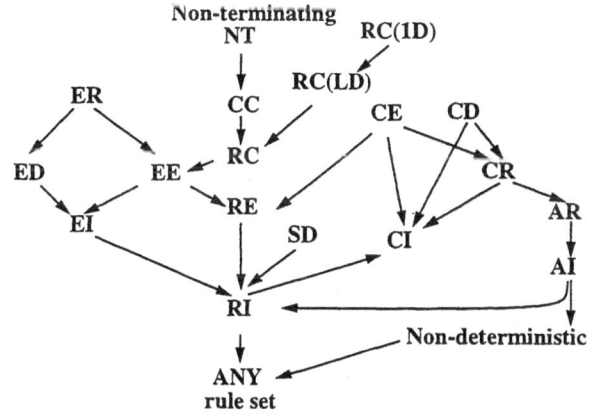

Fig. 1. Taxonomy of rule interactions

Y, *i.e.*, $X \Rightarrow Y$. Obviously, a rule set cannot be non-terminating without having cyclic cascades. It cannot have cyclic cascades without cascades, and it cannot have cascades without event enabling. Similarly there will not exist condition race and action race without action interference. The taxonomy illustrates a way to avoid certain properties. By excluding RC, the rule set will be terminating. The relations can also be formally proven (omitted).

4. HOW TO CONSTRUCT SAFE RULE SETS

What is a safe rule set? What interaction properties should be considered to be harmful? The only reasonable answer is that the distinction between safe and harmful interactions is application dependent. For manufacturing applications where timeliness and determinism are important attributes, the rule sets should satisfy $\neg NT, \neg CC$. AI is acceptable but not $\neg CR$ and $\neg AR$. Given the complexity of the $\neg RC$ criteria, we look for simplifications that are stronger but still sufficient for RC. Among such criteria we find $DFA^{DD}(\Re)$.

$$DFA^{DD}(\Re) \equiv \forall r_1 r_2 q \, [r_1, r_2 \in \Re \wedge r_1 \neq r_2 \wedge$$
$$\wedge q \in \mathcal{Q} \wedge r_1 = \langle \top, sc_1 \wedge nc_1, a_1 \rangle \wedge$$
$$\wedge r_2 = \langle \top, sc_2 \wedge nc_2, a_2 \rangle \Rightarrow \neg (\xi(sc_1, q) \wedge$$
$$\xi(sc_2, q)) \wedge \neg (\xi(nc_1, q) \wedge \xi(nc_2, q))] \wedge$$
$$\wedge \forall r_1 r_2 q \, [r_1, r_2 \in \Re \wedge r_1 \neq r_2 \wedge q \in \mathcal{Q} \wedge$$
$$\wedge r_1 = \langle \top, sc_1 \wedge nc_1, a_1 \rangle \wedge r_2 = \langle \top, sc_2 \wedge nc_2, a_2 \rangle \wedge$$
$$\wedge \xi(sc_1, q) \wedge \xi(nc_2, q) \Rightarrow \xi(nc_2, \delta(a_1, q)) \wedge$$
$$\wedge \neg \xi(sc_1, \delta(a_1, q)) \wedge \xi(sc_2, \delta(a_1, q))]$$

The criterion is a strong version of $\neg RC$ that emulates deterministic finite state automatons (DFAs). Here, the set of conditions are divided into state conditions sc and node conditions nc (Line 2). The state condition is an explicit internal representation of the automata state, whereas the node condition represent the monitored condition. In a DFA constellation, all conditions are

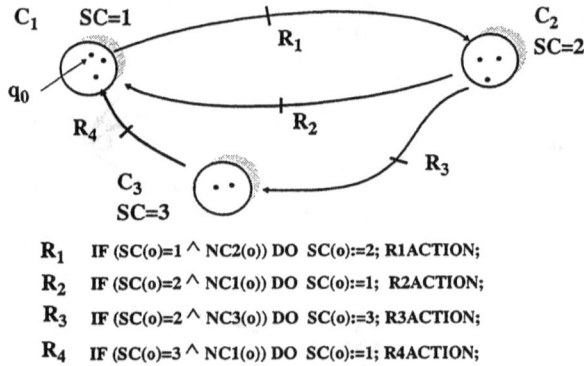

R_1 IF (SC(o)=1 $\wedge$ NC2(o)) DO SC(o):=2; R1ACTION;

R_2 IF (SC(o)=2 $\wedge$ NC1(o)) DO SC(o):=1; R2ACTION;

R_3 IF (SC(o)=2 $\wedge$ NC3(o)) DO SC(o):=3; R3ACTION;

R_4 IF (SC(o)=3 $\wedge$ NC1(o)) DO SC(o):=1; R4ACTION;

Fig. 2. Generated rules in DFA constellation

mutually exclusive (Line 3), which divides the database state into equivalence classes. Next, the set of external updates must be limited. There should be no update in *Sigma* that allows a direct manipulation of the state conditions. When fired, the explicit state is altered by the action and because the state conditions are mutually exclusive, all rules become self-disabling (Line 5). Since $DFA^{DD}(\Re) \Rightarrow \neg RC^{DD}(\Re) \wedge \neg AR^{DD}(\Re)$, any rule set that conforms to the DFA constellation will also fulfill $\neg RC, \neg CC, \neg NT, and \neg AR$. Proof is constructed in two steps. First, SD rules that satisfy $\neg CE$ are shown to satisfy $\neg RC$ with an all-quantification over events (the DFA constellation is not restricted to specific events). Next, DFA^{DD} is shown to be SD and $\neg CE(\Re)$, which concludes the proof.

5. PRACTICAL EXPERIENCE OF THE APPROACH

A feasibility study active database integration was conducted based on a series of four prototypes with increasing capabilities. As a starting point we decided to use the ARAMIS (Holmbom et al., 1992) (Loborg and Törne, 1991) three-layer architecture, initially developed at ABB (formerly ASEA). We designed and implemented the CAMOS (Control Application Mediator Object System). It consisted of four major components: the compiler, an operation manager, an embedded active database, and a real-time server. The manufacturing operations are written in a specialized language, CAMOS(L), that supports common abstractions in manufacturing (resources, operations, cyclic tasks, synchronization, and control algorithms). Using the compiler, manufacturing operations are automatically translated to queries and active rules in the embedded database system. Finally, an operation manager controls and coordinates high-level operations, active rules, and the control algorithms. For the embedded database, we took the view that time-critical processing is normally located in low-level control algorithms rather than in the coordinating software at the mission-level. Therefore, active rules can be used

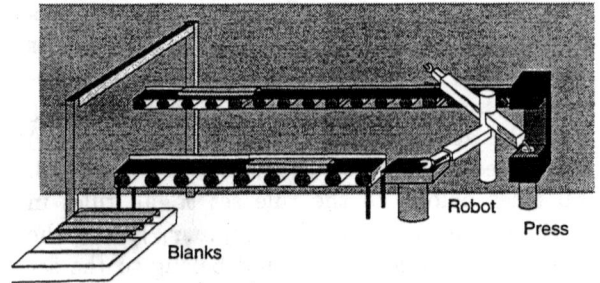

Fig. 3. Production cell in metal processing plant

to initiate and supervise control algorithms, without sacrificing timeliness of closed-loop control. Details about the CAMOS architecture and language can be found in other reports (Falkenroth et al., 1995) (Falkenroth, 1996).

Case study

A case study was conducted as a part of the feasibility study. A small production cell was implemented using CAMOS (Diederich, 1997). The example is taken from a metal processing plant in Karlsruhe, Germany (Lewerentz and Lindner, 1994).

The production cell consists of various transportation systems, a robot, a press, and sensory equipment. The press processes metal parts. The parts are placed in the press by a two-armed robot. Two conveyor belts and a crane transports materials to and from the press. The production cell has several sensors that provide information about the environment and the machinery. A FSM model of the cell consists of about 10^{12} individual states. The corresponding CAMOS(L) program consisted of 12 class definitions and 36 manufacturing operations, together 656 loc including comments. From the source code, the compiler generated a schema with 578 entities of which 79 were active rules, 410 were pre-optimized queries, and 77 were stored functions. Besides the static objects, additional objects are generated during the operation of the production cell resulting in more than 3000 object instances (for equipment, materials, operations, process representation, conditions, and rule instances). In all, it took less than 25 man-hours to write the specification in CAMOS and generate the 79 terminating confluent rules.

Evaluation

The compiler approach worked satisfactorily. The generated rules still contain unnatural state variables for sequences and modes, but the complexity is no longer a problem since the rules are hidden from the programmer. Clearly, issues related to rule-base organization, development and maintenance of rule sets become irrelevant. Instead, they must be addressed in the source language. Fortunately, solutions are easier to find for CAMOS(L) than for active rules. Also verification is easier in the context of CAMOS(L). Several analysis techniques based on petrinets and abstract interpretation have been developed. The compiler relies on several safe constellations (among those are DFA^{DD}), which controls the rule interactions. None of the generated rules may cause nonterminating cascades cycles. The efficiency aspect and the issue of dynamic modes were also addressed in the case study through dynamic rule contexts (Sköld *et al.*, 1995). The integrated system fulfilled the reactivity requirements although the performance was slightly lower compared to a hand-coded controller (Diederich, 1997). A prevailing opinion among researchers is that deferred coupling modes are inappropriate for real-time applications. However, this is a misconception (Stankovic, 1988). Although deferred modes increase the latency, the coupling mode has no bearing on the *predictability* of the response. Nevertheless, the transactions should be short for acceptable performance. In conclusion, the active DBMS was vast improvement from a data management point of view. Portability and timeliness of remains to be addressed.

6. CONCLUSION

Applications with interacting rules are hard to develop. Even a small number of dependencies between rules will make it difficult for a programmer to develop and maintain the system. The scalability of manual coding of rule set in highly dynamic environments is questionable. The approach taken in this paper was to use a compiler to generate correct rule sets and thereby hide the complexity of the active rules.

Contributions

The contributions of the paper are four-fold:

- A formal generic rule model suitable for defining rule-set properties
- A taxonomy of rule interactions
- A compiler technique based on constellations that generates terminating rules for manufacturing control
- Feedback to the research community about development of large rule applications. Particularly, the list of encountered problems should give valuable information.

7. ACKNOWLEDGEMENTS

We would like to thank NUTEK, The Swedish National Board for Industrial and Technical Development for financial support. The local Linkping University funding from CENIIT - The Center for Industrial Information Technology is also gratefully acknowledged. Many thanks to Prof. Tore Risch and Martin Sk"ld for help with the active rules facility, and Peter Loborg for help with the formal definitions. Jörg Diederich and Cinzia Foglietta also deserves thanks for the work with the CAMOS compiler.

8. REFERENCES

Andler, S. F., J. Hanson, J. Eriksson, J. Mellin, M. Berndtsson and B. Eftring (1996). Deeds — towards a distributed active and real-time database system. *ACM SIGMOD Record*.

Baralis, E., S. Ceri and S. Paraboschi (1998). Compile-time and run-time analysis of active behaviors. *IEEE Trans. on Knowledge and Data Eng.*

Buchmann, A. P., H. Branding, T. Kudrass and J. Zimmermann (1992). Reach — a real-time active and heterogeneous mediator system. *ACM SIGMOD Record*.

Dayal, U., B. Blaustein, A. Buchmann, S. Chakravarthy, M. Hsu, R. Ladin, D. McCarthy, A. Rosenthal, S. Sarin, M. J. Carey, M. Livny and R. Jauhari (1988). The hipac project: Combining active databases and timing constraints. *ACM SIGMOD Record*.

Diederich, J. (1997). *Modelling a Production Cell Using the CAMOS Language for Manufacturing Control*. Master Thesis. LiTH-IDA-Ex-97/46. Linköping University, Linköping, Sweden.

Dittrich, K. R., S. Gatziu and A. Geppert (1995). The active database management system manifesto : A rulebase of adbms features. In: *2nd Intl. Workshop on Rules in Database Systems, Athens, Greece*.

Falkenroth, E. (1996). *Data Management in Control Applications - A Proposal Based on Active Database Systems*. Licenciate Thesis No.589. Linkping university, Linkping, Sweden.

Falkenroth, E.T., A. Törne and T. Risch (1995). Using an embedded active database in a control system architecture. In: *Proc. 2nd International Conference on Applications of Databases*. San Jose, CA, USA.

Holmbom, P., P. Loborg, M. Sköld and A. Törne (1992). A model for the execution of task level specifications for intelligent and flexible manufacturing systems. In: *Proc. of the Intl. Symposium on Artificial Intelligence, ISAI92*. Cancun, Mexico.

Lewerentz, C. and T. Lindner (1994). *Case Study "Production Cell" - A Comparative Study in Formal Software Development*. FZI-publication. Forschungszentrum Informatik Haid-und-Neu-Strasse 10-14 Karlsruhe, Germany.

Loborg, P. and A. Törne (1991). A hybrid language for the control of multimachine environments. In: *Proc. EIA/AIE*. Hawaii.

Sköld, M., E. Falkenroth and T. Risch (1995). Rule contexts in active databases - a mechanism for dynamic rule grouping. In: *Proc. Second Intl Workshop on Rules in Database Systems, RIDS'95*. Glyfada, Athens, Greece.

Stankovic, J. A. (1988). Misconceptions about real-time computing. *IEEE Computer*.

Stonebraker, M., M. Hearst and S. Potamianos (1989). A commentary on the postgres rules system. *ACM SIGMOD Record*.

Widom, J. (1994). Research issues in active database systems. *ACM SIGMOD Record*.

Widom, J. and Ceri, S., Eds.) (1996). *Active Database Management Systems - Triggers and Rules for Advanced Database Processing*. Morgan Kaufmann Publ.

Zhou, Y. and M. Hsu (1990). A theory for rule triggering systems. In: *Proc. Intl. Conf. on Extending Data Base Technology*. pp. 407–421.

DISTRIBUTING CONTROL SYSTEMS USING ACTIVE RULES [1]

Sten F. Andler * Joakim Eriksson * Mats Lundin **

* *University of Skövde, Sweden — {sten,joakim} @ida.his.se*
** *Volvo AB, Sweden — mats@morot.nu*

Abstract: Real-time control systems are often distributed, with complex data sharing requirements. A method is proposed using a distributed real-time database, extended with active rules that react to changes in the system. This database is fully replicated and nodes communicate using only the database distribution facilities. Temporal constraints at individual nodes may be met at the expense of temporary inconsistencies in the database. This paper investigates the effects of using a rule-based representation and the replicated database as a distribution mechanism for certain classes of real-time systems. Particular attention is paid to the effect of reduced consistency, assuming eventual consistency. *Copyright © 1999 IFAC*

Keywords: Real-time systems, distributed databases, distributed computer control systems, control system design, database systems, data replication, consistency, distribution control, active database, real-time database

1. INTRODUCTION

Feedback control applications often operate in distributed environments that involve complex data sharing. The development of such applications can be tedious unless tools are available for building these applications on a suitable platform. A distributed database system offers a way of handling distribution and complex sharing of data in a natural way. Thus, such a database can be a useful platform for building distributed control systems. However, the time for replication of data between the nodes may be unpredictable due to network delays.

In this paper it is proposed that such a database, extended with active rules that are used to react to local and remote changes in the database and to perform predefined actions, be used as the basis for distributed real-time control systems. In the database, replication may be delayed, causing

temporary inconsistencies in the database. Issues that are raised when the control application is distributed, by partitioning of the rules to the different nodes in the database, are examined.

2. FEEDBACK CONTROL SYSTEMS

A real-time system for control purposes is used to control functions and processes in the environment, for example in chemical process industries and nuclear power plants. Such a system consists of at least three parts: (1) a sensor function to read necessary data from the environment (examples of such data is current pressure and temperature of the contents of a tank); (2) a control function implementing the algorithm needed to control the environment; and (3) an actuator function for performing the actual control of the system, such as opening and closing of a valve. The sensor function provides the system with feedback on the result of the actuation. Since the control system is interacting with the environment, timing constraints must be taken into account. That is, the time from

[1] This work was supported by the Swedish National Board for Industrial and Technical Development (NUTEK).

sensor reading to actuation must be predictable and short enough, in order not to jeopardize the safety of the environment.

2.1 Distributed Control Systems

A distributed control system is a system in which the functions are located on two or more physically separate nodes. According to Mullender (1993) and Chan and Özgünger (1995), there may be several reasons for building a distributed control system: *Hostile environment* — there may be a severe environment where normal high performance processors cannot work, for example in the reactor of a nuclear plant or in a dusty or hot environment; *Distribution of system* — the system representatives are distributed over a large spatial area; *Distribution of information* — information from one sensor can be used at several locations in the whole system

2.2 Distributed Real-Time Database Systems

A distributed database is a database that has some common data distributed over several sites and allows coordinated sharing and updating of that data. One of the reasons for distribution is that the data must be reachable at several sites at the same time. An advantage compared to a conventional centralized database is the potential for fault tolerance and modular growth. On the other hand, distribution also raises the issue of replication of changes between nodes; there must be synchronization to ensure consistent data at all nodes. The price for this synchronization is the use of a protocol that consumes processor and communication resources and also takes elapsed time to perform. This elapsed time is often hard to predict because of the unpredictability of the communication medium.

In databases, so called ACID properties (Gray and Reuter, 1993) are used to ensure a correct behavior. These properties are *Atomicity* (all or nothing), *Consistency* (correct transformation), *Isolation* (concurrency control), *Durability* (survive failures). There are different protocols proposed for handling data consistency and for ensuring atomic execution of transactions in a database, i.e., to ensure that an operation is executed at all sites or none at all (Lundström, 1997). A common way is to use an *atomicity protocol* that replicates read and write transactions to all other nodes; one such protocol is the *Distributed Two-Phase Commit Protocol*. This protocol and other further developments all use considerable resources for execution.

In this paper, the data in a distributed database is assumed to be available at all nodes, that is, the database is *fully replicated*. If the database can be segmented in such a way that each segment is only accessed within a group of nodes, then the corresponding segment needs only be fully replicated at nodes in that group. For all practical purposes it will appear that the entire database is fully replicated, called *virtual full replication*.

2.2.1. Levels of Consistency
In some cases it is not important to have immediate consistency, e.g., when an application is tolerant to temporal inconsistencies, but instead a more efficient usage of resources is preferred (Lundström, 1997). In such cases an *eventual consistency protocol* can be used, i.e., a protocol that does not guarantee immediate consistency, but consistency at some future point in time. The effect of such a protocol is that the time required to achieve consistency of the database is unpredictable.

In close relation to the level of consistency is the degree of isolation. Gray and Reuter (1993) identify four degrees of isolation where the highest is serializability. In the same way that a higher degree of isolation may be traded for a lower degree in order to increase concurrency in the system, a higher degree of consistency may be traded for a lower degree of consistency to reduce communication delays in a distributed database.

2.2.2. Active Rules for Expressing Control Systems
In this paper, rules are specified using the ECA (*Event—Condition—Action*) paradigm (Dayal *et al.*, 1988). Briefly, a rule specifies that on the occurrence of a certain primitive or composite *Event*, if the *Condition* on the system state is satisfied, then the corresponding *Action* is executed. An event may be a database operation (insert, retrieve, update, delete), but also an external event occurring in the environment, or a periodic or aperiodic temporal event. Conditions can be any expression resulting in true or false. The action can be any database or non-database action.

3. TRANSPARENT DISTRIBUTION

A long-term research problem is to find a simple, transparent, method for transforming a centralized control system or application into a distributed system. Various methods of prepartitioning single-system applications into units of distribution have been discussed, for example, virtual nodes (Burns and Wellings, 1996). This paper investigates whether rules can facilitate distribution of control systems to different nodes in a system, and how database objects must be replicated as needed by the rules.

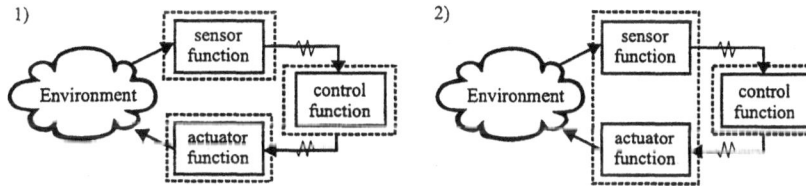

Fig. 1. Two of the four distributed allocation schemas. (A physical node is indicated by a dotted rectangle.)

In a distributed control system there are four static allocation schemas possible for the three main functions of sensor, controller, and actuator. Two of these four allocation schemas are shown in Fig. 1. In the other two schemas, the control function is grouped with either the sensor or the actuator functions.

3.1 *Problems Raised by Distribution*

When a control situation requires distribution, for example, by inherent geographic distribution or demands for fault tolerance, some problems are raised. One of the problems is the delays that come from the need to replicate data. The question is raised of whether precaution must be taken to maintain consistency in the database. This paper addresses the problem of how the level of consistency delivered by a database affects the application, especially in the case where only eventual consistency is guaranteed.

4. ACTIVE RULES AND REPLICATION AS A DISTRIBUTION MECHANISM

This paper proposes the use of a real-time database, with control applications expressed as active rules, with fully replicated data, as a method for simple and transparent distribution of real-time control systems. One of the advantageous features of such a distributed active real-time database platform is the active functionality, which can be used to express the functions of the control system in a location-independent way at different places in the system. The use of different levels of consistency, and their effect on the application, is investigated.

4.1 *Active Rule Model*

In section 2, it was outlined how a control system can be divided into a sensor function, control function and actuator function. Each of these functions can be represented by an ECA rule as follows: The *sensor function* is represented with a rule that is triggered by a periodic event. The periodicity depends on the sampling rate needed

by the application. The action reads the environment and stores the sensor value in the database, which will then be replicated to all the other nodes in the system; A rule representing the *control function* could be triggered by the writing and replication of a sensor value, potentially raised by an event on a node other than where the writing was done. The rule might include a condition part on the sensor value. If the condition is true, then the actual control algorithm is executed in the action. The result is stored in the database, and thereby replicated; If the action results in a database writing and replication, this will in turn raise an event that triggers a rule representing the *actuator function*. The stored data is used by the action part of the rule to perform an action in the environment.

In our model, rules communicate only by means of database operations. Since the database is fully replicated for simplicity and to ensure consistency, updates are performed on all nodes in the system. Thus, rules can be triggered on any node, independently of which node updated the data. To distribute functionality, specific rules can be placed on certain nodes where the functionality is needed.

4.2 *Transient Handling*

Special rules may be introduced to compensate for transients in the system. When such transients are to be handled, an extra rule is used to shorten the delay from the sensor readings to actuation. Consider the case where normally only one rule is used for actuation. With transient handling, both the ordinary control algorithm and the fast transient-handling control algorithm must write data to the same database variable to have it actuated. This may be a source of a write-write conflict (Lundström, 1997). However, a system can detect and resolve such a conflict in an application-dependent way, resulting in a predictable behavior (Andler *et al.*, 1996).

As the control system relies on the database replication functions, delays are introduced. These delays are of importance, as control systems are sensitive to delays (Lundin, 1998). Used in a single node, the proposed transient handling rule would be of little use as the delays introduced are small. In a distributed environment, the delays introduced by database replication are larger, and thus the profit of using this function can be greater.

5. DISTRIBUTION OF CONTROL FUNCTIONS

In section 4.1, the modeling of a control application into active rules was described. The final step of building a distributed control system is the distribution (allocation) of these active rules to the different nodes in the system. This section describes consequences of distributing the rules to different nodes. A discussion is held on where to distribute the rules when building a control system.

5.1 Synchronization of Multiple Sources

A control application where there are several distributed sensors and actuators implies a synchronization problem. It is assumed that sampling in the sensor nodes is synchronized and that the information is written directly to the database. The measurement of interest for a given control function arrives at different points in time. Normally the computation in the control function cannot start until the information from all sensors have arrived. It might, however, be possible to define a rule that uses a certain number of the needed values and infers the others, thus offering fault tolerance. This rule can either be time-triggered, and start the computation at a given point in time, or trigger when enough values are present.

5.2 Cost of Consistency

Consider the levels of consistency discussed in section 2.2.1. Assume that the database can provide three separate levels of consistency, namely:

(1) *Immediate consistency*, where data is replicated to all nodes and committed at the same time in all nodes. To ensure this level of consistency a distributed atomic commitment protocol must be used (Mullender, 1993).
(2) *Eventual consistency*, where data is replicated at a later point in time — this can either be done as soon as possible (ASAP) (Gustavsson, 1995) or within a bounded time (Lundström, 1997).
(3) *Intentional consistency* No guarantee of consistency is given.

The three levels of consistency involve different time delays. Due to control systems sensitiveness to delays, the lowest appropriate level of consistency should be used. To ensure the immediate consistency in a database, a distributed atomic commitment protocol ((Mullender, 1993)) must be used. This protocol has an overhead in communication between the nodes in the system and, thus, takes valuable time to execute. Under the assumption that the underlying network has a broadcast facility, the distributed atomic commitment protocol involves two messages from the coordinator, the node where the information is updated, to the participants, the other nodes in the database. The participants each send one response on the network before the transaction is committed and one more each when committed, resulting in a total of $2n$ messages in 4 rounds of message exchanges.

The second and third levels of consistency both involve propagation of data using the broadcast facility of the underlying network. The second level also involves some overhead in nodes where the data is propagated. The time for the overhead depends on the algorithm. Lundström (1997) describes a version vector algorithm. This algorithm will use $n \cdot m \cdot l$ number of comparisons, where n is the number of nodes, m the number of objects, and l the number of transactions in the log filter, to detect an inconsistency. A conflict resolution mechanism is executed only in the case of an inconsistency.

In choosing an appropriate level of consistency for a control system based on a distributed database it must be examined whether inconsistencies can occur, and if they do, in what situations. Inconsistencies occur when there is a read-write-conflict or write-write-conflict in the database, that is, if one or several control functions use and update the same data in the database.

6. DISCUSSION

In this section, there is a discussion of what to consider when distributing rules over several nodes. Further, some implementation issues are discussed concerning how transients are handled and how the level of consistency affects the performance of the system.

All communication between the functions in the control system uses the database and its replication, and thus, the same information will be available at all nodes in the system. The reasons for using the database replication for all communication are: (1) The database replication mechanism is automatically distributing all information to all the nodes in the database. This means that a given rule will have access to the same information and therefore can be distributed to any node. (2) It is not explicitly defined how the information flows through the network, and thus no reconfiguration of paths is necessary if allocation of the main functions is changed. (3) All information uses the same distribution channel. If more than one channel were used, the order in which information would be perceived could not be guaranteed, which might result in an inconsistent view of system state.

The work presented here is independent of the underlying communication architecture. Although hard real-time constraints can be guaranteed locally, the timeliness of distribution is highly dependent on the network used for replication. A non real-time network can be used if ASAP replication is sufficient. However, normally bounded-delay replication (Lundström, 1997) is needed for control applications, i.e. a real-time communication network is required.

6.1 *Issues in Distribution and Allocation of Rules*

When the control system is modeled using active rules, the rules can easily be distributed to different nodes in the system. However, there are some considerations in distributing the rules. If a control system is expressed in active rules and delays are ignored, it will have the same behavior wherever the rules are distributed. When delays are significant, however, they affect the control system and, thus, should be as small as possible. Control systems are affected, as the control functions are distributed over the different nodes, in the following ways:

- Sensor functions must be executed before other control functions.
- The rules for handling transients should be placed with the actuator functions for best result. These rules improve further when the sensor and actuator are at the same physical node and the control algorithm is at a separate physical node. The reason is that the delay for the control function is significantly larger than than the delay for detecting the transient.

6.2 *Effects of Consistency*

An application using a database may be sensitive to inconsistencies. To avoid an inconsistent state, different algorithms for data replication can be used. In choosing an appropriate level of consistency, one must consider what level is necessary. Most control functions will not need any assurance of consistency, but there are situations where this might be needed. The highest level of consistency implies considerable communication and, thus,

long delays. Long delays must be avoided as much as possible, so this level is usually inappropriate. The second level, with eventual consistency, is less costly when no inconsistencies occur; it has almost the same delays as when no consistency is guaranteed at all (the third level). This algorithm may be inappropriate if the situation is complex and an extensive conflict resolution algorithm must be used. This is however not likely as the system uses simple functions and the worst thing that can happen is that the computations must be done again. It is therefore proposed that the eventual consistency algorithm is used whenever consistency is necessary. Even when consistency is not required, the cost in time for enforcing eventual consistency in replication is usually negligible.

7. RELATED WORK

Rules were introduced into database management systems in the early 1980s as a method for implementing automatic view updates, constraint enforcement, trigger facilities, etc. As a means for specifying rules, the ECA (Event—Condition—Action) paradigm was proposed (Dayal et al., 1988). Introducing an active database in a real-time environment have impact on the predictability of the system (Eriksson, 1997). Several proposals have been presented to handle this increased unpredictability (Chakravarthy et al., 1989; Buchmann et al., 1994; Andler et al., 1996).

Suggestions on using active rules in control applications have been made, for example AMOS (Falkenroth, 1996) but here the sensor and actuator functions are performed by a dedicated real-time server, and the active database is used for mediating data to the control algorithm.

Some research has been performed on distributed active databases, but the emphasis has almost entirely been on distribution of event detection (Jaeger, 1997; Schwidersky et al., 1995; Mellin, 1998). Effort has been made to utilize CORBA in order to distribute reactive mechanisms (Gatziu et al., 1998). However, the aim is to unbundle the reactive mechanism from the database management system rather than to actually distribute it.

Research has been made on using databases for distributed control systems (Törngren and Wikander, 1996). Considerations for building distributed control systems are described. In this work, active rules are not used explicitly for building the control applications but related areas are covered.

8. CONCLUSIONS

In this work, a simple and transparent method for distribution of control system applications using active rules has been introduced. If the delays are not considered the control performance of the distributed system will be the same regardless of placement. It is important to have optimistic replication in the control system, as a faster algorithm for replication can be used when consistency must be ensured. The use of a distributed atomic commitment protocol implies time-consuming replications between the nodes and, thus, degraded performance of the system. Normally, control functions do not need a higher level of consistency than eventual consistency. This distribution method has been demonstrated on a relatively simple application class of real-time control systems. The effects of various degrees of inconsistency (e.g., eventual consistency) on the application have been investigated.

The rise in control performance gained by the use of transient handler functions is insignificant if it executes at the same node as the ordinary control algorithm. The largest advantage of this function is if it can be located at the actuator node and the ordinary control algorithm is at a separate node. In this case the time saving can around one or two replications between physical nodes in the system.

8.1 *Future Work*

In this paper it has been discussed how the distribution of rules affects the behavior of the system. There may be a way to automatically distribute rules to the nodes in the system by taking delays into account. Other considerations to take into account are the processing power, network load and environmental factors. This may also be extended to a tool-set for support of implementing control applications in a distributed database.

The method has only been applied to real-time control systems, but it should be possible to generalize it to other application areas. An open problem is to determine which classes of applications are suitable for distribution using a database that supports eventual consistency.

9. ACKNOWLEDGMENTS

The authors wish to thank Jonas Mellin for his valuable suggestions and comments on this work, and to Ragnar Birgisson for comments on the paper. This work would not have been possible without the exciting research environment of the Distributed Real-Time Systems research group and the DeeDS project.

10. REFERENCES

Andler, S. F. *et al.* (1996). DeeDS towards a distributed active and real-time database system. *ACM SIGMOD Record.*

Buchmann, A. P. *et al.* (1994). Building an integrated active OODBMS: Requirements, architecture and design decisions. Technical Report. Tech. University Darmstadt, Germany.

Burns, A. and A. Wellings (1996). *Real-Time Systems and Programming Languages* (2nd ed.). Addison Wesley.

Chakravarthy, S. *et al.* (1989). HiPAC: A research project in active time-constrained database management. Technical Report XAIT-89-02. Xerox Advanced Information Technology.

Chan, H. and Ü. Özgünger (1995). Closed-loop control of systems over a communication network with queues. *Int'l Journal Control.*

Dayal, U. *et al.* (1988). The HiPAC project: Combining active databases and timing constraints. *ACM SIGMOD Record.*

Eriksson, J. (1997). Real-time and active databases: A survey. In: *Proc. ARTDB '97.* Springer Verlag.

Falkenroth, E. (1996). Data management in control applications: A proposal based on active database systems. Licentiate Thesis, Linköping University, Sweden.

Gatziu, S. *et al.* (1998). Unbundling active functionality. *ACM SIGMOD Record.*

Gray, J. and A. Reuter (1993). *Transaction Processing: Concepts and Techniques.* Morgan Kaufmann Publishers.

Gustavsson, P. M. (1995). How to get predictable updates using lazy replication in a distributed real-time database system. Master's Thesis. University of Skövde, Sweden.

Jaeger, U. (1997). Event Detection in Active Databases. Ph.D. Thesis. Humboldt Universität zu Berlin, Germany.

Lundin, M. (1998). Building distributed control systems using distributed active real-time databases. Master's Thesis. University of Skövde, Sweden.

Lundström, J. (1997). A conflict detection and resolution mechanism for bounded-delay replication. Technical Report HS-IDA-TR-97-10. University of Skövde, Sweden.

Mellin, J. (1998). Multi-purpose predictable event monitoring. Licentiate Thesis, Linköping University, Sweden.

Mullender, S., ed. (1993). *Distributed Systems* (2nd ed.). Addison-Wesley, ACM Press.

Schwidersky, S *et al.* (1995). Composite events for detecting behaviour patterns in distributed environments. In: *Distributed Object Management 95.*

Törngren, M. and J. Wikander (1996). A decentralized methodology for real-time control applications. *Journal of Control Engineers.*

REAL-TIME TREATMENTS AND COMMUNICATIONS
IN SICODI DCS

Miguel A. García *, César de Prada *

* *Department of Systems Engineering and Automatic Control & Sugar Technology Center (CTA)*
University of Valladolid. Spain
e-mail: miguel@autom.uva.es

Abstract: Distributed control systems (DCS) have been in the last years in the front line of the development of industrial communications and system architectures. This is particularly correct when the DCS is integrated in a dynamic simulation system. In such a case, the temporal treatments are given not only for sample periods and discrete-event timers, but also for the time evolution itself if it assumes this role. This paper describes the implementation aspects of the real-time and communication properties, such as the priorization in a multi-timer model, the communication with a remote simulation, or the distributed database for the service of the level of communications between the Local Control Units (LCU) of the DCS. *Copyright © 1999 IFAC*

Keywords: Distributed real-time systems, communication in real-time systems, industrial applications, dynamic Training Simulators, distributed database of a DCS, Windows NT, C++.

1. INTRODUCTION

SICODI (García, 1998) is a complete development of a general purpose configurable DCS (65.000 lines of source code). As an objective of the training policy of the sugar industries in it, the Sugar Technology Center (CTA) is developing in the last four years a complete Training Simulator for the plant operators of beet-sugar factories (Acebes, Achirica, García and Prada, 1995), in which SICODI is integrated and to which provides real-time dynamical evolution. This training system is operative and progressively incorporating sections of sugar factories, degrees of detail and possibilities of particularization.

The results of the control algorithms implemented in SICODI are provided to the simulation through a bi-directional communication service from each one of its LCUs. According to specifications of the project, SICODI has been devoleped under Windows NT operating system and in C++ language, and using sockets as communication primitive, in order to assure the compatibility with auxiliary standard applications, complementary modules of software and wide area communications respectively. Amongst the characteristics of SICODI can be outlined:

- High-priority timer.
- Implementation of the concepts of generical LCU; functional block and control loop based on a predetermined connectivity of functional blocks.
- Integration of discrete-event sequences with bidirectional access to the blocks of the loops.
- Integration of multivariable predictive controllers.
- High-performance Man-Machine Interface.
- System of communications with simulations, pilot plants and between LCUs.

Blocks, loops, predictive controllers and the graphical controls of the MMI are configurable even at execution time. Anyway, most of the characteristics are out of the scope of this article.

The Training Simulator emphasizes the distributed architecture adding a workstation for the simulation and another one for a trainer console from which introduce and evaluate the problems and reactions. Fig. 1 shows a scheme of the whole system.

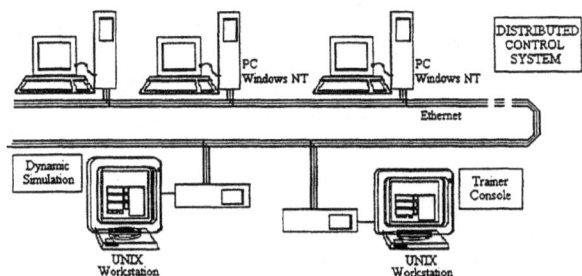

Fig. 1. Architecture of the Training Simulator

In this context, the treatment of the time problem for SICODI working as part of the Training Simulator is presented due to the interest of the implemented solutions, although it can be configured for working with real data -through data-adquisition boards and networks of PLCs- with no extra codification.

The rest of the paper is structured as follows: Section 2 describes the temporization problem and the priorization strategy. Section 3 presents the communication with the simulation from each LCU and the different tools used in order to make it possible, while Section 4 is devoted to some aspects of the communication between LCUs and their relationship with the dynamical distributed database of SICODI. Finally, Section 5 propose some improvements and research lines for the future.

2. TEMPORIZATION

The temporization problems of the system are framed in the event-driven mechanism supported by the operating system and provided by the programming structure of the compiler. Basically, this mechanism consists of an automatical association of a command handler function to each command or event selected from the available message map. The command handler functions are member functions of the class from which the command will be captured. As in the case of interruptions, the events are detected by the operating system deriving the flow control to the functions that handle them, if any. In other case, the treatment is delegated in the default member function of the base class of the given class, again if any, or simply the event is neglected and loosed with no treatment.

Based on this mechanism, the main classes of a C++ application have access to the timer event, in such a way that a member function of them can be coded -from the automatical declaration unchained when is chosen- in order to capture given time instants. The most common way to configure these utilities is by giving a value -tipically in milliseconds- as argument to the handler function, in such a way that the flow control derives to the function with a cadence given by a sample period of this value.

The main problem that appears in a complex system like SICODI, in which a great number of activities and functionalities coexist, is to guarantee optimal conditions of execution to critical tasks in front of another tasks with lower priorities that can demand a great amount of resources. This is the case of the task in charge of adquisition and application of treatments, in front of the task in charge of monitoring and data representation. The characteristics of these tasks are opposite, since the first one must be executed inmediately and its needs of resources are small, while the second one can be delayed and its consumption of resources is massive.

These differences constitute the typical situation for a priorization policy and are propitious for a good design of the accesses to resources.

In spite of its lacks, Windows NT operating system achieves with the needs of soft real time systems in a certain degree. This compliance is based on the capacity of implementation of the basic concepts of multiprocessing and priorization (Richter, 1994). First of all, different processes can be defined, each of them with more than one thread of parallel execution, in such a way that it assures a dynamical parallel multiprocessing based on the load of each processor over available multiprocessor architectures. Besides, a priority from a ranking of 32 levels is associated to each process and its threads. Finally, a group of synchronization primitives (e.g. semaphores, critical sections, mutexes and others) can be used to establish a relative order to the lines of execution. By using appropriately these elements, a good performance in the real time level can be obtained.

Thus, the objective is to assign the highest possible priority to the task in charge of adquisition and application of control algorithms and other treatments. Windows NT assigns a scheduling base priority to each thread determined by the combination of the priority of its process and the priority level of the thread. So, if the highest priority for a thread is needed, not only the TIME_CRITICAL priority must be assigned to it, but also the process to which the thread belongs must own the REALTIME priority. Based on the priority structure, Windows NT performs priority based preemptive scheduling (Ramamritham, Shen, Gonzalez, Sen and Shirgurkar, 1998).

The first step is therefore to get a process with the REALTIME priority. There are different ways to get such a process, but most of them require to make it from a different one. This is because the default priority for a process is NORMAL, and there is no mechanism by means of which a process could change its priority itself. Hence, the best way to get this goal is by using an independent application whose only assignment is to launch SICODI as an application and consequently as a process with the REALTIME priority. This is made by using the *CreateProcess()* function and being part of the arguments the path to SICODI executable and the type of priority to be assigned to its process. This launching application has been implemented without interface, in such a way that it works transparently and the visible effect executing it is the same as executing SICODI directly. However, the base priority assigned to SICODI as a process is manipulated from the external application. Another implementations based on the generation of different processes inside SICODI are more difficult, do not contribute to improve the result and loose the modularity of the proposed solution.

The second part of the problem is to achieve the TIME_CRITICAL priority for a thread of the SICODI process with REALTIME priority. This must be done from the code of SICODI by using the *AfxBeginThread()* function, which has as arguments the callback function that develops the new line of execution and the level of priority of the created thread. Fig. 2 shows the complete operation for obtaining a new thread with the TIME_CRITICAL priority.

Several utilities can help to test the priority assigned to each process and thread based on the configured requests. Amongst them, the one used is Process Viewer, which takes part of the Visual C++ compiler, although the diagnostic is applied to all the processes running and not only to those developed with its language. Fig. 3 is the Process Viewer interface for the thread with TIME_CRITICAL priority of SICODI as a REALTIME process. Note the value 31 in the field "Dynamic Priority".

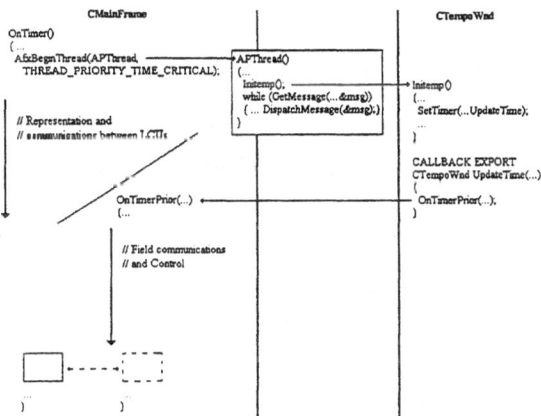

Fig. 2. Multi-timer execution

Hence, the goal is to obtain a new thread with the highest priority as a platform from which to have a new independent timer. For this purpose it implements a system of explicit treatment with the filter for the timer event. Another timer remains configured in the basic thread and using the default event handler treatment, in such a way that SICODI deploys autonomous timers for each priority event at the event capturing level. In fact, given that the two main levels are the adquisition-control and the graphical representation levels, each of them can be configured with the same or different specific sample periods depending on the neccesities in each functionality.

One important aspect of the multithreading model deals with the relative conditions of execution for the implemented threads. In general, the synchronization objects are suitable to withhold certain threads waiting for the arrival of another ones to some concrete points of their evolution, or for resources to be liberated. In the proposed double timer system this kind of treatment is avoided since the purpose is to have both of them running without any mutual interference and completely decoupled. So, if the timer with lower priority is delayed, the other one is

not affected and can execute with no problems. Hence, no synchronization objects are used and once the threads are created, both are allowed to execute without stopping and with the unique restriction of the attention paid by the operating system to each one of them. The fulfilment of the real-time conditions is delegated to the ability of the operating system to preempt the TIME_CRITICAL priority thread in its accesses to resources with respect to the other threads queries.

The model of multiple timer has a great modularity. This is because the different functions for inputs, treatments, outputs, communications, graphics, etc, are located in the handle functions associated to them. From these locations, it is very easy to change the calls and the segments of code related to each functionality from one handle function to the other. This possibility may have no sense for functionalities clearly associated to one of the timers, but can be useful for those with a location policies-dependent, such as certain types of communication between LCUs, and it is obvious when the number of levels and timers is increased. In fact, a set of flags can be declared in order to be used from the user interface to choose the timer in which each functionality is considered, by means of applying or inhibiting the calls to the related functions in alternative locations.

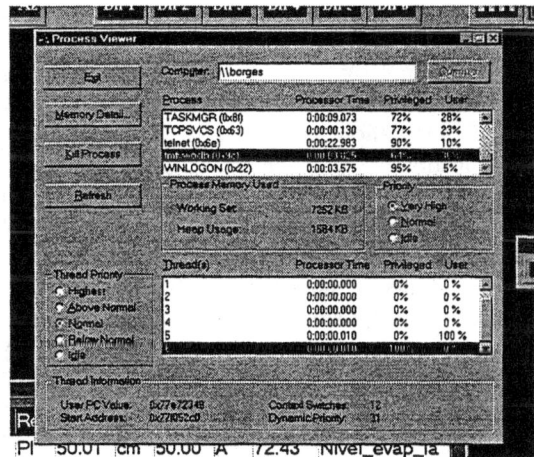

Fig. 3. Checking with Process Viewer

The low priority timer devoted to representation tasks is in charge of the updating of the different types of user interfaces, and it consists on the propagation of the values in the corresponding mainframe variables to the variables of the Views. The system of centralized temporization for updating the Views is based on the recognition of the instances of each kind of representation from the class *CMainFrame* with an exclusive identifier at launching time, together with the sending back of a pointer from the View in order to allow the access to it. All the classes associated to Views of representation have a member function *Actualice()* for the updating of information in it from the corresponding global variables in *CMainFrame*.

Yet another timers are instantiated dynamically associated to the states or actions -as function member of the class that implements the state- of discrete-event sequences and taking as argument the value of minimal stay in each configured state of each configured sequence.

Tests of the high-priority task -adquisition and control- have been made with the function *QueryPerformanceCounter()* and resolution of miliseconds. The average results are:
Real data, 5 loops: 10 ms.
Simulated data, 5 loops: 130 ms.
Simulated data, 52 loops: 460 ms.
As a first explanation, it can be noted the higher speed with real data and the non-linear increase of time with greater amounts of loops assigned, surely due to minimal consumptions in connection and delays of communication even for small packets. These results, obtained even with high graphical and network loads, are completely under specifications taking into account the sample periods of 1 s (most of them of 5 s) in the simulated system.

3. COMMUNICATION WITH THE SIMULATION

Two levels of communications can be considered in the system. The first and most important one is the established with the transmissors ans actuators in the plant, and it is located in the already so called adquisition task. When the LCU is configured as a part of the Simulator, the field elements and devices are replaced by the corresponding variables in the dynamic simulation, in such a way that the communication is simplified and does not depend on any specific details of each component with particular protocols or buses. However, the rest of the characteristics of this kind of communication must be preserved, and specifically the ability of each LCU with assigned control tasks to direct access to the simulation without third-party participation. So, no LCU with special responsabilities must be kept, and each of them is self-sufficient in the accesses to the part of simulation object of its interest. This property is in the base of the expected behaviour of a distributed control system, in such a way that the failure or even absence of any LCU locates its effect in the assigned area, and the rest of the LCUs can operate -with no interactions and well configured subsystems- on their assigned areas without problems derived from it. In fact, successive entries and leavings of LCUs, and the relative order of these operations with respect to other LCUs must be negligible apart from their direct effect on the assigned areas.

The implementation of such a communication system is based on the distribution of controlled variables from the simulation to the LCUs specified in the configured assignments; and in the other way, on the reception of the manipulated variables in a shared memory segment from the different LCUs with rights for applying them, as a previous step to pass them to the simulation. The communication primitive is a standard socket, that allows an effective and wide distribution based on Internet networks and Ethernet architectures. Besides and by means of sampling periods, the timing properties are related to the instant in which the put-in-order manipulated variables are sent to the simulation for a new sample period to run.

All these activities are commended to an intermediate application, so called link application. In the current architecture, the link application executes in a UNIX platform, that can be the same in which the simulation runs or another different one, and in fact the link application launches the configured simulation as a part of its initialization process and establishes the communication channel with the LCUs of the DCS. Anyway, link application is the intermediary to the simulation in both directions. For this purpose and together with the shared memory in charge of collecting the control results of the LCUs, presents a group of synchronization and passing files to access to the simulation. The Fig. 4 represents a complete operation of distribution of the controlled variables to the LCUs and recovery of their manipulated variables for the simulation, using the mechanisms implemented by the link application.

The configuration of link application in terms of delivery of controlled variables to the LCUs of SICODI is based on the absolute order of the total number of controlled and measured variables in the

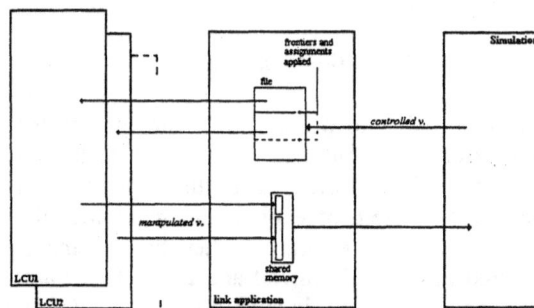

Fig. 4. Comm. between LCUs and simulation

system and on the frontiers applied from these values and related to the assignments to the LCUs. The information of assignments requires an agreement between all the LCUs and with the link application, based on specific fields in the configuration of all of them, and it allows to send each variable from the simulation to only one LCU. This exclusive communication is in fact the assignment, and the receiver LCU mantains the information of the assigned variables for the rest of the system according to the concept of distributed data base.

In the other direction, the shared memory segment allocated by the link application is charged to receive the manipulated variables from the LCUs to which they are assigned. This segment is necessary because the set of variables from each LCU is received through a child process obtained in the spawn process from which the socket attends to each connected LCU. The array in which the values from different LCUs are put in order must therefore be placed in a higher level due to the innaccesibility between the copies on different child processes. The accesses to the shared memory segment from each of these child processes are synchronized by the signaling of a semaphore, as Fig. 5 illustrates.

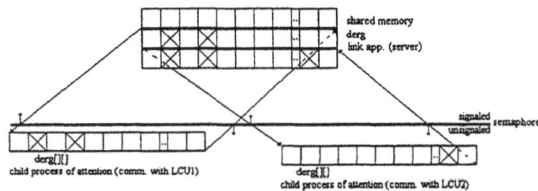

Fig. 5. Accesses and modification in shared memory

The impossibility of implementig advanced mechanisms of the above mentioned type in the language in which the simulation is implemented impose the use of synchronization and passing files for the communication with the simulation. Neither sockets nor semaphores are available in ACSL language, so an extern system of reading and writing in files while the accesses to auxiliary files are blocked must be deployed instead. Each session of the link application owns a set of synchronization and passing files that is shared with the launched simulation. Fig. 6 shows the blocking operations from both sides (link application and simulation) and the exclusive writing and reading operations when the access from each side is monopolized.

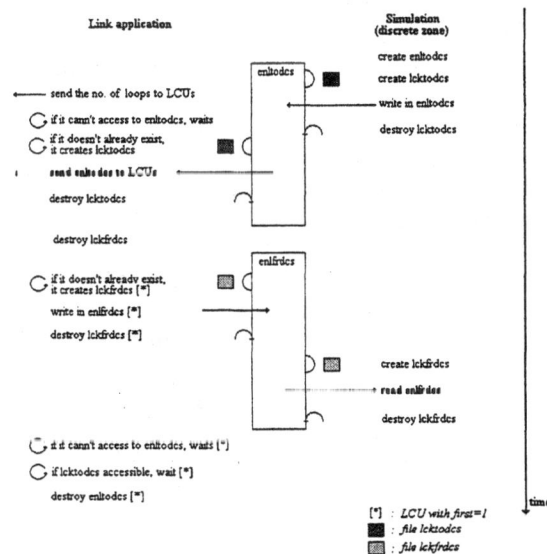

Fig. 6. Comm. simulation - link application

All these treatments are inserted in the link application with the appropriate order between them, which basically consists on calls to the socket functions of server and to the functions of manipulation of the blocking files. Both basic treatments have their corresponding complement in the other sides: the client functions of the socket in the LCUs of SICODI and the complementary reading/writing operation in the simulation. With

respect to the communication via sockets, this is based on a high level protocol with prefixes, numbers of variables to send and their values. The flag *first* is marked by the link application in the child process that attends to the first connected LCU, with the purpose of selecting it as the unique one in charge of some operations and avoiding conflicts between LCUs. As it can be seen in Fig. 7, these operations consist of the writing of manipulated variables in the simulation through the passing file and the sending of the new instant for the simulation to evolve a new sample period.

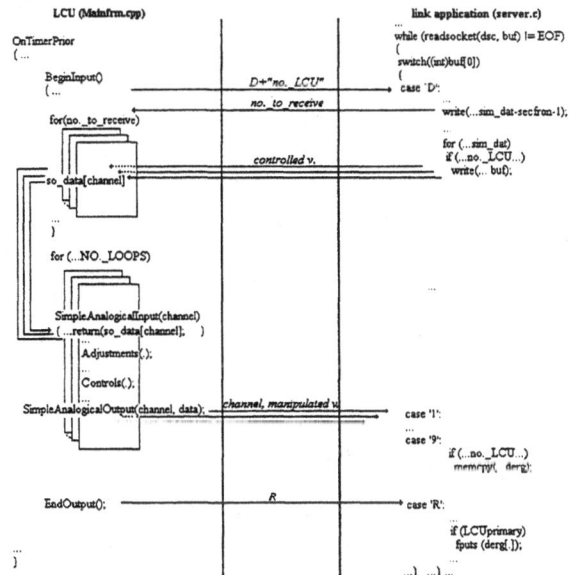

Fig. 7. Comm. LCUs - link application

4. COMMUNICATION BETWEEN LCUs

In a higher level of communication, the LCUs establish a channel between them in order to have the whole information of the system in each one of them transparently and independently of assignments. Although this level of communication works in a lower priority than the communication with the simulation, some aspects of its already operative implementation deals with the optimization of the time requirements and they have some interest from this point of view. Besides, some details of this communication are related to the treatment of the distributed database in SICODI.

Like the communication with the simulation, this new level is supported by a proprietary high level protocol over sockets based on the headers of the message packets and on the order of the values sent in them. The system of communications works on demand, in such a way that the request from a LCU unchains the corresponding answer from the LCU in which the sought information is assigned once the request is received, unpacked and analized. In the same way, the sought information is packed and sent, being recognized as the answer in destiny and suffering the corresponding unpacking for the actualization of local variables. The request remains active while remain the conditions of representation.

The so called transparency consists on the ability of the LCUs of SICODI to represent the information of all the control and process variables assigned to any of them in the same way in which they would be represented if they would be assigned to the given LCU. With inner treatments, each LCU is able to determine whether a request in any of the types of representation corresponds to local information and can be served with it, or on the contrary, it does not belong to the LCU. All the process of recognition of needs and making up of the communication packets is automatical and takes place without the participation of the users beyond the action from the graphical interface asking for information.

Depending on the type of request, the replies are different and contain variable amount of information. Basically, the two main types of request concern instantaneous information alone or together with values in the past. These possibilities are related to the types of representation to be fed. So, the tables and bar diagrams show instantaneous values of the last sample period, whereas the representation of trends needs values of the variables towards the past. Likewise, the answer packets contain the required information, being the packets corresponding to trends much bigger than those in charge of accomplish the needs of tables and bar diagrams.

Because of the direct relationship between the size of the packets and the time neccesary for the treatments, it is convenient to try to optimize this parameter in order to get packets as short as possible. This fact, together with the possibility of maintenance with the last values generated once the massive packet was received and unpacked, is exploited in the case of this type of requests in order to improve the efficiency of the communications. The implemented mechanism can be referred to as rationalization of requests, and consists on a modification of the initial massive request to another of instant values once the answer with past values is received, assigned and considered correct. From this point the system acts automatically over the active requests, removing the massive request already served and substituting it for the corresponding request of instant values. The result is the avoidance of new massive answers and the maintenance of the trend representation in succesive sample periods with the new instant value.

The scheme of communications completes with a system of dynamic generation of storing variables in the LCU that generates the request. Implicitly, this mechanism supposes the existence of a distributed database in the system consisting of the storage in the different LCUs in which they are assigned, and from which the rest of LCUs are fed. A LCU with an active request has dynamic storage not belonging to the distributed database, but serving the local and casual needs. Such a storage is destroyed immediately after the end of the request for which it was created without affecting the permanent storage in the LCUs in which the information is assigned.

One special type of message is the remote modification, which consists on the access to the variables of a loop -typically in order to change its reference, auto/manual mode and the value of the manipulated variable when manual mode is set-assigned to a different LCU from a given one. The message with remote modifications is received in the LCU with the assignment of the loop in which it tries to impose them and the new values are overwritten. The remote modification illustrates the way in which the distributed database works. The new values inserted from a View of the remote LCU are sent directly to the LCUs in charge of the affected loops. The accepted modifications are introduced in the variables of these LCUs that take part of the distributed database. Once the modifications take place, these LCUs propagate them in the system, in such a way that even different Views of the LCU from which the modification came -apart from the View in which was caused- are fed with the values sent from the distributed database, and not with the values of the neighbour View in the same LCU.

5. SUMMARY AND FUTURE WORK

SICODI is open to new additions and functionalities, via ActiveX controls or directly over its classes. Concerning real-time and communications, the use of generic tools (Windows NT, C++ and sockets) does not suppose a lack of rigour. On the contrary, some aspects of real-time systems (high-priority, multithreading) can be obtained, and the rest can be reached progressively in a down-to-top approach.

Windows NT was rejected in a first approach by the developers of control applications, but the last reliable versions and their real-time extensions (HyperKernel, InTime, RTX, 1999), have made it the choice gaining in importance. Some developments (Gergeleit, 1998) have extended the C++ properties of instantiation to the dynamical evaluation of time constraints in real-time systems, showing the suitability of this language for real-time. Finally, modifications of sockets like XTP (Xpress Transport Protocol) (Weaver, 1998) advance in the deterministic aspects of standard communications.

6. REFERENCES

García M.A. (1998). SICODI. Un sistema configurab de control distribuido en tiempo real. Ph. D.

Ramamritham K., Shen C., Gonzalez O., Sen S., Shirgurkar S.(1998). *Using Windows NT for Real-Time App.: Experimental Observ. and Recommend.* 4th IEEE Real-Time Tech.&App.

Acebes F, Achirica J, García MA, Prada C. (1995). *A simulator to train plant operators of a beet-sugar factory.*SAMS'95.Vol.18-19,pp.659-662. Published under license by Gordon & Breach Sci.

Richter J. (1994). *Advanced Windows NT.* McGr-Hill

HyperKernel,InTime,RTX(1999).http://www.imagination.com/, http://www.radisys.com/, http://www.vci.com/.

Gergeleit M. (1998).*Checking timing constraints in anO-O environment.*http://set.gmd.de/RS/Papers/checker

Weaver A.C.(1998).*Distrib Control Using Computer Networks.* IEEE Industrial Electronics SN.

DEVELOPING A TESTBED FOR DISTRIBUTED REAL-TIME APPLICATIONS

P.T.Woolley* W.M.Walker* A.Burns*

* Department of Computer Science, University of York, UK.
e-mail [woolley, billy, burns]@cs.york.ac.uk

Abstract: Fixed priority scheduling is viewed to be a standard technique for the scheduling of tasks in real-time systems on uni-processors. There are, however, also communication protocols which are based upon fixed priority scheduling. This paper discusses a testbed which has been constructed using embedded controller boards connected via a CANbus. The testbed supports experimentation with distributed real-time systems using both the Ada-95 and C programming languages. Support is provided for a higher level CAN communication protocol by a set of interface routines, and for multi-tasking on a single processor through the provision of a kernel of routines based upon the Pthread standards. Results are provided which suggest that the testbed can allow the deterministic scheduling of both tasks and communications, thus facilitating experimentation with distributed real-time systems.
Copyright © 1999 IFAC

Keywords: Ada 95, Real-Time Systems, Testbed, Controller Area Network.

1. INTRODUCTION

Fixed priority preemptive scheduling has become a standard technique for implementing real-time systems (Burns and Wellings, 1996). The majority of attention has been focused on the issues relating to the scheduling of tasks on a single processor, but communication protocols are also available which are based on fixed priority scheduling of messages. It is therefore possible to consider complete distributed systems based upon fixed priority despatching. The theoretical scheduling work has been presented previously e.g. (Liu and Leyland, 1973), (Audsley *et al*, 1993), (Gallmeister, 1995), (Burns and Wellings, 1996).

The modifications made to the Ada programming language during the 9X process has resulted in a language which is suited to programming real-time systems, and so Ada was used as the means of programming fixed priority based applications. To make use of Ada it was first necessary to produce a predictable and effective run-time system, and it is for this reason that the Gnat Ada compiler was ported to a bare processor. The fixed priority communications protocol used was one based upon the CANbus, and thus it was also necessary to develop a programmer's API to the CAN. The way in which Gnat was ported to the hardware architecture means that the testbed also includes support for programs written in C.

The availability of Gnat has significantly improved access to the Ada programming language for general programming, but there is still no easy way to experiment with real-time behaviour, particularly on distributed hardware. The testbed discussed in this paper facilitates such experimentation.

In this paper we address the issues involved in constructing an effective testbed for analysing and evaluating distributed scheduling schemes, and describe our implementation. Data on the performance and predictability of the implementation are presented, and a small demonstration system is also described.

2. HARDWARE ARCHITECTURE OF THE TESTBED

The basic hardware architecture used for the testbed comprises a number of processor nodes connected together via a serial CANbus. The processor nodes are mainly embedded Motorola 68376 based boards. This processor was chosen because it includes an on-board CAN controller; the TouCAN module. The remaining nodes in the network are either standard personal computers (PC's) with PCI CAN hardware, or further Motorola 68332 based processor boards which have access to additional CAN hardware devices; in both cases the CAN hardware is based upon the Intel CAN controller chip.

The CANbus is operated using the maximum 1Mbps data rate, and has an overall length under 10m. The CAN API, discussed in section 4.1, supports up to 10 nodes on the network. The embedded controller boards are used to provide predictable applications, while the PC's provide support for the development of programs. This support is in the form of the Gnat Ada compiler (configured as a cross-compiler), and software tools for downloading programs and data file access via CAN and monitoring the network traffic.

3. PORTING THE ADA RUNTIME

Support for the Ada programming language has been provided by porting the Gnat system to the bare processors used on the embedded controller boards. Apart from the production of a Gnat cross-compiler this has meant the construction of a kernel which implements the functions necessary for multi-tasking and time-based processing.

The kernel supports multi-tasking via the POSIX threads standard (POSIX.1c), and includes support for priority based preemptive scheduling, mutexes, condition variables, and timed delays. The kernel itself is written in a mixture of Motorola assembly language and C, in order to obtain maximum efficiency in terms of execution time and compactness of machine code. This additionally means that it is possible to use POSIX threads from within C programs.

When writing an Ada tasking program, the compiler links to the relevant functions of the pthread-kernel and thus the underlying despatching policy of Ada tasks within this environment is priority based FIFO scheduling by default. This means that the analyses developed for fixed priority scheduling can be used for the applications created. Full details of the porting of Gnat to the processors used within the testbed, and of the pthread kernel are given in (Walker *et al*, 1999).

4. INTEGRATING THE CAN

Integrating CAN into the testbed can be considered to be composed of issues in two areas; technical issues concerned with serial communications via CAN, and provision of a programmer's API (Figure 1) for access to the CANbus.

4.1 *Technical Issues*

The major technical issue to be overcome in setting up a CAN based serial communications scheme is ensuring that all CAN devices on the network are transmitting data at as close to the same rate as possible. A difference in communication rate of as little as 2% can cause messages to go undetected due to loss of synchronisation. Initially the Intel based CAN devices were generating an exact (measured on an oscilloscope) 1Mbps transmission rate, while the TouCAN chips were operating approximately 2% faster. The difference in speed was due to the underlying processor clock rate from which the CAN chip derived its timing. The TouCAN boards were operating with a 16MHz clock rate, while the Intel based chips obtained a 1MHz frequency from dedicated oscillators. This led, after between 15 minutes and 2 hours, to a loss of synchronisation. Additionally some data frames were more likely to fail than others. This effect was caused by the bit-stuffing behaviour of the low level CAN protocol. By changing the processor clock rate on the TouCAN boards to 21MHz the difference in the CAN transmission rates was brought to within 1%, and the problem was alleviated.

4.2 *Programmer's API to CAN*

Once the physical data transfer problems were solved, the problem remained of sending arbitrary length data messages across the CAN bus, while maintaining predictability in message transmissions. The standard and extended low level CAN protocols (2.0A and 2.0B) allow for a maximum of 8 bytes per frame. If more space than this is required for data, then multiple frames must be sent. A further motivation for the provision of a higher level CAN communication protocol was to isolate the programmer from the underlying hardware. This has the advantage that programming communications becomes both easier to achieve, and more reliable (in that direct access to hardware registers is not required).

From the programmer's point of view communication across the CAN network is performed by reading and writing messages of arbitrary length to CAN communication channels. The first step

```
procedure initialise_can_message_system( Node_Number : integer );

procedure reserve_can_channel( channel: integer; mode: unsigned_8);

function open_can_tx_channel(channel: integer; queueing_mode: integer; access_mode: integer )
return integer;

function open_can_rx_channel(channel: integer; queueing_mode: integer; access_mode: integer;
blocking_mode: integer ) return integer;

procedure read_can_message( CAN_handle: integer; data: in out buffer );

procedure send_can_message( CAN_handle: integer; destination: integer; data: in buffer;
byte_count: integer);
```

Fig. 1. Ada Programmer's API to the Distributed Real-Time Execution Environment's CAN protocol

in using the CAN communication protocol developed here, is to initialise the CAN message system. This is required both to activate the Tou-CAN hardware and to assign a unique NODE NUMBER to the board in question. This is important since messages are directed to specific nodes via their node number, and allowing software assignment of node identifiers enables greater flexibility. Step two, is to reserve CAN channels for TRANSMITing and RECEIVEing data. The idea behind reserving channels is to allow protection against accidental use of the CANbus. If a message is directed to a channel which has not been specifically reserved, an error will be generated. The channel number selected is effectively the priority of the CAN message, with higher numbered channels representing higher priorities. Once reserved, a channel may then be used for reading or writing messages to the CANbus. If the message data length is greater than 8 bytes, the message is packaged and sent out in as many frames as required. The reading of CAN messages can be either BLOCKING or NON_BLOCKING, and reading and writing can be performed in QUEUING or NON_QUEUING modes.

Due to the nature of the low level CAN protocol, multi-frame messages of higher priority will be sent in entirety before any lower priority messages sent from another node, even if a multi-frame low priority message has begun transmission. The reason for this behaviour is that priority arbitration is performed for each low level CAN frame sent, not on a per message basis. The multiple parts of the message are queued in 7-frame blocks, within the CAN device's message buffers. This means that as soon as one message has been transmitted, the next (with the same priority) is entered into the arbitration process for the next message frame. This behaviour ensures that the highest priority message, regardless of its length, will be delayed by no more than the time taken to transmit one frame of a lower priority message.

If two tasks on the same node are transmitting messages the lower priority message (i.e. the message on the lower channel number) will be suspended after the currently transmitting frame has been sent. Once the higher priority message has been successfully transmitted the low priority message is resumed. This means that the delay on any message is equal to its message transmission time plus the combined message transmission times of all higher priority messages on any node, plus some overheads for preempting and packaging of data. The traffic on the CANbus can thus be analysed in a similar manner to tasks on a single processor (Tindell et al, 1995).

The aim of the particular communication protocol developed for the testbed was to generate a communication protocol which was simple, and encapsulated the concept of fixed priority preemptive messaging. In this implementation the channel numbers represent message priorities, thus facilitating analysis of the network traffic. Messages may be analysed according to the channel number upon which they are transmitted together with their expected duration on the network (obtained from Figure 2) in order to assess the amount of time required for the message to be transmitted in entirety. In a sense the protocol itself can be considered as a medium level protocol, the purpose of which is to provide a base for predictable scheduling of abritrary length messages.

5. PERFORMANCE EVALUATION OF THE TESTBED

Performance evaluation of the testbed is split into two parts; evaluation of the task switching overheads and evaluation of the data transfer times across the CANbus. The values obtained

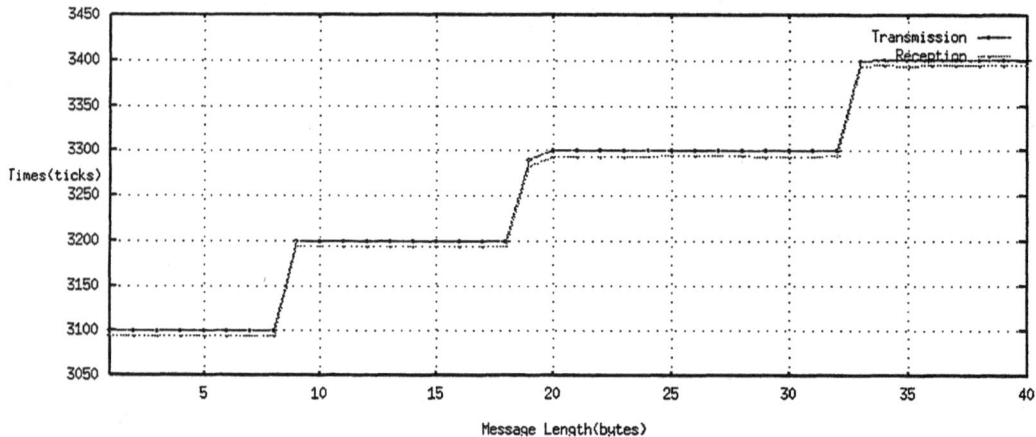

Fig. 2. Transmission and Reception Times for CAN Messages from 1 to 40 bytes.

are timed from the perspective of an application program, so that the measurements represent the timing overheads which are incurred rather than the time taken to perform kernel functions. The time values obtained thus include the overheads of entering and exiting the kernel functions as well as performing the required functions. The timer used was the standard Motorola 68376 real-time clock updating every 244us.

5.1 Task Switching Overhead

Switching between tasks (whether thay are Ada tasks or C-language threads) is governed by the thread_manager within the pthread kernel developed as a part of the run-time environment of the testbed. The major overhead (in terms of execution time) is thus present within this kernel function, since it is here that the thread queues are examined and modified. In order to obtain the execution time of this function, a C program was constructed such that a number of threads with known execution time were executed. A flag within the kernel registered the number of times that the thread_manager was called. The difference in time between the expected execution time of the group of threads (the sum of their individual execution times) and the measured execution time was taken to be the time spent within the thread_manager and was calculated to be 0.5 of a clock tick (122us) per entry.

The timing measurements taken where recorded with the processor system clocks, and this means that there is always a ± 1 clock tick (244us) error on any recorded time value. The experiments were arranged in such a way that these errors were small in comparison to the computations being timed. For example, when recording CAN transmission times the time taken was measured around 100 message transmissions. The error on the clock values was ± 1 and so the error on the duration recorded is ± 2 clock ticks. Once the time

value is divided by the number of messages sent this error is reduced to ± 0.02 clock ticks.

In order to assess the task switching overhead of an Ada tasking program, a similar experiment was set up. A block of Ada code was executed and timed. This code was then used for both the main task and all child tasks in a multi-tasking Ada program. The action of the main task was to record the time and then lower its own priority so that all child tasks were released before the main task could complete its computation. The kernel overhead was then calculated by comparing the measured main task execution time with the combined execution times of all of the tasks in the system. It was expected that, since the Ada compiler adds additional validity checks and code to manage the tasking, that the task switch overhead would be greater than the calculated thread_manager overhead. The results obtained are shown below:

Number of tasks 5 (Main task + 4 child-tasks)

Expected Loop Execution Time = 1410 ticks

Measured Loop Execution Time = 1422 ticks

Number of calls to thread_manager = 7

Task Switching Overhead = (1422 - 1410)/7 = 1.71 ticks (approx. 418us)

The task switching overhead can be broken down by tracing the kernel function calls which the Ada program makes. It was found that for a change of priority the Gnat Ada compiler checks the priority range against the maximum and minimum allowbale priority values returned by the kernel, creates and locks a mutex, performs the priority change, unlocks the mutex, and then enters the thread_manager to select which task to make active next. This means that the thread_manager is entered more than once per call to the set priority function in an Ada program, thus accounting for the increased task switching overhead.

104

5.2 CAN Communication Overheads

The overhead of using the CANbus for communications can be split into three considerations; the first is the time between issuing the send command and the time at which the full message has actually been sent onto the CANbus. The second is the time required to receive a message from the CANbus and make it available to the application, and the third is the impact of on-node and off-node preemption on multi-frame messages.

To assess the transmission time characteristics of the CAN protocol developed for this testbed the time taken to transmit 100 messages of varying data length (from 1 byte to 200 bytes) was measured. To ensure that the complete message had been sent, immediately after the send command the program entered a loop continually polling a kernel flag (via a function call) which was set once the message had completed its transmission. Figure 2 shows the resulting transmission times to be step-wise linear with respect to message length. The steps in the plot are consistent with increasing frame sizes of the messages, and can be expected from the transmission behaviour (i.e. a 16-byte message requires 2 frames, and thus has an additional overhead involved in initialising 2 CAN buffers compared to a message which requires only 1 CAN buffer). After every 7th frame there is also an additional delay caused by the re-filling of buffers, because messages are placed into the buffers 7-frames at a time (the TouCAN chip has space for 16 CAN buffers, and so 7 of these are used for transmission leaving the rest free for reception). Since the message transmission times are broadly speaking linear and repeated experimentation gives the same results, they are also predictable.

In order to assess the reception time for CAN messages, two nodes were employed. The first node transmitted 100 messages each of sizes from 1 byte to 200 bytes. The second node polled for these messages in NON_BLOCKING mode, and recorded the amount of time taken from the start of transmission until the complete messages had arrived. Synchronisation was achieved by having the reception node BLOCK on a CAN channel before commencing timing, and having the transmitting node send a *header* for each block of messages. The reception time was found to closely follow transmission time. The activity of receiving messages is slighlty faster than the complete transmission of a message, and in Figure 2 it appears that messages are received before they have been completely sent. It must be remembered that the transmission program was timing until the time at which confirmation of the transmission was recorded, and will include time taken to change the TouCAN hardware registers and update the

kernel status data. If BLOCKING mode had been used for reception, there could have been an additional 1 tick (244us) on all reception time values due to the need for the kernel to wake-up the task when a CAN frame was detected. This must be considered when using BLOCKING mode reads, but represents an additional delay on the basic function of reception.

The impact of message preemption was estimated by arranging for two tasks to write messages to the CANbus at a regular interval. The priority of one message was much higher than the other, with the lower priority message also having the longer data length and lower task priority. Thus when the lower priority task is running and transmitting its long message, the higher priority task can become active and preempt that message. By timing the transmission time of the lower priority message the impact of the preemption of messages can be assessed. This was performed with both on-node preemption of messages and off-node preemption and the results are shown below.

Time to transmit 200 byte messages = 17667 ticks

Time to transmit 100 byte messages = 974 ticks

Estimated time for preempted 200 byte messages = 17667 + 974 = 18641 clock ticks

Measured time for preempted 200 byte messages = 18639 clock ticks

Accounting for a $\pm$ 2 clock tick error on all timings, these results indicate that there is negligible cost involved in the on-node preemption of CAN messages. Additionally, the low-priority message did not interfere with the transmission of the high priority message.

5.3 Implementation of the CAN Driver

The memory model of the boards used for this testbed is simple linear addressing with no memory protection. An application running on one of the boards has access to all available memory at any one time. The CAN "driver" consists of kernel functions for transmitting and receiving messages. This means that there is no task or daemon running to serve the CAN and thus no impact is made upon schedulability unless the CANbus functions are actually used.

For transmission the overheads described in the previous section are all that are incurred. The Transmission function can be treated from a timing perspective like any other function, since it simply copies the messages into the TouCAN hardware buffers. Reception requires more consideration, as it is interrupt driven. If a CAN frame arrives at the node, an interrupt is generated and the message is copied from the TouCAN hardware into a kernel data structure. Control then returns to whatever task was previously running, unless a higher priority task is blocked awaiting a message on the appropriate channel number. Only CAN frames which have been expressly directed to the node have any impact on timing, because the TouCAN hardware is used to mask out messages which do not have the correct node mask in their CAN identifier.

The additional overhead which needs to be considered in schedulabilty analysis is therefore related to the expected arrival rate of messages at a particular node. The overhead in terms of schedulability analysis can be treated as a highest priority task with period equal to message arrival rate. Where there are multiple messages directed to one node, each message has a priority with respect to other messages related to its channel number.

6. A DEMONSTRATION SYSTEM

The controlled system for the demonstration was an electric slot-racing set, modified such that there were infrared sensors around the lanes of the track (784 per lane), and the cars were fitted with infrared emitters. The sensors were connected to a Motorola 68030 processor board in such a way that when the emitter from a car triggered a sensor, an interrupt was generated on the processor board and the sensor number and current processor clock counter value were recorded. in memory locations. This processor board was connected via a VMEbus to a Motorola 68332 based processor board with an Intel CAN controller chip. Track sensor data was broadcast on the CANbus every 8ms, and this was sufficient to locate the position of cars on either lane to the closest sensor just passed. The lane voltages were set in response to a CAN message containing a lane number and requested voltage value. An interesting feature of the track layout was the chicanes, where it was possible for two cars to collide. The aim of the demonstration controller was to allow two cars to complete the circuit with fast laps times, while ensuring that they did not collide on any of the chicanes. The control software was arranged such that two MC68376 boards were used, one for each lane, and co-ordination was performed by direct observation of the position of both cars (by sensor number). By ensuring that, when in chicanes, the cars were never closer than 5 sensors apart (just over one car length in this case), the cars were able to complete circuits of the track without colliding on the chicanes. In addition the trailing car was not held up for the full length of the chicane, but only for sufficient time to void collision with the lead car.

7. CONCLUSION

The software and hardware of the testbed provide a suitable platform for experimentation with real-time distributed applications. The use of the Gnat Ada compiler in conjunction with the pthread kernel which has been developed, enables the predictable and timely execution of multi-tasking programs on individual nodes in the network, and the CANbus facilitates the analysis of network traffic. The measurements undertaken thus far indicate that sufficient data is available on the performance of the testbed to enable the accurate assessment of analysis techniques, and to facilitate experimentation with real-time distributed software.

8. REFERENCES

Audsley, N. Tindell, K. and Burns, A. (1993). The end of the line for static cyclic scheduling ? *In Proc. Fifth Euromicro Workshop on Real Time Systems* 36-41.

Burns, A. and Wellings, A. (1996). *Real-Time Systems and Programming Languages.* 2nd ed.. Addison-Wesley. Harlow.

Gallmeister, B.O. (1995). *POSIX.4: Programming for the Real-World.* O'Reilly & Associates. Sebastapol CA.

Liu, C. and Layland, J. (1973). Scheduling Algorithms for Multiprogramming in a hard Real Time Environment. *JACM* **20(1).** 46-61.

Tindell K, Burns A, and Wellings, A. (1995). Calculating Controller Area Network (CAN) Message Response Times. *Control Engineering Practice.* **3(8).** 1163-1169.

Walker, W.M. Woolley, P.T. and Burns, A (1999). An Experimental Testbed for Embedded Real-Time Ada95. To Appear in *Ada Letters, 1999.*

EXPERIMENTAL EVALUATION OF HIGH-ACCURACY TIME DISTRIBUTION IN A COTS-BASED ETHERNET LAN

Ulrich Schmid [*,1] Herbert Nachtnebel [**,2]

* Technische Universität Wien, Department of Automation,
Treitlstraße 1, A-1040 Vienna, Austria. Email:
s@auto.tuwien.ac.at
** Technische Universität Wien, Department of General Electrical
Engineering and Electronics, Gußhausstraße 25–29, A-1040
Vienna, Austria. Email: nachtneb@iaee.tuwien.ac.at

Abstract: This paper reports on the experimental evaluation of the *Network Time Interface* (NTI) M Module developed in the SynUTC project at TU Vienna. Designed for COTS-based distributed systems, the NTI provides all hardware-related features for high-accuracy interval-based external clock synchronization, like a high-resolution rate- and state-adjustable clock, local accuracy intervals, interfaces to GPS receivers, and various timestamping features. The evaluation results show that the NTI enables GPS time distribution with μs-accuracy even in Ethernet-based systems. However, the available configuration parameters must be carefully chosen —and experimentally verified— in order to cope with the various hidden sources of timing uncertainty. *Copyright © 1999 IFAC*

Keywords: GPS time distribution, Ethernet, experimental evaluation, interval-based external clock synchronization, COTS hardware support, M-Modules

1. INTRODUCTION

It is well-known from (Lundelius-Welch and Lynch, 1984) that the synchronization tightness achieved by any clock synchronization scheme depends primarily on the uncertainty (= variability) ε in the end-to-end transmission delay. For typical LANs, ε lies in the ms-range, which makes it impossible to use a simple packet data exchange to disseminate time with high accuracy. Additional techniques are required for this purpose, which,

however, should be compatible with existing network controller technology to be useful in practice.

One of the achievements of the authors' research project SynUTC (Schmid, 1994) is a suitable add-on hardware (Horauer et al., 1998) for high-accuracy time distribution in COTS-based distributed systems. A prototype implementation of this *Network Time Interface* (NTI) is available as an industry-standard M-Module (Mandl et al., 1999). It was thoroughly evaluated recently in an Ethernet-coupled distributed system made up of several VMEbus-CPUs running ISI's multiprocessor real-time kernel pSOS^{+m}.

This paper reports on the major[3] results and insights gathered during this evaluation. It is organized as follows: After a brief overview of ar-

---

[1] The SynUTC-project received support from the Austrian Science Foundation (FWF) grant P10244-ÖMA, the OeNB "Jubiläumsfonds-Projekt" 6454, the BMfWV research contract Zl.601.577/2-IV/B/9/96, and the Austrian START programme Y41-MAT. For further reference see *http://www.auto.tuwien.ac.at/Projects/SynUTC/*
[2] Supported by the Austrian *Gesellschaft für Mikroelektronik* (GMe).

---

[3] A forthcoming technical report (Schmid et al., 1999) will contain all the details.

chitecture and features of the NTI in Section 2, the hard- and software of the evaluation system is outlined in Section 3. Section 4 presents two issues that were found to have a major impact upon ε, along with supporting measurement data. In Section 5, it is shown how the obtained results plug into a very simple interval-based time distribution algorithm. A relation to existing work in Section 6 and some conclusions in Section 7 eventually complete the paper.

2. NTI FEATURES AND ARCHITECTURE

The *Network Time Interface* (NTI) (Horauer *et al.*, 1998) has been designed for high-accuracy fault-tolerant external clock synchronization in LAN-based[4] distributed systems, cf. Section 6. Apart from advanced clock synchronization algorithms, this goal primarily requires hardware support for exact timestamping of data packets and a sophisticated rate and state adjustable clock at each node. Fig. 1 shows the basic architecture of a 2-node system.

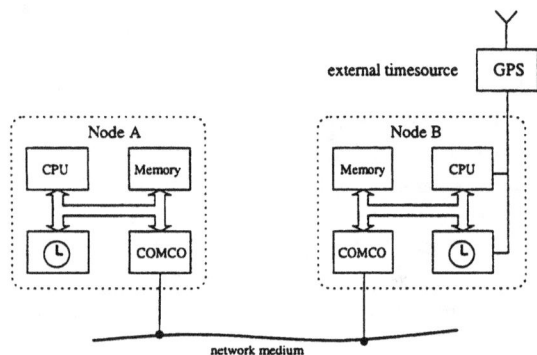

Fig. 1. *Basic architecture of a 2-node system*

Accordingly, each node must be equipped with a clock device (the UTCSU, see below), a CPU responsible for executing the clock synchronization algorithm, and a *Communications Coprocessor* (COMCO) providing access to the network by reading/writing data packets from/to (shared) memory. At least one node must also be connected to a GPS timing receiver (Höchtl and Schmid, 1997).

In purely software-based clock synchronization, timestamping of *Clock Synchronization Packets* (CSPs) at the sending resp. receiving side is done by reading the clock when assembling the CSP for transmission resp. in the packet reception interrupt service routine. However, as explained in detail in (Horauer *et al.*, 1998), this implies

that the transmission delay uncertainty ε includes both the network channel access uncertainty and the reception interrupt latency, which can be quite large. To get rid of them, a refinement of the DMA-based coupling method originally proposed in (Kopetz and Ochsenreiter, 1987) can be used. The key idea is to insert a timestamp on-the-fly into the memory holding a CSP in a way that minimizes ε, as outlined in Fig. 2.

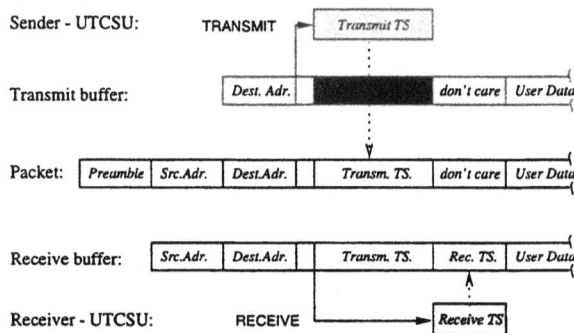

Fig. 2. *Packet timestamping*

Whenever the COMCO fetches data from the transmit buffer holding the CSP, it has to read across the particular offset that causes a special decoding logic to generate the trigger signal TRANSMIT. Upon its occurrence, the UTCSU puts a *transmit timestamp* (TxTS) into a dedicated sample register, which is transparently mapped into a certain portion of the transmit buffer and hence automatically inserted into the outgoing packet. Note that the trigger address and the mapping address are configurable parameters that may be different. By the same token, when the receiving COMCO writes a certain offset in the receive buffer, the trigger signal RECEIVE is generated by the decoding logic, which causes the UTCSU to sample the *receive timestamp* (RxTS) into a dedicated register. This timestamp is saved subsequently in an unused portion of the receive buffer.

The proposed approach works for any COMCO that accesses CSP data in memory. Suitable chipsets are available for a wide variety of networks, ranging from fieldbusses like Profibus over Ethernet up to ATM networks. It must be stressed, however, that it is usually impossible to decide *a priori* whether a particular COMCO is suitable for a certain application. In fact, the many hidden architectural intricacies make it very difficult to accurately assess ε without actual measurements.

The *Network Time Interface* (NTI), see (Mandl *et al.*, 1999) for the comprehensive user manual, provides all the hardware support required for high-accuracy clock synchronization on a single-height

---

[4] Note that the approach can also be adopted to more general topologies commonly known as WANs-of-LANs, provided that all gateway nodes are also equipped with the NTI, cf. (Schmid, 1994) and (Schossmaier and Schmid, 1995).

(146 x 53 mm) MA-Module[5]. Fig. 3 shows the major components on-board the NTI, which can be accessed from any COTS CPU/COMCO with MA-interface via ordinary memory and memory-mapped registers.

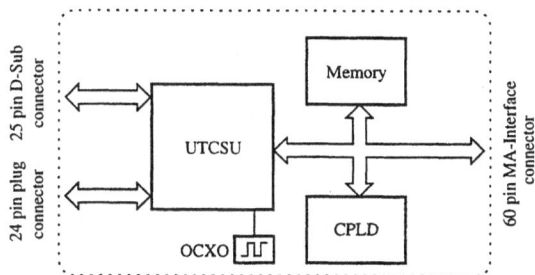

Fig. 3. *NTI block diagram*

All required decoding and glue logic of the NTI is assembled in a single, in-circuit programmable *complex programmable logic device* (CPLD) designed using VHDL (Nachtnebel *et al.*, 1998). It adapts the UTCSU and the memory to the MA-Module interface, forwards interrupt requests from the UTCSU to the carrier-board, generates the acknowledgement signal terminating a bus cycle, etc.

The *memory* serves as control and data interface between the CPU and the COMCO, providing the special functionality for COMCO accesses as outlined before. It consists of up to four 64k x 16 bit SRAM chips and supports byte, word, and longword read/write accesses.

The UTCSU-ASIC (*Universal Time Coordinated Synchronization Unit*) described in (Schossmaier *et al.*, 1997), (Loy, 1996) contains most of the dedicated hardware support for clock synchronization. Manufactured as an ASIC in 0.7 μm CMOS technology, the UTCSU accommodates about $80,000$ gates on a 100 mm^2 die packed into a 208-pin PQFP case. Fig. 4 gives an overview of the major functional blocks.

The centrepiece of the UTCSU is a *local clock unit* (LTU) utilizing a 56 bit NTP-time format. Clock time can be read atomically as a 32 bit *timestamp* with resolution $2^{-24} \approx 60$ ns and a 32 bit *macrostamp* containing the remaining 24 most-significant bits + an 8 bit checksum protecting the entire time information. The LTU employs an unconventional *adder-based* clock design, which uses a 91-bit high-speed adder instead of a simple counter for summing up the elapsed time between succeeding oscillator ticks. Owing to this, the UTCSU can be paced by a quartz oscillator

---

[5] *M-Modules* (MUMM, 1996) are an open, simple, and robust mezzanine bus interface primarily designed for VME carrier boards, which are commonly used in Europe. *MA-Modules* are enhanced M-Modules, providing a 32 bit data bus and enhanced addressing capabilities.

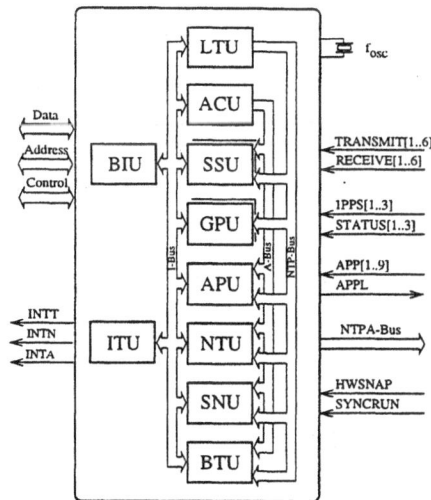

unit	full name
ACU	Accuracy Unit
APU	Application Unit
BIU	Bus Interface Unit
BTU	Built-In Test Unit
GPU	GPS Unit
ITU	Interrupt Unit
LTU	Local Time Unit
NTU	Network Time Interface Unit
SNU	Snapshot Unit
SSU	Synchronization Subnet Unit

Fig. 4. *Interior of the UTCSU*

of arbitrary frequency $f_{osc} \in 1 \ldots 20$ MHz; alternatively, an external frequency source like the 10 MHz output of a high-end GPS receiver can be used. Moreover, the local clock is fine-grained rate adjustable in steps of about 10 ns/s and supports state adjustment via continuous amortization as well as (optional) leap second corrections in hardware.

To support interval-based clock synchronization (see Section 5), the UTCSU contains two more adder-based "clocks" in the ACU that are also driven by the oscillator frequency f_{osc}. They are responsible for holding and automatically deteriorating the 16 bit *accuracies* α^- and α^+, thereby maintaining a bound on the local clock's instantaneous deviation w.r.t. real-time.

A number of external events, supplied to the UTCSU via buffered/opto-decoupled, polarity programmable input lines, can be *time/accuracy-stamped* with local time and accuracy upon the appropriate input transition. Optionally, an interrupt can be raised on such an event as well. Three different functional blocks in the UTCSU utilize this feature: First, trigger signals generated by the decoding logic at CSP transmission and reception sample the current local time/accuracy into dedicated UTCSU registers in one of the six available SSUs. Second, three independent GPUs are provided for timestamping the *one pulse per second* (1pps) signal —indicating the exact beginning of a UTC second— from up to three GPS receivers. Finally, nine independent application time/accuracy-stamping inputs are provided by the APU.

Those timestamping features are complemented by several 48 bit programmable *duty timers*: Whenever local time reaches the programmed time of an armed duty timer, an internal or external signal changes state and an optional interrupt is raised. Duty timers are required for triggering activity of the clock synchronization algorithm, for controlling continuous amortization, inserting/deleting leap seconds, and generating application-related events.

All UTCSU application time/accuracy-stamp inputs and duty timer outputs as well as all interfaces to GPS receivers are available via the NTI M-Module's front-panel 25-pin D-sub connector. In addition, all receive and transmit time/accuracy-stamp signals are fed to the carrier board via the 24-pin plug connector. High-speed opto-couplers or transceivers are provided for all inputs to ensure a decoupled and reliable interface.

3. EVALUATION SYSTEM

Aiming at the support of existing technology, it was only natural to use COTS components for building up the evaluation system as well. Each node thus consists of a Motorola MVME-162 CPU (M68040 CPU + Intel i82596CA Ethernet-controller) and a AcQ i6360 or, alternatively, a MEN A203 VMEbus carrier-board hosting the NTI MA-Module, which are plugged into a dedicated A32/D32 VMEbus [6] backplane. All nodes are interconnected via the CPU's Ethernet port using thin-wire technology. Fig. 5 outlines the basic hardware architecture.

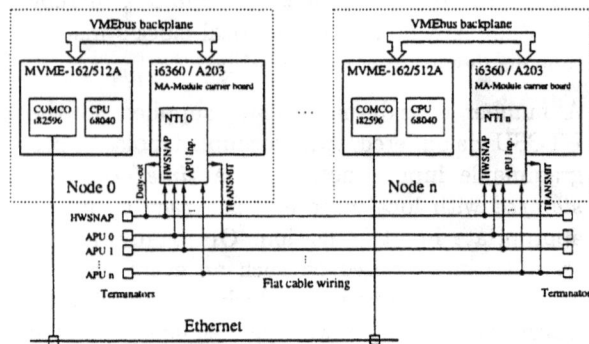

Fig. 5. *Evaluation system hardware architecture*

---

[6] Actually, VMEbus equipment from another research project was re-used for this purpose. Whereas the resulting architecture is unlikely to be chosen in practice, it nevertheless constitutes an suitable "worst case environment" for evaluation.

Dedicated flat cable wiring —the only add-on to Fig. 1 required for evaluation purposes— is used for measuring the (one-way) transmission delay between any two nodes in the system. More specifically, the TRANSMIT-signal (recall Section 2) triggering the transmit timestamp on node i's NTI is exported and fed to the application timestamp input A_i on each NTI in the system. This way, any node that receives a CSP from node i just has to compute the difference of the —locally available— timestamps RxTS $- A_i$TS to obtain the actual transmission delay. In addition, the HWSNAP-timestamp inputs of all NTIs are also tied together and driven by a duty-timer output of NTI 0. This enables the generation of simultaneous interrupts at all nodes, a feature required for initiating coordinated activity.

The evaluation system's software consists of two layers. The lower-level one is provided by the *i82596 NTI-driver* (Richter *et al.*, 1999), which integrates the NTI and the i82596 Ethernet-COMCO into ISI's pSOS^{+m} multiprocessor real-time kernel.

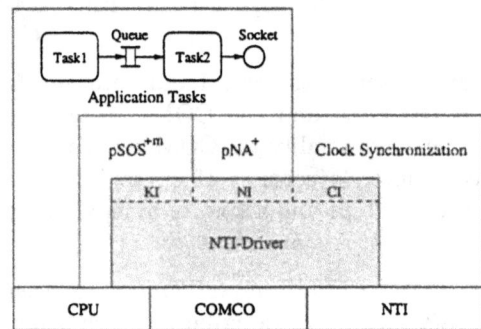

Fig. 6. *A pSOS^{+m} node using the NTI-driver*

From Fig. 6, it is apparent that the NTI-driver multiplexes three different interfaces to the i82596 and the NTI:

(1) *Kernel Interface* (KI): pSOS^{+m} supports multiprocessing by means of remote objects (tasks, queues, semaphores, etc.), which are implemented atop of RPCs. To keep the kernel reasonably independent of the particular communications network, a user-supplied KI is required that maps a simple message-passing interface to the particular COMCO.

(2) *Network Interface* (NI): In addition to kernel services, application tasks can use TCP/IP sockets for communication with remote sites if the additional software component pNA$^+$is present. Like the pSOS^{+m} kernel, pNA$^+$ is kept hardware-independent by means of a user-supplied NI, which is similar to the KI but plugs into a different message-passing interface.

(3) *Clock Interface* (CI): The third component that requires network services is the clock

synchronization algorithm. Again, a simple message-passing interface CI is sufficient here; it is the only one that uses the time-stamping feature of the NTI.

Since the primary intention of the NTI-driver was transparent integration of clock synchronization into pSOS^{+m}, it also simplified the development of the evaluation system's software considerably. In fact, using the powerful features of pSOS^{+m}, the multitasking system shown in Fig. 7 was developed without much difficulty.

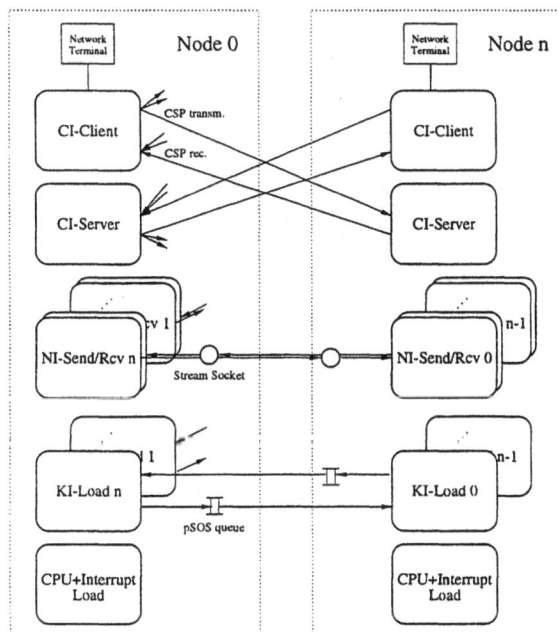

Fig. 7. *Evaluation system software architecture*

It consists of a number of concurrently executing tasks (the same at each node), which can be parameterized online by means of a custom configuration dialogue.

- The *CI-Client* is the major task of the evaluation system. It periodically broadcasts CSPs to the *CI-server* at any other node and collects the reply messages. After a specific number of rounds has been reached, the CI-client computes all transmission delay statistics and prints them on the network terminal. Among the configurable parameters are e.g. the number of rounds, broadcast period, CSP size, TxTS and RxTS trigger offsets, etc.
- Any pair of *NI-Send* and *NI-Receive* tasks at a remote node exchange data via a TCP/IP stream socket, thereby generating network traffic of configurable load and type.
- Similarly, any pair of *KI-Load* tasks at different nodes use two global pSOS^{+m}-queues to exchange data over this IPC mechanism.
- Finally, there is a task responsible for generating a configurable CPU and/or interrupt load with optional accesses to NTI memory.

Relying upon the widespread belief that using COTS components makes development easy, 6 months were projected for crafting the above evaluation system. In reality, however, about two years were spent on putting the system into full operation, and certain problems almost caused the project to fail at all. In fact, given the particular pitfalls listed below, *"COTS Integration: Plug and Pray?"* (Boehm and Abts, 1999) would have been a more appropriate belief for planning the development work:

(1) When encountering a certain misbehavior of the NTI prototype on a COTS M-Module carrier board, the NTI was obviously blamed for it. It took some time to recognize that the real cause might be the carrier board as well: Both the i6360 and the A203 had at least two serious bugs, ranging from ringing at critical signals over specification violations up to tricky race conditions.

(2) Finding a bug on a COTS module is one thing; convincing the manufacturer of its existence and pressing him to fix it is a completely different matter. In fact, if custom hardware like the NTI is involved, suppliers usually divert the problem to these components. Hence, one is forced to develop a handy hardware+software system that clearly reproduces the bug, and ship it to the manufacturer. Even if this can be managed, however, it depends upon the severity of the error —and the importance of the customer— whether and when a bug fix is provided.

(3) Even the best documentation of a COTS product usually lacks an in-depth description of certain non-standard features. For example, when the CSP memory was moved from the on-board RAM of the MVME-162 to the NTI, the NTI-driver showed completely irregular behavior. The first guess was that the i82596 on the MVME might be unable to access memory on the VMEbus, contradicting both hints in the user manual and pre-purchase statements of the supplier. Several weeks after issuing the problem to the customer support, this guess was confirmed by Motorola. Accidentally, however, a clue to an i82596 test that allows to specify the RAM address was found in the MVME diagnostics utilities manual. Trying it out with the NTI-RAM, it worked fine! (The problem was eventually tracked down to the fact that i82596 locked access cycles had been enabled, which are not supported on the VMEbus.)

In view of this experience, we are not sure whether it was indeed a good idea to design the NTI prototype without an on-board CPU and COMCO. W.r.t. overall development costs, it might well have been the case that improved control over all components had amply brought in the increased design complexity.

4. MEASUREMENT RESULTS

Given the wealth of measurement data gathered during the various experiments, only the two most interesting findings can be presented in this paper. Further details can be found in (Schmid *et al.*, 1999).

It had been clear right from the beginning that the COMCO's large FIFOs used for tolerating varying bus latencies when accessing memory would impair ε. The i82596 provides a 64 bytes *Tx-FIFO* and a 128 bytes *Rx-FIFO*, which are filled from/emptied to memory by two on-chip DMA channels. Elaborate prefetching policies ensure that the data consumed/produced by the serial side at the constant rate of 10 Mbit/s can be handled even when the bus access grant is delayed. Given the NTI's flexibility w.r.t. programming the TxTS and RxTS timestamp offsets (recall Section 2), however, it was to be hoped that a gross impairment could be circumvented. Fortunately, this proved to be true at the end, although it turned out that "theoretical" knowledge from the —comprehensive and quite detailed— i82596 user manual was not sufficient to avoid certain tricky pitfalls revealed by experimental evaluation.

For example, the i82596 manual says that the serial side initiates transmission when Tx-DMA has loaded the first two lwords (=8 bytes, destination address and length field) from CSP data. Therefore, it was concluded that transmission is in progress, i.e., not deferred due to carrier sense or collision resolution, when the succeeding lwords are eventually fetched. Since CSP data is preceeded by 4 additional lwords containing control information for the i82596, this led to the decision to trigger TxTS at offset 28 [bytes] in the CSP buffer. Experimental evaluation, however, revealed an unacceptable $\varepsilon \approx 1.2$ ms in this case, which is the duration of a maximally-sized Ethernet-packet.

Fig. 8 outlines the reason for this unpleasant effect. The shown curves relate the instant of accessing a certain offset in the CSP buffer with the outgoing transmission at the serial side. Since the latter is performed at a constant rate of 10 Mbit/s, the x-axis can be interpreted as a time axis. The time origin is start of channel activity, which coincides with reading offset 24, as said before. The fat solid curve (1) represents the case of channel idle.

What was not anticipated correctly is the fact that the Tx-FIFO is also completely filled when transmission is deferred, as shown in the fat dashed curve (2). By choosing the TxTS trigger offset 28 as above, the packet could be delayed by the maximum channel access time *after its TxTS was drawn*! Hence, the TxTS trigger offset has to be

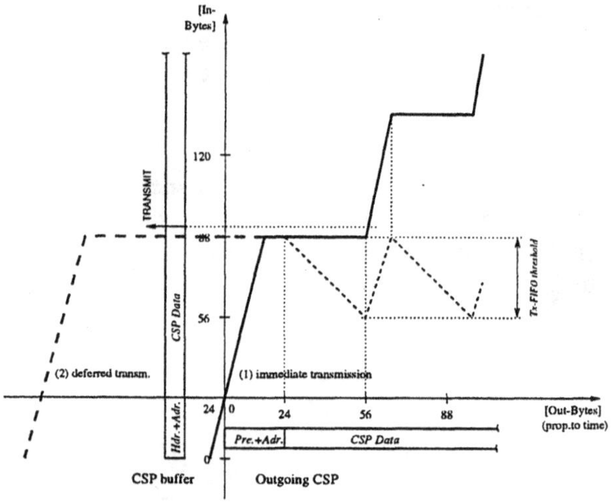

Fig. 8. *Buffer accesses during CSP transmission*

moved beyond offset 24+64=88, so that the 64 byte Tx-FIFO can be completely filled without triggering TRANSMIT. TxTS is eventually drawn during the next Tx-FIFO fill, which takes place after 32 bytes have been read off the Tx-FIFO and sent to the channel, cf. the dashed line in the figure above. The appropriate experiments confirmed that any TxTS offset > 92 performs almost equally well.

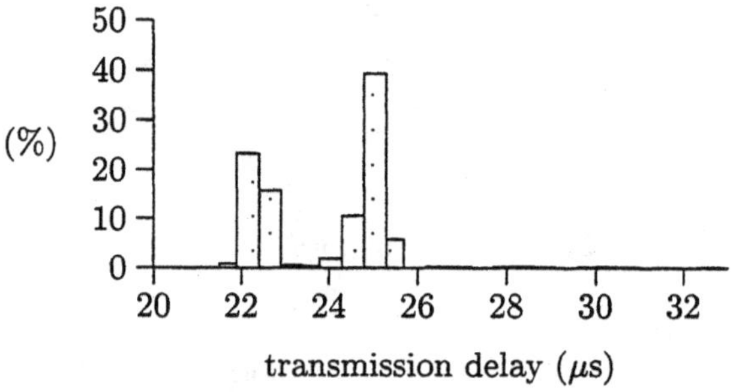

Fig. 9. *Histogram (rel. frequency) of transmission delays for MEN A203 carrier board*

Nevertheless, the resulting histograms like the one in Fig. 9 showed a disturbing long tail of rare excessive transmission delays, which needed explanation. It was soon recognized that the length of the tail is different for the two carrier-boards available, which are functionally equivalent but have different bus speeds. In fact, the ratio of the encountered maximum tail excess of about 12 μs for the A203 vs. 20 μs for the i6360 was found to match the bus speed ratio of about 600 ns/lword vs. 900 ns/lword, suggesting an interfering bus activity of at most 20 lword accesses.

This is eventually explained by the fact that the Tx-FIFO fill of a (deferred) transmit command can interfere with CSP reception. More specifically, upon reception, the Rx-FIFO is filled with incoming data, which are not transfered to memory by Rx-DMA until the Rx-FIFO threshold of

64 bytes has been reached. Now, if it happens that a transmit command is issued to the i82596 at a time that causes the Tx-FIFO fill to start right before the Rx-threshold is reached, Rx-DMA can be delayed (at most) for the time t_{88} required to fill 88 bytes, i.e., 22 lwords, into Tx-FIFO. Note that this is obviously a rare event, since Tx-DMA must start in the time window $[t_{Rx} - t_{88}, t_{Rx}]$ to be interfering at all. However, some experiments revealed that coordinated activity (simultaneous CSP sends at all nodes) can increase this probability considerably at particular nodes, like at the one that produced Fig. 9.

Unfortunately, since the CPU issuing the transmit command does usually not know about ongoing CSP receptions, there is no easy way of avoiding [7] this pitfall. Hence, the only remedy is increasing the bus speed, which both decreases the maximum disturbance of ε and the probability of its occurrence. For example, using the maximum speed of the NTI's M-Module interface (100 ns/lword), one could achieve a quite reasonable $t_{88} \leq 2.2$ μs.

Fig. 9 finally reveals a remaining worst case timing uncertainty $\varepsilon \approx 5$ μs (with a standard deviation of about 1.5 μs), which is almost the same for the A203 and the i6360. Therefore, it must be caused by bus arbitration delays and/or by i82596-internal uncertainties that do not depend upon VMEbus speed. A possible explanation of the former are MVME-162 CPU-activities that delay granting the local bus to the on-board i82596; the latter uncertainty might be caused by a different transmission initiation at the serial side in case of channel idle resp. channel busy. Unfortunately, those effects cannot be distinguished easily by means of the present evaluation setup, since MVME-162 CPUs with different local bus speeds are not available.

Anyway — comparing the above result to $\varepsilon >$ 10 ms measured for pure software-based time-stamping reveals an improvement of a factor of 1000, which seems to be remarkably enough.

5. TIME DISTRIBUTION ALGORITHM

A unique feature of the NTI/UTCSU is its support of the interval-based paradigm (Marzullo, 1984), where real-time t (e.g. GPS time or UTC)

is represented by an *accuracy interval* $\boldsymbol{A}(t)$ satisfying $t \in \boldsymbol{A}(t)$. Referring to Section 2, a node p's accuracy interval $\boldsymbol{A}_p(t) = [C_p(t) - \alpha_p^-(t), C_p(t) + \alpha_p^+(t)]$ is formed by adding its UTCSU's clock value $C_p(t)$ and its interval of accuracies $\alpha_p(t) = [-\alpha_p^-(t), \alpha_p^+(t)]$. Exploiting this on-line bound on the local clock's deviation from real-time, a clock synchronization application can judge whether the instantaneous accuracy is sufficient for a certain goal, a feature that is particularly interesting for multi-clustered [8] applications. The price to be paid for this additional information, however, are explicit bounds on certain system parameters like transmission delay uncertainty — and this is where the results of the experimental evaluation come into play.

During the last few years of working on the SynUTC project, a thorough theoretical framework for fault-tolerant interval-based clock synchronization was established, see (Schmid, 1994), (Schmid and Schossmaier, 1997), (Schossmaier, 1997), (Schmid, 1997), etc.; a comprehensive collection of those results can also be found in (Schossmaier, 1998). To link this research with the NTI and the evaluation results, it is sufficient to consider a very simple (non-fault-tolerant) algorithm for external synchronization: In a system consisting of a single node g equipped with a GPS receiver and one or more ordinary nodes, let the UTCSU-clock of node g be continuously locked [9] to GPS time. Then, the following trivial algorithm can be used for time distribution:

(G) *GPS node g*: Periodically, at $C_g(t) = kP$, $k \geq 1$, node g uni- or broadcasts a CSP containing $\boldsymbol{A}_g(t)$ ($=$ NTI g's transmit time/accuracystamp) to all other nodes in the system.

(O) *Ordinary node p*: If a CSP from node g (containing $\boldsymbol{A}_{gp} = [T_{gp} \pm \boldsymbol{\alpha}_{gp}]$) arrives at local time T_p^g ($=$ the timestamp-part of NTI p's receive time/accuracystamp $\boldsymbol{A}_p^g$), compute

$$\boldsymbol{I}_p^g = \boldsymbol{A}_{gp} + [\delta_{gp} \pm \varepsilon_{gp}] + \boldsymbol{u} + \overline{\boldsymbol{G}}$$
$$+ T^R - T_p^g + (T^R - T_p^g)\rho_p$$

and setup the UTCSU clock correction duty timer for time $T^R = kP + \Delta$, $\Delta > 0$ sufficiently large, to initiate local clock correction towards $\boldsymbol{I}_p^g$. Note that $\Delta > 0$ must secure that message

---

[7] It should be mentioned that the idea of triggering RxTS by the reception interrupt signal INT, which is the method used in all other existing approaches (see Section 6) would perform even worse here. In fact, there are technical problems that rule out this method for the i82596 at all: Apart from the difficulty of tapping the interrupt line on a COTS module like the MVME-162, INT signals various conditions other than packet reception as well.

[8] Accuracy can be used to secure precision among clusters that do not participate in a common clock synchronization algorithm: If $\boldsymbol{A}_p$, $\boldsymbol{A}_q$ at nodes p, q located in different clusters are both non-faulty, inclusion of t implies that their clock values $C_p(t)$, $C_q(t)$ cannot be further apart than $-\left(\alpha_p^-(t) + \alpha_q^+(t)\right) \leq C_q(t) - C_p(t) \leq \alpha_q^-(t) + \alpha_p^+(t)$.

[9] This is easily accomplished by adjusting node g's UTCSU-clock according to the difference GPS time vs. sampled 1pps UTCSU-timestamp upon every 1pps-pulse from the GPS receiver.

reception and computation of the above interval are completed before clock correction takes place.

As explained in detail in (Schmid and Schossmaier, 1997), the apparently complicated formula above expresses a few basic operations only: First, the received accuracy interval must be enlarged by ε_{gp} to account for the variable transmission delay $\delta'_{gp} \in [\delta_{gp} \pm \varepsilon_{gp}]$ (delay compensation). Second, when shifting the resulting interval from T_p^g to resynchronization time T^R, a sufficient enlargement ("deterioration") of the shifted interval is required to compensate the non-zero drift $\rho'_p \in \rho_p$ of the local clock $C_p(t)$. Note that this drift compensation is performed continuously by the UTCSU in hardware during the whole round as well. Finally, to cope with clock granularity resp. rate adjustment uncertainty, the intervals $\overline{G} = [-G, 0]$ resp. $u = [-u, u]$ are provided; using an oscillator with $f_o = 10$ MHz as in the evaluation setup, $u = 1/f_o = 100$ ns and $G = 2^{-23} \approx 120$ ns.

Clearly, I_p^g also gives node p's interval of accuracies α_p immediately after resynchronization. Assuming identical ρ_p and ε_{gp} for any p for simplicity, one obtains $\alpha_p \subseteq \alpha = \alpha_g + \varepsilon + \Delta\rho + u + \overline{G}$. Moreover, the interval of accuracies immediately before the next resynchronization is bounded by $\alpha' = \alpha + P\rho$, so that the worst case precision of any two nodes in the system evaluates to $\pi = |\alpha'|$.

Assuming an OCXO with $\rho_p = [-10^{-7}, 10^{-7}]$, $\Delta = 100$ ms, $\alpha_g = [-370\text{ns}, 370\text{ns}]$ and choosing $\delta_{gp} = \delta = 22.5 \ \mu s$, $\varepsilon = [-3\mu s, 12\mu s]$ according to the A203's evaluation results in Section 4, the following worst case accuracy and precision can be guaranteed in the evaluated setting:

P [s]	α [μs,μs]	α' [μs,μs]	π [μs]
10	[-3.6,12.5]	[-4.6,13.5]	18.1
50	[-3.6,12.5]	[-8.6,17.5]	26.1
100	[-3.6,12.5]	[-13.6,22.5]	36.1

The above choice of ε was guided by securing correctness for any CSP, which requires incorporating the rare cases where the transmission delay is extremely large. The small standard deviation shows, however, that the average accuracy/precision is about 10 times better than the worst case. When more advanced (namely, fault-tolerant) versions of the above algorithm are employed, the long tail of ε can be cut. Note that there are several directions for such an improvement, ranging from multiple CSPs over multiple GPS receivers up to advanced clock validation algorithms (Schmid, 1994).

6. RELATION TO OTHER WORK

Since the transmission delay uncertainty ε is one of the key factors determining the worst-case synchronization tightness π of any clock synchronization scheme, the following classification of communication subsystems is commonly employed in the literature:

(I) Nodes within a few 10 meters of each other can be connected by a dedicated clocking "network". Phase-locked-loop clocks with clock voting for increased fault-tolerance can be used here to provide π in the 10 ns-range (Ramanathan et al., 1990b).

(II) Nodes within a few 100 meters of each other are usually interconnected by a packet-oriented communications subsystem. Sending data packets is the only means for exchanging (time) information here. Purely software-based solutions typically achieve π in the ms-range, which can be brought down to μs with moderate hardware support.

(III) World-wide distributed systems connected via long haul networks exhibit highly variable and even potentially unbounded end-to-end transmission delays. This effectively rules out any bounded π. Nevertheless, an accuracy in the 10 ms-range can be achieved by means of the prominent Network Time Protocol NTP (Mills, 1995) under "reasonable" conditions, see (Troxel, 1994).

There is a large body of research in clock synchronization, which primarily addresses type (II) systems. However, only a few papers deal with high-accuracy clock synchronization, i.e., hardware support for decreasing ε.

In their pioneering paper (Kopetz and Ochsenreiter, 1987), the authors describe a Clock Synchronization Unit (CSU) chip that ensures ε in the few 10 μs-range if used in a collision-free Ethernet. A downsized successor of the CSU targeted to the TTP fieldbus for automotive applications is briefly outlined in (Kopetz et al., 1995) (see below). No implementation details are available for the hardware-assisted clock synchronization scheme (Ramanathan et al., 1990a), which claims ε in the 100 μs-range for not necessarily fully connected point-to-point networks.

An even better ε in the 1 μs-range can be achieved with any hardware support if the network controller provides the required transmit and receive timestamp trigger signals directly. Although this cannot be expected from COTS COMCOs, there are appropriate research prototypes for TTP (Kopetz et al., 1995) and LON (Horauer and Loy, 1998) fieldbusses. For example, $\varepsilon \approx 1.9 \ \mu$s is claimed for the (collision-free) TTP running at a speed of 100 Kbit/s.

The NTI evaluated in this paper also leans on the general hardware architecture and the (DMA-based) method of packet timestamping proposed in (Kopetz and Ochsenreiter, 1987). Still, several "uncertainty-saving" engineering improvements have been added that guarantee ε in the 10 μs-range even for standard (= non-collision-free) Ethernet. Note also that the NTI's UTCSU-ASIC does not have much in common with the original CSU chip and its successor: The UTCSU not only relies upon the interval-based paradigm but employs fundamentally different building blocks internally as well, like the high-resolution adder-based clocks.

A completely different approach that does not need (much) hardware support but still achieves π in the 10 μs-range is the remarkable *a posteriori agreement technique* used in (Veríssimo *et al.*, 1997). Basically, it exploits the simultaneity of reception in broadcast networks for ruling out the channel access uncertainty. However, unlike the NTI, it is only applicable to networks with hardware broadcasting capabilities and generates considerable network and CPU load.

Finally, in view of the negligible costs of GPS receivers, it is tempting to solve the clock synchronization problem simply by equipping each node with a modular GPS receiver. However, apart from fault-tolerance considerations (Höchtl and Schmid, 1997), there are practical problems with this approach. First of all, one has to consider the effort of accommodating and connecting the "forest" of antennas required for a, say, distributed factory automation system with 100 nodes. In addition, the large time-to-fix of GPS receivers implies that it may take 30 seconds or more until correct timing information is available. This in turn implies a large node join delay in case of re-integrating a newly powered-up node.

The NTI in conjunction with interval-based clock validation provides a way to escape from the abovementioned problems by simultaneously increasing the fault-tolerance degree and decreasing the number of GPS receivers required in the system. This is basically done without additional (cabling) costs, since it uses the existing data network only. The price to be paid is additional hardware support and decreased precision/accuracy, which is hopefully acceptable for typical applications.

7. CONCLUSIONS

In this paper, some results and insights gained during the experimental evaluation of the NTI M-Module supporting the distribution of GPS time over LANs have been presented. As opposed to the typical worst case transmission delay uncertainty $\varepsilon \approx 10$ ms of purely software-based solutions, the NTI provides $\varepsilon \approx 10$ μs even in "bad" system architectures, incorporating Ethernet, slow buses, and COTS network controllers with large FIFOs. Proper choice of TxTS triggering and increasing bus speed have been identified as key issues in reducing ε, which can be brought down to the 1 μs-range if the full speed of the NTI's M-Module interface is exploited.

A simple time distribution algorithm has been sketched to link the evaluation results with interval-based clock synchronization. Although it does not incorporate any advanced feature, it still provides an accuracy/precision in the 10 μs range with negligible system overhead. More advanced algorithms are available, which use multiple GPS nodes and elaborate clock validation techniques for improving accuracy/precision and fault-tolerance degree. A forthcoming paper will be devoted to an in-depth experimental evaluation of those algorithms.

ACKNOWLEDGEMENTS

This work could not have been done without vital support from all the other members of the NTI development team, namely, Thomas Mandl, Michael Schmidt, Gerhard R. Cadek and Nikolaus Kerö. The invaluable assistance of Johann Klasek is also gratefully acknowledged.

8. REFERENCES

Boehm, Barry and Chris Abts (1999). COTS integration: Plug and pray?. *IEEE Computer* **32**(1), 135–138.

Höchtl, Dieter and Ulrich Schmid (1997). Long-term evaluation of GPS timing receiver failures. In: *Proc. of the 29th IEEE PTTI Systems and Application Meeting*. Long Beach, California. pp. 165–180.

Horauer, Martin and Dietmar Loy (1998). Hardware-unterstützte Uhrensynchronisation in Verteilten Systemen. In: *Proc. Austrochip 1998*. Wiener Neustadt, Austria. pp. 67–72. (ISBN 3-901578-03-X).

Horauer, Martin, Ulrich Schmid and Klaus Schossmaier (1998). NTI: A Network Time Interface M-Module for high-accuracy clock synchronization. In: *Proc. Workshop on Parallel and Distributed Real-Time Systems (WPDRTS'98)*. Orlando, Florida. pp. 1067–1076.

Kopetz, Hermann and Wilhelm Ochsenreiter (1987). Clock synchronization in distributed real-time systems. *IEEE Transactions on Computers* **C-36**(8), 933–939.

Kopetz, Hermann, Andreas Krüger, Dietmar Millinger and Anton Schedl (1995). A synchronization strategy for a time-triggered multicluster real-time system. In: *Proc. Reliable Distributed Systems (RDS'95)*. Bad Neuenahr, Germany.

Loy, Dietmar (1996). GPS-Linked High Accuracy NTP Time Processor for Distributed Fault-Tolerant Real-Time Systems. Dissertation. Technische Universität Wien. Faculty of Electrical Engineering.

Lundelius-Welch, Jennifer and Nancy A. Lynch (1984). An upper and lower bound for clock synchronization. *Information and Control* **62**, 190–204.

Mandl, Thomas, Herbert Nachtnebel and Ulrich Schmid (1999). Network Time Interface user manual. Technical Report 183/1-87. Department of Automation, TU Vienna.

Marzullo, Keith A. (1984). Maintaining the Time in a Distributed System: An Example of a Loosely-Coupled Distributed Service. PhD dissertation. Stanford University. Department of Electrical Engineering.

Mills, David L. (1995). Improved algorithms for synchronizing computer network clocks. *IEEE Transactions on Networks* pp. 245–254.

MUMM (1996). *ANSI/VITA 12-1996, M-Module Specification*. Manufacturers and Users of M-Modules e.V.

Nachtnebel, Herbert, Nikolaus Kerö, Gerhard R. Cadek, Thomas Mandl and Ulrich Schmid (1998). Rapid Prototyping mit programmierbarer Logik: Ein Fallbeispiel. In: *Proc. Austrochip 1998*. Wiener Neustadt, Austria. pp. 99–104. (ISBN 3-901578-03-X).

Ramanathan, Parameswaran, Dilip D. Kandlur and Kang G. Shin (1990*a*). Hardware-assisted software clock synchronization for homogeneous distributed systems. *IEEE Transactions on Computers* **39**(4), 514–524.

Ramanathan, Parameswaran, Kang G. Shin and Ricky W. Butler (1990*b*). Fault-tolerant clock synchronization in distributed systems. *IEEE Computer* **23**(10), 33–42.

Richter, Gerda, Michael Schmidt and Ulrich Schmid (1999). i82596 NTI device-driver software documentation. Technical Report 183/1-90. Department of Automation, TU Vienna.

Schmid, Ulrich (1994). Synchronized UTC for distributed real-time systems. In: *Proc. IFAC Workshop on Real-Time Programming (WRTP'94)*. Lake Reichenau, Germany. pp. 101–107.

Schmid, Ulrich (1997). Interval-based clock synchronization with optimal precision. Technical Report 183/1-78. Department of Automation, Technische Universität Wien. (submitted).

Schmid, Ulrich and Klaus Schossmaier (1997). Interval-based clock synchronization. *J. Real-Time Systems* **12**(2), 173–228.

Schmid, Ulrich, Johann Klasek, Thomas Mandl, Herbert Nachtnebel, Gerhard R. Cadek and Nikolaus Kerö (1999). A Network Time Interface M-Module for distributing GPS-time over LANs. Technical report. Technische Universität Wien. Department of Automation. (forthcoming).

Schossmaier, Klaus (1997). An interval-based framework for clock rate synchronization algorithms. In: *Proc. 16th ACM Symposium on Principles of Distributed Computing*. St. Barbara, USA. pp. 169–178.

Schossmaier, Klaus (1998). Interval-based Clock State and Rate Synchronization. Dissertation. Technische Universität Wien. Faculty of Technical and Natural Sciences.

Schossmaier, Klaus and Ulrich Schmid (1995). UTCSU functional specification. Technical Report 183/1-56. Technische Universität Wien. Department of Automation.

Schossmaier, Klaus, Ulrich Schmid, Martin Horauer and Dietmar Loy (1997). Specification and implementation of the Universal Time Coordinated Synchronization Unit (UTCSU). *J. Real-Time Systems* **12**(3), 295–327.

Troxel, G. D. (1994). Time Surveying: Clock Synchronization over Packet Networks. PhD thesis. Department of Electrical Engineering and Computer Science, Massachusetts Institut of Technology.

Veríssimo, Paulo, Luís Rodrigues and Antonio Casimiro (1997). Cesiumspray: a precise and accurate global clock service for large-scale systems. *J. Real-Time Systems* **12**(3), 243–294.

SUPPORTING COMMUNICATING REAL-TIME STATE MACHINES BY A CUSTOMISABLE ACTOR KERNEL

Giancarlo Fortino, Libero Nigro, Francesco Pupo

Dipartimento di Elettronica Informatica e Sistemistica
Università della Calabria, I-87036 Rende (CS) - Italy
E-mail: {g.fortino, l.nigro, f.pupo}@unical.it

Abstract: This paper proposes an approach to the development of real-time systems which relies on Communicating Real-Time State Machines (CRSM's) as the specification language, and on a customisable actor kernel for prototyping, analysis and implementation of a specified system. CRSM's were chosen since they offer an intuitive and distributed specification of a system in terms of a collection of co-operating state machines interacting through timed CSP-like IO commands. On the other hand, the underlying actor framework is designed to provide a time-sensitive scheduling structure which can be tuned to CSRM's in order to support temporal validation through assertions on the recorded time-stamped event histories. The actor kernel lends itself to be easily hosted, e.g., into Java. *Copyright © 1999 IFAC*

1. INTRODUCTION

It is recognised that the design of real-time systems strongly demands the adoption of notations and mechanisms which enable formal specification of and reasoning about system behaviour (Ghezzi, et al., 1991; Gerber and Lee, 1992; Shaw, 1992). Formal tools are required in order to capture timing constraints existing on the functional responses the system generate in coping with external environment originated stimuli. The usefulness of a formal notation is enhanced by the possibility of supporting validation activities which can help guaranteeing the correctness of the timing behaviour. Besides, the specification language should be clearly linked to subsequently design and implementation phases in order for it not to remain too abstract and far from the final system construction.

It was this perspective which stimulated the work described in this paper. Starting from the definition of Communicating Real-Time State Machines (Shaw, 1992) as a specification language which centres on a timed extension of finite state machines, a well-known formal tool also from an industrial point of view, the aim is to support CRSM's on the basis of a customisable actor model (Agha, 1986) especially designed for real-time (Kirk, et al., 1997; Nigro and Pupo, 1998a). The actor kernel provides the prototyping environment used for analysis and simulation of system properties, and can be used as the concrete representation of CRSM's in the final target architecture. The actor framework naturally supports validation activities through assertions specification and verification on the dynamically generated time stamped event histories (Shaw, 1997). The proposed transformation of CRSM's on to actors ensures that a complete development life cycle is defined. A real-time system gets specified, analysed and implemented according to a common set of concepts which propagate unchanged from modelling down to implementation. Currently, the Java programming language is used for experimentation.

Section 2 summarises the CRSM's language and gives an example of a control system specification. Section 3 highlights the basic concepts of the adopted customisable actor framework. Section 4 describes a mapping from CRSM's to actors together with a programming style in Java. Section 5 discusses the scheduling issues and the management of time under both the prototyping and the real execution operating environments. Finally, the conclusions are presented with a summary of the implementation status and an indication of some research directions which deserve further work.

2. COMMUNICATING REAL-TIME STATE MACHINES

CRSM's were designed as an executable specification language (Shaw, 1992), which rely on a distributed model of concurrency and synchronous one-to-one communications. A system specification is closed, i.e., machines are introduced to model both the computer system and its controlled environment. The CRSM's of a system are linked to one another by unidirectional typed channels where IO is modelled directly after CSP (Hoare, 1985).

A CRSM is as a timed state machine with one start state and one or more halt states, where timing constraints can be involved in every state transition. A state transition is a guarded command with a timing constraint: $G \rightarrow C[\tau]$. The guard G is made up of state variables of the CRSM. It must be true (the default value if the guard is missing) for enabling the command. The command C can be computational (e.g., expressing an internal processing or a physical activity performed on an interface object like a sensor or an actuator) or an IO. An IO command is an input, e.g., Ch(v)?, or output, e.g., Ch(expr)!, operation on a channel. When both the CRSM's involved in a communication over a channel are ready (rendezvous), the couple of IO commands is satisfied with a simultaneous state change in both machines, the resultant assignment v:=expr, and the two CRSM's which resume concurrent execution. Command execution is atomic.

The timing constraint is typically a timing interval $[t_{min}..t_{max}]$, $t_{min} \leq t_{max}$, which expresses the possible rendezvous times for an IO or the possible duration times for an internal command. The time interval can degenerate to a single time value. The default timing interval for an IO command is $[0, \infty]$ which means that a CRSM is ready for a communication at any time the partner is. A rendezvous is only possible at the times corresponding to the intersection of the intervals associated to the two IO commands. An impossible rendezvous can introduce a deadlock for a CRSM in the case the IO command is the only command out of the current state.

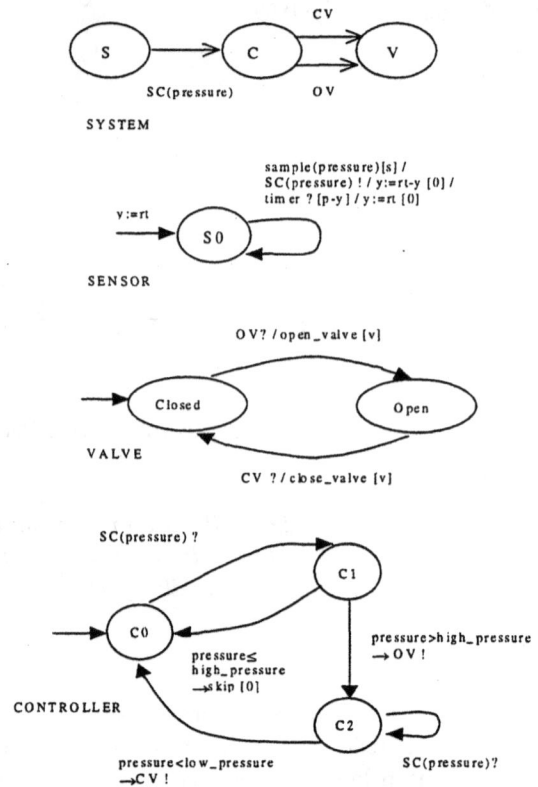

Fig. 1. A simple CRSM control system

A global time notion is provided in a system of CRSM's and is available through the *rt* variable. Each machine M has a companion real-time clock machine, with a *timer* channel, which acts as a time server. A command like timer?[y] allows M to set a timeout to occur after y time units from the current time. These facilities can be used to program a periodic behaviour. Figure 1 portrays a simple steam boiler process control system. A sensor S periodically samples the pressure and sends it to the controller C. When the controller detects the pressure exceeds a given high_pressure value, it commands the opening of a valve V which remains open until the pressure decreases under a low_pressure value. The CRSM's S, C and V are linked by three channels: Sensor Channel (SC), Open Valve (OV) and Close Valve (CV). SC carries as data a sampled pressure value. OV and CV are pure synchronisation event channels.

Timing attributes of the system are: the sensor period p, the reading time s of the sensor, the open/closing time v of valve. The controller is provided of the *high_pressure* and *low_pressure* values. For simplicity, some transitions are inscribed by chains of consecutive commands (separated by /) which have associated hidden states. The sensor machine takes the real time value at the beginning of each cycle (state S_0). Then it samples the pressure, engages a communication with the controller, takes again the current time, waits for the completion of the period and, finally, returns

118

to the home state S_0. Other details of the three machines should be self-explanatory.

Timing requirements are associated to sensor periodicity and valve deadline. For the system to behave correctly, the sensor period should not be overflowed (with some tolerance). Moreover, from the detection of a critical pressure the control system should open the valve within a deadline d. These properties must be verified through analysis of the system behaviour.

The evolution of a CRSM system is represented by the history of its transition firings. From time to time one transition is chosen from the set of the enabled and ready transitions on the basis of the Earliest-Time-First strategy. Ties are broken non deterministically. Transition firing advances the system time of the minimum quantity. For each channel event a copy of the channel transmitted data and the occurrence time of the event are registered.

The history of transition firings can be generated by executing the specification and explored through assertions. An assertion captures a desired system property. For instance, the deadline upon the opening of the valve could be expressed by the assertion:

when OV {
    assert(time(OV, -1) - time(SC, -1) ≤ d)
}

where *time* returns the occurrence time of an instance of a channel event. In the example the latest events of OV and SC channels are checked. Assertion specification can exploit the operations upon time stamped event histories proposed in (Raju and Shaw, 1994; Shaw, 1997).

A prototyping environment for specifying and testing properties of a system of CRSM's has been previously described in (Raju and Shaw, 1994). In this paper a mapping of CRSM's on to actors is proposed as an alternative method. The goal is to use an actor kernel which supports prototyping and testing of timing properties and can serve also as the target implementation level, thus linking the specification to concrete development.

3. A CUSTOMISABLE ACTOR KERNEL

In (Kirk, et al., 1997; Nigro and Pupo, 1998a) an actor framework is proposed for the development of distributed and predictable real-time systems. The approach favours the fulfilment of application requirements through a holistic integration of an application and its operating software, i.e., a programmer-defined scheduling algorithm.

An actor system with timing constraints can be formalised by Coloured Petri Nets and checked using simulation and occurrence graphs (Nigro and Pupo, 1998b).

Actors are modelled as finite state machines. Asynchronous message passing represents the basic communication paradigm. The arrival of a message triggers a state transition and the execution of an action (message processing). Actions are atomic. The execution of an action, besides the modifications to the internal data state of the actor, can generate new messages toward acquaintances (including the sender actor). At the end of current message processing the actor is ready again to receive the next message.

Predictable time behaviour rests on the lightweight nature of actors. They have no internal threads. Pre-emption and conventional mutual exclusion mechanisms are avoided. Within a community of actors allocated to the same physical processor, concurrency is achieved by action interleaving, i.e., by orchestrating message scheduling and dispatching among the various actors.

For modularity reasons, a *control machine* concept is introduced with the following sub-components: Scheduler, Message Plan, Controller and Real-Time Clock. Messages are transparently filtered by the Scheduler which applies to them a suitable control policy which is driven by timing constraints enforcement. The Scheduler schedules "just sent messages" on the Message Plan. The Controller is responsible of message selection from the Message Plan and message dispatching to the relevant actor. The Real-Time Clock provides the *time* notion to the control machine and application actors. Time can also be virtual time for prototyping purposes.

The key point is that Scheduler and Controller are actors themselves. They can be customised through programming. A specific architecture of the control machine directly follows from the adopted timing model. In this paper an organisation based on CRSM's is proposed.

4. MAPPING CRSM'S ON ACTORS

CRSM's can be mapped on actors in a straightforward way by means of a *self-driving* model. The following summarises the mapping and assumes Java as the target programming language.

The base class RT provides a discrete time model and the member function *rt()* which returns the current system time (a long).

The Actor base class provides the common services for actors, e.g., the non-blocking *send* operation, the *become* operation for changing the actor state, the (abstract) *handler* for responding to an incoming message. A CRSM-actor class extends

```
public class SENSOR extends Actor{
  final byte S0=0, S1=1, S2=2, S3=3, S4=4; //states
  float pressure;
  long y, p, s;
  Channel SC;
  TimedMessage timer, ic;
  public void initialise(long p, long s, Channel SC){
    this.p=p; this.s=s; this.SC=SC;
    timer=new TimedMessage(this);
    ic=new InternalCommand(this);
    y=RT.rt(); become(S0); send(ic.set(s));
  }//initialise
  protected void handler(Message m){
    switch(currentStatus()){
      case S0: if(m==ic){
                 pressure=pressure_reading(); become(S1);
                 send(SC.out(new Float(pressure)));
               }; break;
      case S1: if(m==SC.outRendezvous){
                 become(S2); send(ic.set(0));
               }; break;
      case S2: if(m==ic){
                 y=RT.rt()-y; become(S3);
                 send(timer.set(p-y));
               }; break;
      case S3: if(m==timer){
                 become(S4); send(ic.set(0));
               }; break;
      case S4: if(m==ic){
                 y=RT.rt(); become(S0);
                 send(ic.set(s));
               }
    }//switch
  }//handler
}//SENSOR
```

Fig. 2. A Java SENSOR class

Actor and captures, through a redefinition of the handler method, the CRSM life cycle, i.e., the machine transition diagram.

Internal commands, timers and channels are uniformly transformed into messages. A machine-actor sends to itself all the messages corresponding to the enabled guarded commands for the transitions out of current state. One of this message will be (possibly) chosen and delivered by the control machine and the others de-scheduled.

Figure 2 shows a Java class for the SENSOR machine in Figure 1. The SENSOR class has the method *initialise* which is used instead of the constructor to initialise the actor state. *initialise* creates the internal command *ic* and the timer *timer* by specifying itself (this) as the destination actor. Then sets the initial state (S_0) and sends to itself the internal command for beginning a reading period. States from S_1 to S_4 are hidden states in Figure 1. They serve the purposes of supporting the execution of chained commands. As one can see from Figure 2, a state change is immediately followed by the self-sending of all the messages which can bring the machine out to a next state and so forth.

The collection of messages an actor understands is pre-allocated and reused in order to avoid dynamic memory allocation/deallocation operations.

The Message base class provides common attributes for messages: the destination (target) actor, the carried data (an Object), the oc-

currence time, and bookeeping pointers useful to insert/extract the message into/from one or more message lists. The TimedMessage class extends Message by a time window $[t_{min}, t_{max}]$. A timer can directly be programmed as an instance of a TimedMessage. The InternalCommand class extends TimedMessage by a duration interval $[d_{min}, d_{max}]$. The Channel class extends TimedMessage and provides the channel abstraction. It hosts the following information: (a) the identity of the channel sender/receiver actors (b) two time windows associated respectively to the sender and the receiver (c) status information for indicating the readiness of the sender/receiver to a rendezvous (d) a Rendezvous inner class, two instances of which are pre-allocated in the channel for communicating the rendezvous respectively to the sender and the receiver. The two Rendezvous instances have a link to the Channel object (e) the channel transmitted data (Object) (f) the time stamped event history, i.e., an instance of the TEH class, and (g) the size of the time stamped event history.

The *set* method allows to set the time window of an internal command or a timer. It is overloaded to allow the specification of single time intervals. The *in/out* methods of the Channel class correspond respectively to the input (?) and output (!) operations of a channel. Since Java doesn't directly support passing a parameter by-reference, the *in* method has a parameter which is an array of Object. All the channels are capable of transmitting Object (and derived classes) items. The runtime type information coming with objects allows a safe access to transmitted information.

The configuration of a CRSM system, e.g., performed in the *main* method, consists in building the machine-actors, then making the channel objects with bindings, for each channel, to the sender/receiver. After that, actors are initialised by invoking the *initialise* method. The constructor of the Channel class allows the transmission of the initial binding, i.e., the identity of channel sender/receiver. The binding operation can be re-executed dynamically by using the Channel *rebind* method. Finally, the Controller method of the control machine is launched (see the next subsection).

5. OPERATING ENVIRONMENTS FOR CRSM'S

Figure 3 illustrates the architecture of a control machine tailored to prototyping and testing of a system of CRSM's. The organisation can slightly be refined to become the runtime control machine for real execution.

The Scheduler actor takes as input the "just sent messages" generated by the last activated machine and schedules them on the Message Plan which is ranked by the $t_{\min}$ attribute and then by the $t_{\max}$ attribute of the message time intervals. An internal command is scheduled according to the timing window $[0, \infty]$. A ready for rendezvous channel has the scheduling window $[t_1, t_2] = [t_{1sender}, t_{2sender}] \cap [t_{1receiver}, t_{2receiver}]$. A timer is scheduled according to its expire time, i.e., the time interval is $[t_{expire\_time}, t_{expire\_time}]$. In reality all the above intervals are relative to the time the machine which generated the associated command entered its current state. If t_E denotes the entering time of such a state, the relative scheduling window $[t_1, t_2]$ is converted into the absolute window $[t_E + t_1 + \delta, t_E + t_2 + \delta]$ where δ is the *dwell-time* (Shaw, 1992), i.e., a minimum time a machine (under simulation) is supposed to remain in a state.

The organisation of the Message Plan purposely avoids dynamic memory allocation/deallocation operations by having that messages act themselves as the nodes of the Plan list. Insert/remove operations reduce to pointer adjustments. The set of proposed messages by a machine are linked to one another in a ring to facilitate de-schedule operations (see below).

The controller selects from the Plan the next message according to the Earliest-Time-First strategy. The message with the minimum $t_{\min}$ is chosen. However, to be selected, the time interval $[t_1, t_2]$ of a ready-to-commit rendezvous message must be active (rt is within the interval) and its margin, $t_2 - rt$, the minimum. When multiple choices are possible, the controller selects non deterministically. The chosen message is time stamped with the current value of rt (occurrence time). Pending messages in the Message Plan directed to the same selected machine are de-scheduled. As a side effect, the de-schedule operations can inactivate a previously set timer. Finally, the message is dispatched. If the chosen message is a channel, the transmitted value along with the message time stamp are copied into the channel time stamped event history (TEH). The two contained rendezvous messages are then delivered in two control cycles.

Prototyping execution of CRSM's is usefully complemented by the Assertion Checkers. An assertion checker is an actor which is capable of capturing the under dispatching message and applying to it a given assertion. Assertion clauses have access to the TEH of channels. Moreover, since assertion programming isn't different from actor programming, the assertion checker can have a state and keep data useful for the expression of general assertions. Assertion checkers are initialised with

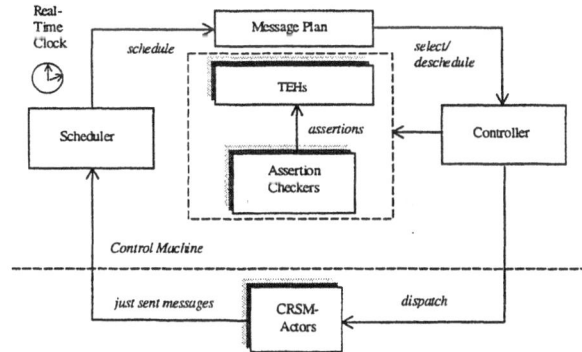

Fig. 3. Control machine architecture for CRSMs

the timing requirements (e.g., deadlines) which the system behaviour must fulfil. For example, the assertion concerning the correct operation of the valve since a critical level of pressure can be coded in Java as follows:

```
if (m==OV.outRendezvous && m.target==Valve)
    Assert.assert("Valve deadline",
        OV.teh.time(-1)-SC.teh.time(-1)<=d);
```

The implementation of the TEH class exports the operations time(-i) and value(-i) which return respectively the time and value of the i-th most recent event occurrence registered on the channel time stamped event history.

Figure 3 summarises the message handling from scheduling to selection and dispatching. However, from the control point of view, it is the Controller which carries the steps of the control loop. At each iteration, the Controller gives the Scheduler the chance to schedule the "just sent messages". Then it proceeds with the selection of the next message and the de-schedule of deactivated messages. After that the Controller activates the assertion checkers with the selected message for the verification of assertions. Finally, the chosen message is dispatched to the destination actor. At the end of the message handler in the actor, the control loop is resumed for a new iteration.

It is worth noting that the control machine concept enables a smooth transition from the prototyping environment to the real execution environment. The machine-actors can be operated under prototyping or real execution with minimal changes. Differences are concerned only with the time management.

During prototyping the system is handled under simulation. The "real time" rt is advanced of a discrete amount by the controller just after the selection process. In particular, if the selected message corresponds to an internal command with duration interval $[d_1, d_2]$, then rt is incremented of an amount d, $d_1 \leq d \leq d_2$. For worst-case-analysis, rt is augmented of d_2. If the message is a rendezvous with absolute time interval $[t_1, t_2]$, then rt is incremented of a quantity t, $t_1 \leq t \leq t_2$.

On the other hand, in the real operation of the system on the target architecture rt advances automatically and there is no need to take in account a dwell-time in states. If required, the Controller waits for the rt to be increased up to the t_{min} value of the most imminent message. However, in both operating environments, the selection strategy of the Controller remains the same.

6. CONCLUSIONS

This paper proposes a support for Communicating Real-time State Machines (CRSM's) specification method in terms of a customisable actor framework which is adequate for predictable real-time systems. The approach enables prototyping and verification of a CRSM system and provides a concrete representation for final implementation which faithfully mirrors the specification. Thus the risks of distortions during the transformation from analysis to design and implementation are minimised.

The resultant approach is useful for the development of hard-real-time systems and can be followed also in a language like Java where the use of non-predictable features (e.g., dynamic memory management, unfair thread scheduling ...) is avoided.

An implementation in Java of the base classes supporting CRSM's was achieved. The translation from CRSM's to actors is currently accomplished manually. However, since the established semantic equivalence between CRSM's and actors it is possible to automate the translation. It is under development a Java tool which will permit the graphical editing of CRSM's and the automatic generation of the Java actor code which is integrated with the control machine and other runtime classes to compose the prototype executing environment. The tool will accommodate for the interactive specification of assertions which are verified during prototyping and testing. The goal is to support also the generation of final target code of a verified system.

Prosecution of the research aims also at extending the CRSM language towards a better separation of functionality from timing constraints (Nielsen, et al., 1998). The goal is to favour machine compositionality by having that the timing issues are handled by meta-level synchronisers which control scheduling. A benefit of this approach would be that of possibly dynamically changing timing constraints (for example the reading period of a sensor when more critical behaviour occurs) transparently to the application CRSMs.

Another direction which deserves further work concerns a more comprehensive support for time stamped event histories according to the proposals contained in (Shaw and Rupp, 1996; Shaw, 1997). The aim is to enable a programming style directly based on TEH's both for assertion and CRSM's specification.

7. ACKNOWLEDGMENTS

Work carried out under MURST Project "Methodologies and Tools of High Performance Systems for Distributed Applications".

8. REFERENCES

Agha, G. (1986). *Actors: a model for concurrent computation in distributed systems.* MIT Press.

Gerber, R. and Lee, I. (1992). A layered approach to automating the verification of real-time systems. *IEEE Trans. on Software Engineering* **18(9)**, 768-784.

Ghezzi, C., Mandrioli, D., Morasca, S., Pezzè, M. (1991). A unified high-level Petri net formalism for time-critical systems. *IEEE Trans. on Software Engineering* **17(2)**, 160-172.

Hoare, C.A.R. (1985). *Communicating Sequential Processes.* Prentice-Hall International.

Kirk, B., Nigro, L. and Pupo, F. (1997). Using real time constraints for modularisation. *Lecture Notes in Computer Science* **1204**, 236-251, Springer-Verlag.

Nielsen, B., Ren, S. and Agha, G. (1998). Specification of real-time interaction constraints. Proc. of 1st IEEE Int. Symp. on OORT Computing.

Nigro, L. and Pupo, F. (1998a). A modular approach to real-time programming using actors and Java. *Control Engineering Practice*, **6(12)**, 1485-1491.

Nigro, L. and Pupo, F. (1998b). Using Design/CPN for the schedulability analysis of actor systems with timing constraints. http://www.daimi.aau.dk/CPnets/workshop/

Raju, S. and Shaw, A. (1994). A Prototyping Environment for Specifying, Executing and Checking Communicating Real-Time State Machines. *Software-Practice and Experience* **24(2)**, 175-195.

Shaw, A. (1992). Communicating Real-Time State Machines. *IEEE Trans. on Software Eng.* **18(9)**, 805-816.

Shaw, A. (1997). Time-stamped event histories: a real-time programming object. Proc. of WRTP'97, 97-100.

Shaw, A. and Rupp, A.(1996). Real-time programming with time-stamped histories. TR#UW-CSE-96-05-02, Dept. of Computer Science and Eng., U. of Washington.

Achieving Internal Synchronization Accuracy of 30 μs Under Message Delays Varying More Than 3 msec

Horst F. Wedde,
Jon A. Lind,
Guido Segbert

Informatik III
University of Dortmund
44221 Dortmund / Germany
wedde@ls3.informa tik.uni-dortmund.de

Abstract

In order to support real-time distributed experiments, in particular for safety-critical applications such as studied in our MELODY project, featuring task execution times of 20 – 50 msec, we could tolerate a clock synchronization accuracy of ca. 100 μsec between a distinguished (master) clock and any other system clock. Due to the lack of driver software in AIX for running GPS on IBM RS/6000 machines (models 570 and 580) – which would have resulted in a maximum deviation of 200 nsec - we investigated the message delay behavior in the Token Ring and Ethernet architectures in order to determine and compare clock drifts between different machines. Based on the insights gained we defined a novel two-step distributed clock synchronization protocol. First a clock at a master site initializes a synchronization procedure affecting the remainder (slave) clocks, and ensuingly only local updates are at times performed during a distributed experiment adding to making our internal synchronization non-intrusive. As a key feature a novel masking technique is employed through which messages with "undue" delays have no effect. Thus, in spite of a variation of message delays ≥ 3 msec the protocol guarantees a maximum deviation of 30 μsec between the master and any slave clock in the system. The approach and experimental findings are discussed in detail.
Copyright © 1999 IFAC

Keywords: *Global Positioning System (GPS), internal clock synchronization, communication delays, distributed operating systems, real-time systems, safety-critical systems*

INTRODUCTION

The hardware environment. In the MELODY project, an innovative distributed operating system for supporting real-time and safety-critical applications has been stepwise developed since 1988 (Wedde, Lind(1997)). Given that measures for achieving dependability (e.g. replication of files) and for improving the real-time performance (in particular *survivability)* are in conflict there cannot be a closed-form design approach. Instead a novel incremental development procedure was established which is heuristic yet systematic. It is experiment-based and has been termed *Incremental*

Experimentation. (Details can be found in (Wedde, Lind(1997))). After 5 subsequent design phases for which simulation served as the main tool we did a distributed implementation of the fully developed MELODY model. This was done on a Token Ring of 8 IBM RS/6000 machines (models 370, 570 and 580). In order to evaluate the real-time performance under real communication delays (instead of preset values in simulations) the task deadlines – which are meant as values of the *system environment time –* were to be checked against clocks anywhere in the system which in turn had to be in good synchrony with the environmental time. In order to avoid synchronization influences on messages on the Token Ring, or from other network traffic on time synchronization, we operate a separate Ethernet LAN between the nodes which is completely dedicated to time synchronization. *The protocol used is UDP.* It is commonly used in workstation and even PC LANs.

The synchronization problem. Our idea had been to use GPS (Hough(1992), Ano92, Wannemacher, Halang(1995)) facilities which guarantee an accuracy of 100 nsec between the system clocks. Unfortunately IBM did not develop the driver software for the local system buses used in the models 570 and 580. With this in mind we gave up the idea of external clock synchronization. Looking at previous work on internal synchronization (one clock being considered as the master after which the others were to be synchronized) the AIX Timed Protocol (Stevens(1992)) had to be ruled out, due to its inappropriate accuracy (10's of milliseconds). NTP (Mills(1991)) does better (accuracy about 1 msec) but this is still inadequate since it is notably intrusive on the performance of tasks that execute in the range of 20-50 msec. In contrast, we considered a maximum deviation of 100 μsec as imperative for a non-intrusive synchronization policy.

We are aware that for micro-controller networks operated under the CAN bus a clock accuracy 20 μsec has been achieved (Gergeleit, Streich(1994)), a consequence of the very accurate broadcast function used in this protocol. However, for typical workstation environments this could not be utilized. The paper describes our timing studies and a resulting distributed synchronization protocol which guarantees, in spite of a variation of message delays ≥ 3 msec, a maximum deviation of 30 μsec between the master and any slave clock in the system. We termed this inexpensive novel method *Real-Time Network Protocol (RTNP)*.

Organization of the paper. After describing the experimental set-up for measuring the slave clock deviations from the master in section 1, we present our synchronization protocol RTNP in section 2, and we prove it correct to the problem specification. Experiments under normal system workloads confirm the non-intrusiveness of the protocol. At the end we discuss the results and some extensions.

1. MEASURING CLOCK DEVIATIONS AND DRIFTS

When we purchased our IBM RS/6000 machines (models 370, 570 and 580) a little over 5 years ago they ran AIX version 3.2. This operating system was then the only one allowing absolute priorities for user-level commands. For all our experiments in the MELODY project we used this capability to ensure fixed response times under real-time constraints. Since priority levels between 1 and 10 are meant for hardware-related operations (interrupts) it is quite risky to assign such priority values to user processes (unpredictable system crashes). In all experiments reported here (as well as in the time synchronization implemented in MELODY) we mostly used priority level 12 for very high priority operations.

The physical basis for clock drifts is that the crystals used, which oscillate with a very high frequency, cannot be manufactured with arbitrary precision. Given that the frequency is *constant* for each crystal (with variations well below nanoseconds) this implies that the drift between two clocks is a *linear* function over time. The practical problem which we had to face was how to determine this function by using messages for which we experienced delays varying more than 3 msec while the needed accuracy was to be at least 100 μsec.

In our first experiments we determined the following communication delays between two machines: a message M_1 was sent from site A to site B. It was immediately acknowledged at B by sending a message M_2 from B to A. The total communication delay T was measured in site A's time. No other user processes were active. There is no measurable overhead during this communication procedure (below 25 nsec).

delay but by the argument above M_1 would then not be minimal. Let the minimal message delay be denoted by D_0.

The number of measurements in fig. 1 is 10000. We also studied the patterns achieved by looking at 1000 and 100000 measurements. The results were as follows (see also fig. 1):

1.1 *Property*: For the master site A and any given slave site B there is a minimum communication delay T which is independent of the number of measurements (up to 2 μsec).

1.2 *Property* : There is a communication delay value (peak) which is assumed in a maximum number of measurements. Summarizing over all slave sites B, 70 – 80% of all measured values differ at most 10 μsec from the corresponding peak. This peak is termed the *standard communication delay*.

1.3 *Property*: For every site B, the difference between minimal and standard communication delay is at most 40 μsec.

1.4 *Property (Symmetry):* If the communication procedures described above originate and end at B (instead of A) the corresponding measurements have the same properties 1.1, 1.2, 1.3. Furthermore, for each B the minimum and standard delays are mostly identical (up to 1 – 4 μsec) with those corresponding values from the communication procedures originating and terminating at A.

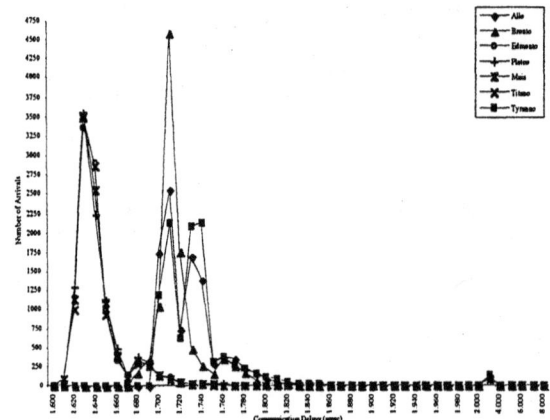

Figure 1: Communication Delays (10,000 measurements)

From these remarkable properties it follows – if T_0 denotes the minimum communication delay – that *the delays of M_1 and M_2 – denoted by D_1, D_2, respectively – to be minimal themselves, even $D_1 = D_2$. Due to the symmetry property (1.4) for every minimal M_1 a message M_2 (from B to A) with the same delay can be expected.* If M_2 was not minimal then there existed a message M_2^* with a smaller

Using a similar argument for the standard delay T', its constituting values, the delays of messages M_1'

124

and M_2', can be assumed equal, namely 1/2 T'. Let they be denoted by D_1', D_2', respectively. *D_1' and D_2' differ from D_0 by at most 20 μsec (see 1.3).* The dependencies discussed can be found in fig.2 assuming that there is no timestamp used.

1.5 *Property*: If a timestamp is placed on M_2 before it is sent to the slave (see fig. 2) *a constant overhead Ov of at most 43 μsec is encountered at the master site.* (This is due to reading M_1 from the physical buffer, generating M_2 and placing the timestamp on it, and writing M_2 into the physical buffer.)

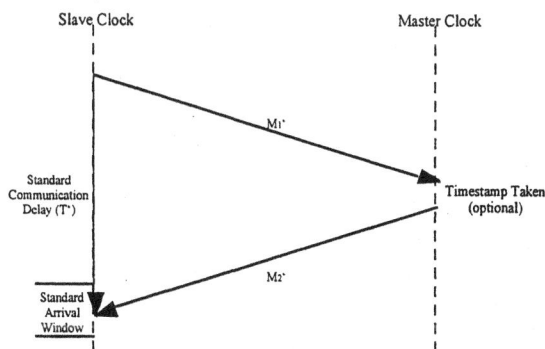

Figure 2: Model for Determining Clock Synchronization/Drift

The key for measuring clock deviations is:

1.6 *Proposition*: Let the size for a **standard time arrival window** (a window around the standard arrival time (see fig. 2)) be set to ±x μsec. Assume that M_2 arrives within the window. Let Ov denote the overhead at the master site described in 1.5, and let D_1, D_2 be the delays of M_1, M_2, respectively. Under the assumptions of 1.5, if 1/2 T' (see fig. 2) is added to the timestamp when M_2 arrives at the slave site, this value will be compared to the local time. The difference is the current deviation between master and slave clock (in terms of the slave time). *The clock deviation y is bound by $-20 \le y \le 20 + x$ μsec.*

Proof.

(0) The drift between the clocks during the transmission of M_2 (which takes at most 4 μsec) is neglectible. It is in the low range of nanoseconds.

If D_1' = D_2', i.e. D_i' + 1/2 Ov = 1/2 T'; i = 1,2 (see fig. 2 and the discussion after 1.4) then M_2 (= M_2') arrives just within the standard delay,

and the difference between the local slave time and (timestamp + 1/2 T') is the current deviation, according to (0).

(1) If $D_1 > D_1$' we may assume that in the worst case D_1 could deviate from 1/2 (T' – Ov) by at most (20 + x) μsec since for M_2 arriving within the arrival time window, D_2 has to be off by 20 μsec into the opposite direction. Formally speaking: If D_1 := D_1' + (20 + x) we need D_2 := D_2' – 20 such that

$$D_1 + D_2 + Ov = D_1' + (20 + x) + D_2' - 20 + Ov = T' - Ov + x + Ov = T' + x .$$

But D_2' – 20 ≤ D_0, the minimal delay of M_2.

(2) Also, D_1 cannot be more than 20 μsec lower than D_i'. - (1) and (2) conclude the proof.

In the sequel we will use 5 and 10 as values for x.

Measuring the clock drift. In order to determine the (linear) clock drift between the master B and a given slave A (see fig. 2) the communication described in fig. 2 and 1.6 was done 1000 times in the following way: The standard arrival window was set to ± 5 μsec. The clock deviation according to 1.6 was applied as frequently as possible, i.e. immediately after the deviation had been recorded, at the end of the communication the next procedure was started. If M_2 did not fit into the arrival window the corresponding procedure was discarded. The results are depicted in fig. 3. There is one curve for every master/slave pair.

We observed that:

- Rarely did messages not fit into the (small) standard arrival window (less than 20% failures). These accidents were quite evenly distributed over the time range of execution.

  Although 1.6 guarantees an accuracy of ± 25 μsec the measured deviation from the median linear curves are not more than 5 μsec! This suggests that the standard delays T', D_1', D_2' (see 1.6), or values very close to them, were found very frequently. This is to be expected since a large majority of messages experience delays that differ at most 5 μsec from the standard delay (see previous bullet and fig. 1, 1.2).

125

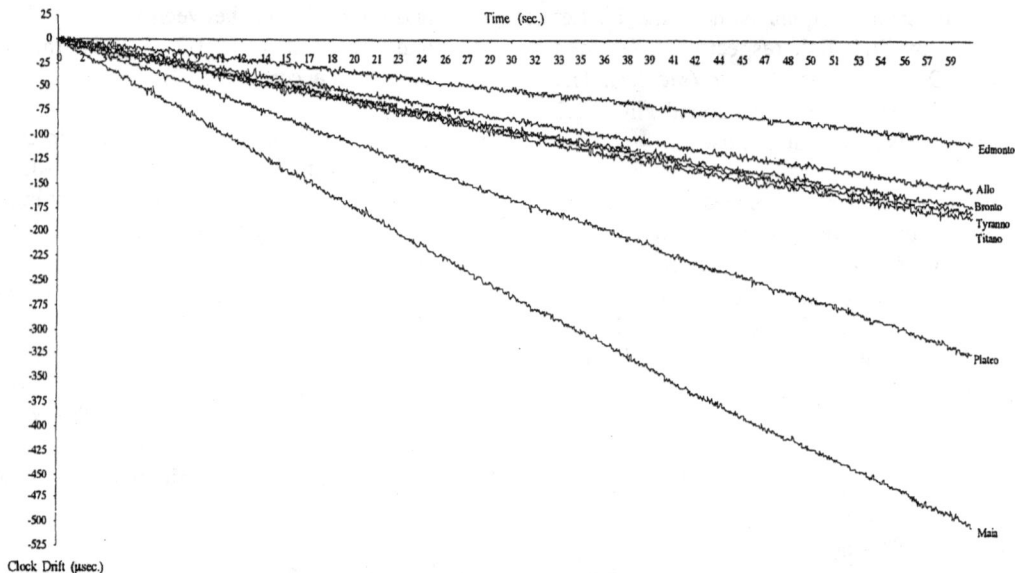

Figure 3: Linear Clock Drift

2. THE REAL-TIME NETWORK PROTOCOL

Our novel protocol for synchronizing the slave clocks starts with a broadcast message from the master to all slaves. This invokes them to start an *initial synchronization procedure* which is essentially the same as the one described in 1.6 except that the local slave time is not *compared* to the master time + 1/2 T' (the latter having been determined off-line, according to 1.2) but *replaced* by it. The window size is set to ± 10 μsec in order to catch a large majority of arriving messages (see 1.2). According to 1.6 the worst-case deviation between master and slave clock could be between – 20 and 30 μsec. If M_2 fails to fit into the standard arrival window *another attempt has to be made until the synchronization has been successful.* After the initial synchronization has been completed each slave updates its clock after every second. The update value is the drift value with respect to the master clock. This value is determined off-line, and is extremely accurate (see section 1). This completes RTNP. A pseudo-code specification is in the appendix.

In order to test the actual accuracy of the synchronization we extended the RTNP protocol. After each local update the deviation between master and slave was measured according to 1.6 (with x := 10). The results are depicted in fig. 4. Despite a few missing points (messages had not arrived within the standard arrival window) the deviation – which is guaranteed to be between -20 and 30 μsec – is measured throughout to be below 5 μsec. As in case

of the linear clock drift (see fig. 3) these results suggest that the message delays for testing the effect of the accuracy of the local update are typically (see 1.2) very close to the standard delay values. Thus the synchronization is much better than under the worst-case reasoning.

Finally we invoked MELODY some 30 seconds after the initial synchronization under RTNP had been started. We terminated MELODY after ca. 120 seconds. During the whole time we continued with the local update part of RTNP until we stopped after 150 seconds total experiment duration time. As shown in fig. 5 the pattern of describing the local updates is essentially unchanged, the measured deviations again in the range of 5 μsec. This proves very convincingly that MELODY processes have no effect on the clock synchronization. Since the local updates take only 25 nsec or less (one update per second) we conclude that RTNP is non-intrusive, and it is not affected by MELODY.

Conclusion

Based on detailed educated experimental insights into the AIX communication behavior under UDP, we were able to develop the simple distributed protocol RTNP for internal synchronization in a LAN. It is extremely inexpensive and non-intrusive. Meanwhile it is part of our MELODY environment.

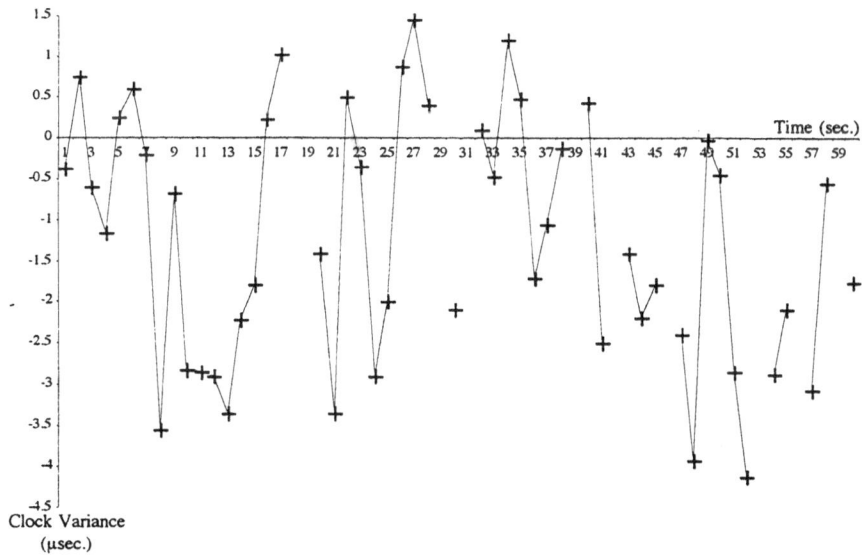

Figure 4: Test of Clock Deviation without MELODY Operations

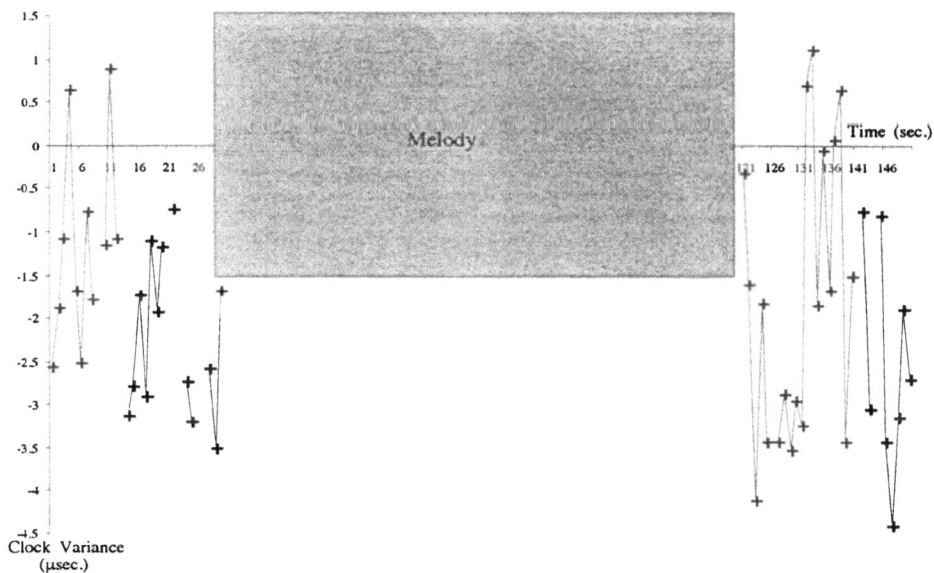

Figure 5: Test of Clock Deviation with MELODY Operations

Through our experiments (see fig. 4 and figure 5) RTNP has turned out quite satisfactory for safety-critical environments with distributed task execution times between 20 and 50 msec.

While the worst-case accuracy is between -20 and 30 μsec under the parameter settings selected – which has not been achieved yet, up to our knowledge – the results suggest already that the choice of the window sizes was overly pessimistic. The local update could easily be done every 100 msec. Finally, instead of the standard delay, one could choose a delay time much closer to the minimum delay (see 1.2 and its implications on 1.6). While then less message

delays were to be found in the narrow vicinity of such a "standard" delay time (see fig. 1) this might cause more attempts for some slaves to reach the end of the initial synchronization. On the other hand the discussion of fig. 3,4,5 revealed that the experimental results were much better than expected under worst-case reasoning. Besides message delays being close to the standard delay we detected - while performing the measurements of communication times - that generally subsequent messages in the experimental stream tend to have nearly identical delays. We are currently studying this phenomenon further.

All measures combined the worst-case accuracy could be lowered to 10 µsec and below as current studies show. However, this is beyond the scope of this paper.

LIST OF REFERENCES

Ano, "Global Positioning System Information Center (GPSIC), Users Manual"; Alexandria, VA, USA Department of Transportation, U.S. Coast Guard 1992

M. Gergeleit, H. Streich, "Implementing a Distributed High-Resolution Real-Time Clock Using the CAN Bus"; Proc. of the First International CAN Conference, Erlangen 1994

H. Hough, "A GPS Precise Timing Sampler"; *GPS World;* October 1991

D.H.. Mills, "Internet Time Synchronization: The Network Time Protocol"; *IEEE-TC* **Vol. 39** (1991)

W.R. Stevens, "Advanced Programming in the Unix Environment"; Addison Wesley; June 1992.

M. Wannemacher, W.A. Halang, "GPS-basierte Zeitgeber: Realzeitsysteme werden endlich echtzeitfähig" (German) Proc. of PEARL'95, in: P. Holleczek (ed.): Series "Informatik aktuell", Springer 1995

H.F. Wedde, J.A. Lind, "Building Large, Complex, Distributed Safety-Critical Operating Systems"; *Real-Time Systems* **Vol. 13 No. 3 (1997)**

APPENDIX

Real-Time Network Protocol:

```
void RTNP_slave_clock(void)
{

receive_startup_broadcast(Master_Sit
e_ID);

  do {
    Start_Time = CURRENT_TIME();

send_timestamp_request(Master_Site_I
D);
    Master_Timestamp =
receive_timestamp_response();
    Comm_Delay = CURRENT_TIME() —
Start_Time;
    }
  while ((Comm_Delay < (STANDARD_DELAY
- 0.000010))
        OR ((STANDARD_DELAY +
0.000010) < Comm_Delay));

  set_current_time((Master_Timestamp
                + (Standard_Delay /
2)));
```

```
  while (TRUE) {
    sleep(1);

adj_current_time(Local_Clock_Drift);
    }
}
void RTNP_master_clock(void)
{
  send_startup_broadcast();

  while (TRUE) {

receive_timestamp_request(Slave_Site_ID
);

build_timestamp_response(CURRENT_TIME()
);

send_timestamp_response(Slave_Site_ID);
    }
}
```

CONGESTION CONTROL FOR ATM REAL-TIME UPSURGE SERVICE

Peijiang Yuan Lichen Zhang*

*Institute of AI & PR, Science center, Shantou University, 515063 Shantou, Guangdong province,
China E_mail: ypjj@106.com lczhang@mailserv.stu.edu.cn

Abstract. ATM traffic control mechanic is one of the important problems that the vendors are forced to face. ATM network traffic control model is usually built according to the type and usage of services. In this paper, a common problem about transport traffic and cell discard/accept is addressed and a revised buffer threshold based method is discussed. With this method, the demand of real time transportation, the order of cells to be sent and rate of loss cells can be guaranteed. Because of authority coefficient, Cells of low priorities but strict deadlines could be sent before they are discarded *Copyright © 1999 IFAC*

Key Words. traffic control, buffer based mode, ATM network, deadline

1. INTRODUCTION

The ITU-T, ANSI, and ATM Forum have selected ATM as part of the B-ISDN specification to provide the convergence, multiplexing, and switching operations(See in Rec.I.365.1 (1993),Rec.I.555 (1993)). With the development of B-ISDN, the demand for ATM to provide a high speed, low delay multiplexing and switching networks to support any type of user traffic grows great rapidly. ATM and sonet are considered to be the single solutions to transport and proprietary implementations that the industry has fostered to face. One component of ATM allows ATM nodes to support multiapplication systems such as voice, video, data, music image, data and FAX, etc. What is more important, ATM network is expected to serve those kinds of services whose bandwidth differs from several thousands bits/per second to several thousands mega bits/per second. After adaptation, ATM switch or other exchange equipment accepts all sorts of services. Then the ATM switch deals with those services according to their types and sends them to the networks. ATM networks ought to cope with those demands freely.

ATM segments and multiplexes user traffic into small, fix-length units called cells. Each cell is identified with virtual circuit identifiers contained in the cell header. An ATM network uses these identifiers to relay the traffic through high-speed switches from the sending customer premises equipment (CPE) to the receiving CPE. ATM provides no error detection operations on the user payload inside the cell. It provides no retransmission service and few operations are performed on the small header. The intention of this approach is to implement a network fast enough to support multimegabit transfer rates. According to the cell header, an ATM Switch identifies a cell and exchanges it to the correct port. Then the route table is filled and the cell header is changed. So the main function of an ATM switch is routing. Considering the arrival of upsurge services, cells from different in ports need to be exchanged to the same out port at the same time. There will happen a competition. If the congestion happens in the internal of a switch, internal congestion is occured. Since the transport speed of ATM networks is very high, the congestion tends to be deteriorating and the network is forced to paralyze. It's a big problem for vendors to think of it.

The switching technologies proposed concentrated on this point.During the recent years there are many methods that have been addressed such as Zhang (1994b),K.Bala (1990), Zhang (1994a), Shared memory switch provides a common memory for the storage of the cell and the switching fabric. Because other equipment can share the memory, it is cost-effective and can be designed to large capacity with the development of VLSI. Yet this kind of control system is very complex and less efficient.

Shared bus switch (See in A.E.Eckberg (1989) and F.Bernaber (1993) and S.Saunders (1994)) is implemented with two buses for fully redundant operations. Each in port module is directly to the bus when a cell from one module needs to be exchanges. It sends a request to the bus arbiter and the arbiter decides whether the cell can be delivered. The structure of this kind of switch is simple and it is easy to implement broadcast, but its capacity is limited because of the bandwidth of the bus.

Multistage switching and Banyan and Delta switching networks have proved to be highly efficient and provide very high throughout.(See in Zhang (1992),Zhang (1996b)) But they can't deal with those traffics with low priority but strict time limit. Here a revised buffer threshold based method is addressed. In that case, deliver of cells with time limit and low priority could be guaranteed. So the safety and possibility of a real time service system can be implemented.

The following describes the structure of this paper: section 2 outlines the characteristic of common upsurge services; section 3 describes the model of the method; the conclusion is made at section 4.

2. ANALYSIS OF ATM UPSURGE SERVICES

The ITU-T recommendation I.211 describes the services offered by B-ISDN. The services are classified as either interactive or distribution services.

ATM services are classified by ITU-T for the transport capacity of those services from Constant Bit Rate and Variable Bit Rate(See in T.M.Chen (1996)) , Request of port to port time limit and without request of port to port time limit, Connect oriented and Connect without oriented. For the sake of traffic control, ATM Forum keeps the first two standards of that of ITU-T, and divided ATM services into 2 classes: Constant Bit Rate, Variable Bit Rate, Available Bit Rate and Undefined Bit Rate. Because of the time sensibility, VBR services are divided into real time VBR services (VBR-rt) and non real-time VBR services (VBR-nrt). Those kinds of services can be divided according two parts:

1.   Services character Those parameters such as Peak Cell Rate (PCR), Sustained Cell Rate (SCR), Minimize Cell Rate (MCR) and Maximum Bursitis Sequence (MBS) that describe the flow character of traffic are usually called Flow Parameters.

2.   Quality of service Such as Peak-to-Peak Cell Delay Variance (Peak-to-Peak CDV), Max Cell Transfer Delay (Max CDV), Cell Loss Rate (CLR), Cell Insertion Rate (CIR), Bit Error Rate (BER), etc.

The following figure illustrates the upsurge degrade of these services:

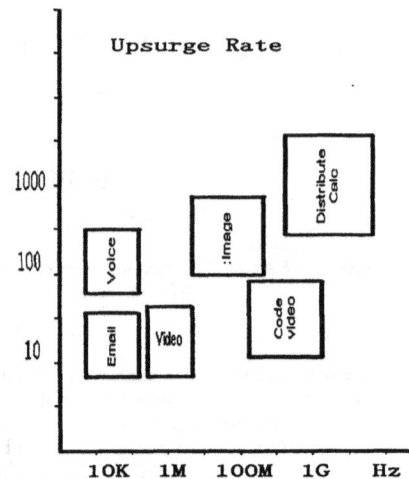

Fig. 1. Upsurge Rate of some services

According to their upsurge rate, these services need different QoS. The following tables show the different demand of different services.

1. Bit Error Rate BER is used to illustrate the performance of the transfer mechanic. Since Fiber is widely used in ATM networks, the BER that ATM networks offer is usually very low.

Table 1   Different ATM Services

Application	A.B. R	AAL BER	N BER
CBR	2M b/s	3e-11	1.3e-6
VBR	5M b/s	1e-11	1.8e-6
CATV	20M b/s	3e-13	6e-7
MPEG1	1.5M b/s	4e-11	2.5e-6
MPEG2	10M b/s	6e-10	1.5e-6

A.B.R: Average Bit Rate
AAL BER: AAL Layer BER
N BER: Non AAL Layer BER

2. Cell Loss Rate The reason for Cell Loss Rate can be describe as follows:

• Cell loss in time of transfer because of buffer

overflow
- Header of the cell error
- VPI/VCI error
- Out of time deadline

Table 2   Different ATM Services

App.	A.B. R	AAL BER	N BER
CBR	64K 2M b/s	1e-8	8e-6
VBR	5 Mb/s	4e-9	8e-6
CATV	20 Mb/s	1e-10	5e-6
MPEG1	1.5Mb/s	1e-8	9.5e-6
MPEG2	10 Mb/s	2e-9	4e-6

A.B.R: Average Bit Rate
AAL BER: AAL Layer BER
N BER: Non AAL Layer BER

3. Cell Insertion Rate Cell insertion rate means that Cell header error causes not only cell loss but also cell mis-insertion. If cell header error occurs and is not detected by the switch, this cell will be sent to the wrong destination through PVC. Obviously, this cell will cause synchronization loss and impair the performance the network. In the worst case, it could cause congestion. The common services' demand of CIR list as following:

Table 3   Different ATM Services

Services	CIR
CBR(Visible telephone)	1e-3
VBR(Data transfer)	1e-6
CATV	1e-6
MPEG1	1e-6
MPEG2	1e-7

A.B.R: Average Bit Rate
AAL BER: AAL Layer BER
N BER: Non AAL Layer BER

Congestion is defined as a condition that exists at the ATM layer in the network element such as switches, transmission links or the cross connects where the network is not able to meet a stated and negotiated performance objective.

Traffic control defines a set of actions taken by the network to avoid congestion. Traffic controls measure to adapt to unpredictable fluctuations in traffic flows and other problems within the network. Traffic control mechanics allot the limit resource according to the QoS of services.

To meet the objective of traffic control and congestion control, the ATM networks must:

- 1. Perform a set of actions called connection admission control (CAC) during a call setup to determine if a user connection will be accepted or rejected. These actions may include acquiring routes for the connection.
- 2. Establish controls to monitor, and regulate traffic at the UNI; These actions are called usage parameter control (UPC).
- 3. Accept user input to establish priorities for different types of traffic through the use of the cell loss priority (CLP) bit.
- 4. Establish traffic shaping mechanism to obtain a stated goal for managing all traffic (with differing characteristics) at the UNI.

3. BUFFER MECHNACHIC AND MODEL STRUCTURE

The main purpose of common traffic control is to allot resource and reject excessive connects request. ATM traffic control mechanic assigns the resource to the request connects after it was accepted. The resource can't be shared till the connection is deleted. If there is no resource left, new connect request will be rejected no matter how important it is. It has been frequently claimed that the priority scheduling approach has superior "stability" compared with other approaches, because "essential" processes can be assigned high priorities in order to ensure that they meet their deadline.

ATM implementations usually adopt the leaky bucket approach to avoid congestion.More details can be seen in M.Butto (1991). The bucket is a number of counters that are maintained at the UNI network side for each connection. A token generator periodically issues values named tokens that are placed into a token pool. Each cell should get a token before it was connected. When cells are sent to the network, the token pool is reduced. When excessive traffic comes, the token pool is exhausted.Figure 2 shows the principle of leaky bucket mechanic.

In such a situation, the cell will be discarded or dropped to the UNI without further processing.

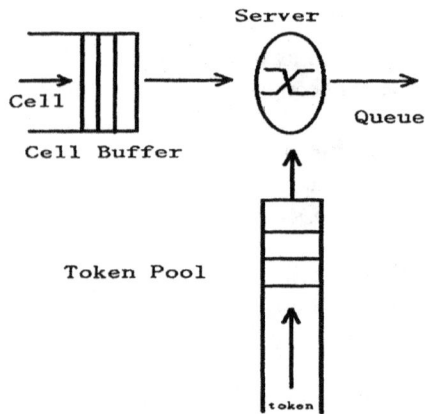

Fig. 2. Principle of leaky bucket mechanic.

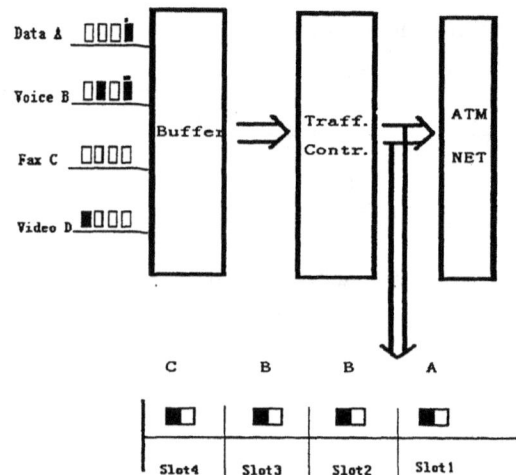

Fig. 3. the basic principle of this method

Consider real time upsurge traffic arrives when the token pool is exhausted, it is obviously that the cell will be discarded. If the cell is important and can't be retransmitted because of time limit, a disaster would happen.So in this paper a method is proposed which could deal with those kinds of cells well. Figure 3 illustrates the basic principle of this method.

ATM network traffic control model is usually built according to the type and usage of services. In general, the model of non real-time services is based on Poisson Model. Real time services and data flow is built as MMPP Method or Multi MMPP method. Considering the cells of the same connection can have different priorities but same demand of discard/accept sequence. This paper define vector $Ci(t0, td, pi)$ as the request priority of cell i, t0 indicates the start time, td means the time deadline, so td-to means time left for a cell to be sent; and take pi as the priority of this cell. So the time limit and the priority can be considered at the same time.

The model of this method can be described as follows:

Ci indicates the sequence order of cell i, ei means the given priority, $\Phi(i)$ illustrates the offset to each cell , matrix A is the authority coefficient matrix that rely on the type of services. Matrix A depends on the model of the system. Usually, In this paper the model is built according to MMPP Method or Multi MMPP method. According to this curriculum, all the cells that waiting for service can be ordered .

4. EXPERIMENTS AND CONCLUSIONS

To test the practicality of this method, the following environment is built:

Different services are adopted by service emulation equipment. After clock recovery and SAR (segment and resemble, clock synchronization method can been seen in H.Aghili and R.Strong (1986)), they are sent to cell process equipment called MUX that can assemble the cells, where all the cells are sent through PVC(Permanent Virtual Channel) that exists in the NET (Network exchange Board). This channel is built by equipment called main process unit (MPU) which fulfill the switching function. E1 service, CATV, MPEG1 and MPEG2 are sent to this kit.

The structure of this system is built according to the following algorithm: (Relevant method can be seen in Gilbert H.Young (1998),Zhang (1996c), Zhang (1996.a)):

Start of procedure
Schedule H0, H1, H2 jobs in their deadline slots;
Initialize the ATM Switch Units (MPU, MUX NET);
Detect the kinds of services;
Create exchange table for cell header;
Add Permanent Virtual Channel for cells;
Save PVC table;
Input cell ;
Get cell head and set priority in Cell head bit CLP;
Set deadline in cell head GFC (usually not used in UNI and NNI);

$$\begin{bmatrix} C1 \\ C2 \\ C3 \\ \ldots \\ Cn \end{bmatrix} = A * \begin{bmatrix} p1-e1 \\ p2-e2 \\ p3-e3 \\ \ldots \\ pn-en \end{bmatrix} + A' * \begin{bmatrix} T1-T0 \\ T2-T0 \\ T3-T0 \\ \ldots \\ T4-T0 \end{bmatrix}^{-1} + \begin{bmatrix} \varphi1 \\ \varphi2 \\ \varphi3 \\ \ldots \\ \varphi n \end{bmatrix}$$

$$A = \begin{bmatrix} a_{11} & \cdots & a_{1n} \\ & a_{22} & \\ & & \ddots \\ & & & a_{nn} \end{bmatrix}$$

$$0 \le a_{i,j} \le 1$$

$$\sum_{i}^{n} a_{i,j} = 1$$

Fig. 4. The model of this method

Initialize the tail job queue empty;

For i:=1 to n do
If I is occupied by Hi
Then Hi in its deadline slot: Enquenue;
ElseI is empty
Iftail job queue is empty
Then
Begin do
Input cell head ;
Search for next head job Hi with the least Ci;
Case Hi of:
H0: reschedule H0 in slot i;
H1: reschedule H1 in slot i;
H2: if there is another empty slot;
then
reschedule H2 at i; Enqueue;
else
if slot d(H2)+C2 is empty and there is a H0 job in [i,i+C2]
then reschedule the first H0 in slot i;
else output "Stop";
End case;
End do;
else tail job is not empty
Begin do
Schedule the first tail job in the tail job queue in slot i; Dequeue;
if the temporal distance constraint is not satisfied,
then output" Stop";
end do;
End if;

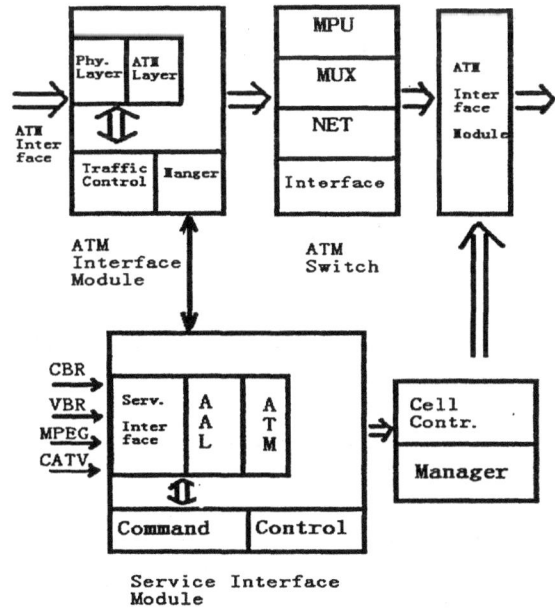

Fig. 5. Hardware of the Experiments

End if;
End of procedure;

Sunset equipment E10 is used and tested for 24 hours. From the experiments it can conclude, when the priority and time deadline at the cell headers are set , sometimes cells with low priority were discarded if large quantity of packets were sent. After the method addressed in this paper is adopted, no such cell was ever discarded.

This method proved to be effective in cell loss at the cost of a tolerant degradation of the behavior of the network.

In this paper, a common problem about transport traffic and cell discard/accept is addressed. On the base of the proposed model, a revised buffer threshold based method is discussed. With this method, the demand of real time transportation, the order of cells to be sent and rate of loss cells can be guaranteed. Because of authority coefficient, Cells of low priorities but strict deadlines could be sent before they are discarded. Experiment shows that this method is efficient in ATM real time transport.

Table 4   Different ATM Services

Kinds	BER		CIR		SLIP	
	B.	A.	B.	A.	B.	A.
E1	-9	-11	-9	-9	-7	-9
TV	-9	-11	-7	-8	-7	-8
M1	-10	-11	-9	-9	-7	-8
M2	-8	-11	-9	-10	-6	-8

BER means Bit Error Rate
CIR illustrates Cell Insertion Rate
SLIP shows the loss of frame rate
B: before this method is adapted
A: After using this method

5. REFERENCES

A.E.Eckberg (1989). 'Meeting the challenge: Congestion and flow control strategies for broad band information transport'. *GLOBER-COM'89* pp. 1769–1773.

F.Bernaber (1993). 'Analysis of on-off source shaping for atm multiplexing'. *INFOCOM'93* pp. 1330–1336.

Gilbert H.Young, C. W. N. (1998). 'Scheduling real-time jobs with chain tree temporal distance constraints'. *IFAC/IFIP Workshop on Real-Time Programming.*

H.Aghili, F. and R.Strong (1986). 'Clock synchronization in the presence of omission and performance faults, and processor joins'. *Proc. FTCS 16* pp. 218–223.

K.Bala (1990). 'Congestion control for high-speed packet switched networks'. *INFO-COM'90* pp. 520–526.

M.Butto (1991). 'Effectiveness of the "leaky bucket's policing mechanism in atm networks'. *IRRR J-SAC* **9**, 335–342.

Rec.I.365.1, I.-T. D. (1993). *Frame Relaying Service Specific Convergence Sublayer.* Geneba.

Rec.I.555, I.-T. D. (1993). *Frame Relay Bearer Service Interworking.* Geneba.

S.Saunders (1994). 'Atm forum ponders congestion control options'. *Data Communications* pp. 55–60.

T.M.Chen, S.S.Liu, V. (1996). 'The available bit rate service for data in atm networks'. *IEEE Communications Magazine* pp. 56–71.

Zhang, L. (1992). 'Generating physical channel access protocol data unit of ccsds recommendation by use of the transputers and occam'. *Lecture Notes in Computer Science* p. No.634.

Zhang, L. (1994a). 'Applying software engineering principles in train control systems'. *Proc. of IEEE Int. Conf. on Real-time Systems* pp. 123–129.

Zhang, L. (1994b). 'A design methodology for industrial real-time control systems'. *Proc. of Third Int. Conf. on Control, Automation, Robotics and Vision* pp. 125–132.

Zhang, L. (1996.a). 'Architecture supports for predictable real-time systems'. *the proceedings of the Eighth IASTED International conference on parallel and distributed computing and systems.*

Zhang, L. (1996b). 'A design methodology for real-time to be implemented on multiprocessors'. *International Journal, the Journal of system and software* pp. 37–56.

Zhang, L. (1996c). 'Integrating different views to develop complex real-time systems'. *the proceedings of 21st IFAC/IFIP workshop on real-time programming* pp. 113–117.

A TOTAL ORDERING SCHEME FOR REAL-TIME MULTICASTS IN CAN

Mohammad Ali Livani* Jörg Kaiser*

*University of Ulm, Department of Computer Structures, 89069
Ulm, Germany*

Abstract: The Controller Area Network (CAN) is a broadcast medium providing advanced features, which make it suitable for many real-time applications. Common application layer protocols designed for CAN (e.g. CAL, SDS, DeviceNet) exploit these features in order to provide reliable real-time communication. However, they do not provide a consistent global message ordering in certain fault situations, nor do they consider temporal properties of the application.

This paper presents a multicast protocol which supports timely delivery of messages and guarantees atomic order, under anticipated fault conditions. In order to support causal ordering of events, the paper provides a discussion on establishing Lamport's precedence relation between events by appropriate usage of real-time multicasts.

In a CAN-based distributed system, the restricted communication bandwidth constitutes a serious bottle-neck. Therefore an important feature of this multicast protocol is to achieve optimal protocol termination time while requiring minimum communication overhead. *Copyright © 1999 IFAC*

Keywords: Real-time communication; fault-tolerance; multicasting; total order; CAN

1. INTRODUCTION

Future computer systems will, to a large extent, monitor and control real-world processes. This results in an inevitable demand for timeliness and reliability. Distributed systems which inherently provide extensibility and immunity against single failures, are an adequate architecture to cope with spatially distributed real world applications. Moreover, the availability of inexpensive, powerful micro-controllers promotes distributed solutions.

Group communication is the basic mechanism to coordinate distributed activities. In order to control the competition and cooperation among distributed processes, and to allow for object replication, atomic delivery of multicast messages is inevitable. In this paper, the following definition of atomic multicast delivery is used.

Definition 1. Atomic multicast delivery – Let G be a multicast group, i.e. a set of objects which must receive a multicast message. Let $del_i(m)$ denote the delivery of a message m to an object i and '$\longrightarrow$' define the precedence relation. Then two messages m and m' are delivered atomically to the group G, if and only if the following holds:

$$\forall i,j : i,j \in G \land i,j \text{ non-faulty}$$
$$\Rightarrow (del_i(m) \longrightarrow del_i(m')$$
$$\Leftrightarrow (del_j(m) \longrightarrow del_j(m'))$$

In other words, atomic delivery of multicast messages implies two properties:

- *Consistent Delivery*: if a multicast message is received by a non-faulty group member, then it is received by all non-faulty group members.

- *Consistent Ordering*: messages received by different non-faulty group members, are received in the same order.

Achieving the consistent delivery requires either a two-phased commit protocol with a considerable acknowledgment overhead (Babaoglu and Drummond, 1985; Birman and Joseph, 1987; Chang and Maxemchuk, 1984), or a mechanism to retransmit each message so many times that the probability of losing all copies is negligible (Cristian, 1990; Livani, 1998; Rufino et al., 1998).

Consensus on the ordering of the messages comes free with the two-phased commit protocol. But if the multiple transmission approach is applied, an additional mechanism for establishing a unique order of the messages must be used. (Schneider, 1990) has described an approach to achieve consensus on the message ordering among a group of receivers, using a time-dependent unique identifier of messages. An approach which exploits knowledge about the delivery deadlines of the messages, has been introduced by (Cristian et al., 1985). A similar mechanism has been proposed for embedded applications by (Zuberi and Shin, 1996). These approaches, however, rely on the timely and reliable delivery of all messages.

The ordering scheme presented in this paper considers also the late transmission of soft real-time messages, which is a timing failure of the message transmission, but does not lead to a system failure. As already shown in (Livani et al., 1998) it is possible to schedule the real-time communication in a CAN-bus system so that hard real-time messages are always transmitted timely while soft transmission deadlines are missed in overload situations. In such a system it is not possible to guarantee a total delivery order among all hard and soft real-time messages.

To see the reason of this restriction consider a hard real-time message H, a soft real-time message S, and two nodes N_1 and N_2. Since no latest delivery time can be guaranteed for S, its consistent delivery must depend on the reception of a signal σ from the bus (e.g. S itself, a commit packet, or another packet depending on the protocol). Assume that N_1 receives σ at time t_1, and N_2 does not because of a single communication error. If N_1 and N_2 receive H after t_1 and N_2 does not receive σ until d_H (i.e. the deadline of H) due to bus overload, then at d_H, N_1 must deliver S before H, and N_2 must deliver H before S.

Due to the restriction mentioned above, this paper proposes an ordering algorithm, which establishes a total order for the messages of each class (i.e. hard and soft real-time) separately.

The paper is organized as follows: section 2 introduces some properties of the CAN bus which are essential to understand the approach. Section 3 presents the deadline-based total ordering scheme. Section 4 discusses how to reflect Lamport's precedence relation using the deadline-based total order. A summary concludes the paper.

2. SOME PROPERTIES OF CAN

The CAN-bus (BOSCH, 1991) is a priority bus targeted to operate in a noisy environment with speeds of up to 1 Mbit/s, exchanging small real-time control messages. The priority-based arbitration mechanism of the CAN bus can be exploited to guarantee the timely transmission of hard real-time messages under anticipated load and fault conditions (Tindell and Burns, 1994). In order to provide best effort scheduling of soft real-time communication in a CAN bus while guaranteeing hard deadlines, different approaches may be applied (Davis, 1994; Livani et al., 1998) which are based on multiple priority classes.

The CAN protocol provides efficient hardware-implemented error handling, which is based on various error detection mechanisms with a very high total coverage, and an approach to immediately signaling the error condition. These features of the CAN protocol ensure atomic broadcast delivery in most fault situations. However, if an error occurs within the transmission time of the two last bits of a data frame, some receivers might accept and other receivers reject the frame. Although the sender (or in case of immediate sender crash an alternative mechanism like Eager Diffusion (Rufino et al., 1998) or Shadow Retransmitter (Livani, 1998)) will retransmit the frame, the consistent ordering of the incoming messages in different sites is not trivial due to the possible transmission of other (high-priority) messages between the first and the second transmission of a frame. Another problem caused by such errors is the duplicate frame reception by some receivers.

In such situations, commercially available application level protocols for CAN, like CAL (CiA, 1993), SDS (Crovella, 1994), and DeviceNet (Noonen et al., 1994) manage to discard frame duplicates, but they fail to provide consistent message ordering among a group of receivers.

Relying on a feasible and flexible real-time scheduling policy (Davis, 1994; Livani et al., 1998) combined with a reliable broadcast delivery mechanism (Livani, 1998; Rufino et al., 1998), following assumptions can be made about the communication system in a CAN bus:

(A1) Hard real-time messages are transmitted timely (i.e. until their transmission deadline) under anticipated fault scenarios, even in overload situations.

(A2) Soft real-time messages are scheduled by static or dynamic priorities, but their deadlines may be missed in overload situations.

(A3) If a non-faulty receiver receives a message, then eventually all non-faulty receivers receive the message.

Given these assumptions, atomic multicasting can be achieved by establishing a consistent global delivery order for multicast messages.

3. THE DEADLINE-BASED TOTAL ORDERING SCHEME

This section introduces a scheme to achieve globally consistent ordering of multicast messages in a CAN-based system. The algorithm presented here relies on the assumptions A1 through A3 in order to achieve atomic multicast with minimum communication overhead, based on the knowledge about the message transmission deadlines. The transmission deadline of a message denotes the time, at which the message must be successfully transmitted to all non-faulty destination sites. Since the transmission deadline is tightly related to the time, where the sender expects the message delivery to all receiving application objects, this ordering mechanism allows an application specific ordering, which is related to temporal requirements. Although this approach is similar to some other ones known from much previous work (Cristian *et al.*, 1985; Schneider, 1990), it considers the late transmission of soft real-time messages in overload situations. Another advantage of this scheme is that it needs no additional communication to establish a globally consistent ordering decision.

Let m and m' be two real-time messages with transmission deadlines d_m and $d_{m'}$. Then the deadline-based ordering algorithm implements the following rule:

(O1) $d_m < d_{m'} \Rightarrow del_j(m) \longrightarrow del_j(m')$

This means that if the deadline of the message m is before the deadline of the message m', then m must be delivered to destination objects before m'. If the underlying protocol layer adjusts the transmission deadlines of hard real-time messages to reserved time-slots (Livani *et al.*, 1998), then the delivery order of hard real-time messages can be always established by this rule.

However, when the deadlines of different messages are equal, the following rule must be used to ensure the globally consistent decision on delivery order of messages.

(O2) $d_m = d_{m'} \Rightarrow (del_j(m) \longrightarrow del_j(m')$
$$\Leftrightarrow header_m < header_{m'})$$

Where $header_m$ consists of a tuple (d_m, $sender_m$, $subject_m$) and identifies a message uniquely. This rule requires that different messages sent by the same sender with the same subject have different transmission deadlines, otherwise an additional sequence number (or "toggle-bit") must be provided in the header. Note that this requirement is also necessary for distinguishing duplicate frames from successive frames.

3.1 *Protocol Termination and End-to-End Delivery*

An atomic multicast protocol may terminate and deliver a message to destination objects, if and only if a) the message is received and accepted by all non-faulty destination sites, and b) the message can be consistently ordered, which means that no other message will arrive later, which must precede the current message according to the ordering criteria.

a) Consistent Message Reception in CAN

In order to minimize the communication overhead, the nodes must not exchange explicit information about the status of their message queues. Thus they have to conclude this information from implicit knowledge. Since the underlying message transfer protocol guarantees the timely transmission of hard real-time messages under anticipated fault conditions, the protocol can assume that a hard real-time message is consistently transmitted to all receivers at its deadline. Hence a receiver can deliver the message to the destination objects at the transmission deadline.

For soft real-time messages, however, the transmission deadline may be missed because of bus overload, and hence, late transmission of the message is still possible. In these situations, the deadline cannot be used to specify the time when the message may be delivered. However, in such situations the following considerations lead to a decision criterion:

Claim 2. If a node has received a message m, and then another message with lower priority is observed on the CAN bus, or the bus is idle, then the sender of m will not retransmit it in future.

PROOF. Assume that a message m is transmitted at least once on the CAN bus. Further, assume that the sending CAN controller still attempts to retransmit m due to the inconsistent transmission. According to the CAN specification (BOSCH, 1991), the sender will try to retransmit m "immediately", thus no bus idle period will be observed before the retransmission of m. Furthermore, no lower-priority message will be transmitted before the retransmission of m, because m will win the arbitration process against any lower-priority message. $\square$

As a consequence, the receivers of a soft real-time message m can be sure that either the sender is crashed, or the transmission was successful, as soon as they detect either a bus idle period or a lower-priority message after receiving m.

If a frame is accepted by a subset of the receivers, and rejected by other receivers, the immediate sender crash may lead to inconsistent message reception. This failure must be tolerated by having some nodes retransmit the frame. The Eager Diffusion Protocol (Rufino et al., 1998) retransmits every frame at least by one receiver. However, due to processing delay the Eager Diffusion Protocol cannot guarantee the immediate retransmission. The ordering scheme presented in this paper relies on dedicated Shadow Retransmitters (Livani, 1998), which guarantee immediate retransmission whenever an inconsistent message reception is possible. The Shadow retransmitters also transmit an ultra-low priority frame (called trailer) at the end of each non-broken sequence of CAN frames. So an idle bus can be detected by observing a trailer on the bus.

From the previous discussion, following properties are derived:

(P1) Any receiver of a hard real-time message m with deadline d_m, can assume that m will be received by all sites until d_m.

(P2) Any receiver of a soft real-time message m can assume that the message has been received by all sites, if it observes a frame with a lower priority on the bus after receiving m.

b) Achieving a Consistent Message Order in CAN

In case of hard real-time messages, the knowledge of time is sufficient for stabilizing the ordering decision: after a deadline d_m, no hard real-time message with an earlier deadline will arrive. But this statement is not true in case of soft real-time messages. Here, the following assumption constitutes the constraint necessary to find a stable ordering decision criterion:

(A4) Every real-time message m is ready to transmit before $d_m - \Delta T_{\max}$, with $\Delta T_{\max}$ being the maximum time required for a single transmission of an arbitrary frame.

The assumption (A4) concludes the following:

Claim 3. Assume a deadline-based message priority scheme e.g. (Livani et al., 1998), where a soft real-time message with a higher priority has an earlier deadline than a soft real-time message with a lower priority. If after the transmission deadline d_m of a soft real-time message m another message with lower priority is transmitted on the CAN bus, or the bus is idle, then no other soft real-time message m' with a deadline $d_{m'} \leq d_m$, will be transmitted later.

PROOF. Assume that a message m' with deadline $d_{m'} \leq d_m$ is pending for transmission at the time $t > d_m$. Because of the deadline-based priority assignment the priority of m' is not lower than the priority of m. Due to (A4), m' has been pending for transmission at least since $d_{m'} - \Delta T_{\max}$, hence at least since $d_m - \Delta T_{\max}$. Hence the sender of m' must have been trying to transmit m' at the beginning of every bus-idle period since $d_m - \Delta T_{\max}$. Therefore no idle bus can be observed between d_m and the successful transmission of m'. Also, no lower-priority message can be transmitted on the bus before m', because m' wins the arbitration process against any lower-priority message. Thus, if after d_m a message with a lower priority than m is transmitted on the CAN bus, or the bus is idle, then no message m' with $d_{m'} \leq d_m$ will be transmitted later. $\square$

From the previous discussion, following property is derived:

(P3) For any soft real-time message m, no preceding soft real-time message will arrive later, if after the transmission deadline d_m a lower-priority message is observed on the bus.

3.2 Atomic Multicast Delivery Algorithm

The proposed atomic multicast delivery algorithm is based on the following rules:

Order: Messages are ordered by their deadlines. In case of equal deadlines, the value of the message header is used for the order decision.

Time1: every received hard real-time message is delivered at its deadline. (This realizes the deadline-based order implicitly)

Time2: Every received soft real-time message must be delivered as soon as either another message with later deadline, or a bus idle time is observed after its transmission deadline.

Due to the rule (Time2), the delivery of a soft real-time message m may be delayed until $d_m + \Delta T_{\max}$. This must be considered when calculating the transmission deadline of any soft real-time message.

In the following, the atomic multicast delivery algorithm is presented. The algorithm assumes the availability of a real-time clock, which is synchronized globally with a bounded inaccuracy.

Atomic_multicast_delivery
initialization:
```
    SRT_q ← empty_q;
    HRT_q ← empty_q;
    next_HRT_delivery ← eternity;
receive(m,t): /* m is sent or received at t */
    if m.dest_group ∈ my_groups
        and m.category = HRT then
    HRT_q.insert (m);
    if m.dl < next_HRT_delivery then
        next_HRT_delivery ← m.dl;
        set_wakeup (next_HRT_delivery);
    if m.dest_group ∈ my_groups
        and m.category = SRT then
    SRT_q.insert (m);
    if m.dl ≤ t then
    /* overload! Deliver preceding SRTM */
        while SRT_q.head.dl < m.dl
            or SRT_q.head < m do
            mx ← SRT_q.gethead;
            SRT_deliver (mx, mx.dest_group);
    else
    /* Deliver SRTM with passed DL */
        while SRT_q.head.dl < t do
            mx ← SRT_q.gethead;
            SRT_deliver (mx, mx.dest_group);
end receive;
wakeup(): /* invoked when the wake-up
                time is reached */
    mx ← HRT_q.gethead;
    HRT_deliver (mx, mx.dest_group);
    if HRT_q ≠ empty_q then
        next_HRT_delivery ← HRT_q.head.dl;
        set_wakeup (next_HRT_delivery);
    else
        next_HRT_delivery ← eternity;
end wakeup;
bus_idle(t): /* invoked if at time t a
                trailer frame is received */
    while SRT_q.head.dl < t do
    /* Deliver all SRTM with passed DL */
        mx ← SRT_q.gethead;
        SRT_deliver (mx, mx.dest_group);
end bus_idle;
end Atomic_multicast_delivery
```

4. GLOBAL ORDERING OF EVENTS

The deadline-based ordering of multicast messages enables a consistent global order between different events observed in the distributed system. In this paper an event is called a global event, if the object which observes it, sends an atomic multicast to enforce a global observation of the event by all concerning objects. As seen by application objects, the order of global events is the same as the globally consistent delivery order of their notification messages. This globally consistent delivery order of messages is established by the atomic multicast protocol described in section 3.2. Note that if a set of events has to be globally ordered, then all multicast messages propagating those events must belong to the same class (hard or soft real-time).

4.1 *Achieving Causal Ordering of Events*

If global events have to be ordered according to their causality, following cases can be observed:

A: Let e and e' be two global events locally observed by an object i, with a causal precedence relation $e \longrightarrow e'$, and let m and m' be messages, which propagate the events e and e' in the system. Let d_m denote the deadline of m. In order to globally agree on the order $e \longrightarrow e'$, for every object j, the relation $del_j(m) \longrightarrow del_j(m')$ is sufficient. Hence, due to the rule (O1), the relation $d_m < d_{m'}$ is sufficient. This means, that in order to reflect a precedence order $e \longrightarrow e'$ globally, i must assign deadlines $d_m < d_{m'}$ to the messages propagating e and e' in the system.

B: Let e and e' be two events observed by two different objects i and j, and $e \longrightarrow e'$, then (according to Lamport's definition of precedence relation (Lamport, 1978), and the discussion in section 3.1) following conditions hold:

(1) Event e is observed by i at local time $T^i(e)$.

(2) Event e' is observed by j at local time $T^j(e')$.

(3) There is a message m sent by i at the local time $T^i(send_i(m))$, which propagates the event e, where $T^i(e) \leq T^i(send_i(m)) < d_m$.

(4) There is a message m' sent by j at the local time $T^j(send_j(m'))$, which propagates the event e', where $T^j(e') \leq T^j(send_j(m')) < d_{m'}$.

(5) If j receives m, then it receives m at the local time $T^j(del_j(m))$, where $d_m \leq T^j(del_j(m))$, and $T^j(del_j(m)) \leq T^j(e')$. Thus due to (4) it follows that $d_m < d_{m'}$.

(6) If j does not receive m, then there is a set of messages $m_1, m_2, m_3, \cdots, m_n$, and a set of objects $k_1, k_2, k_3, \cdots, k_n$, where firstly, k_1 receives m and sends m_1, k_2 receives m_1 and sends $m_2, \cdots, k_n$ receives m_{n-1} and sends m_n, and j receives m_n. Secondly, $del_{k_1}(m) \longrightarrow send_{k_1}(m_1)$, $del_{k_2}(m_1) \longrightarrow send_{k_2}(m_2)$, $\cdots$, and $del_{k_n}(m_{n-1}) \longrightarrow send_{k_n}(m_n)$. And thirdly, $T^j(del_j(m_n)) \leq T^j(e')$. In this case, $d_m \leq T^{k_1}(del_{k_1}(m)) < T^{k_1}(send_{k_1}(m_1)) < d_{m_1} \leq T^{k_2}(del_{k_2}(m_1)) < T^{k_2}(send_{k_2}(m_2)) < d_{m_2} \cdots < d_{m_n} \leq T^j(del_j(m_n)) \leq T^j(e')$. Again, due to (4) it follows that $d_m < d_{m'}$.

In either case (i.e. A; B–5; B–6), every object l in the system which receives m and m', will establish the delivery order $del_l(m) \longrightarrow del_l(m')$ due to the relation $d_m < d_{m'}$. Consequently, l will establish the precedence order $e \longrightarrow e'$. Thus, if e and e' are two global events observed in the distributed system, and $e \longrightarrow e'$ according to Lamport's definition of precedence relation, then all non-faulty objects in the system – which are addressed by messages propagating e and e' – will be notified about e before e', and hence they will establish consistently the precedence order $e \longrightarrow e'$.

5. CONCLUSION

The paper introduced a real-time multicast ordering scheme, which achieves consistent delivery ordering of multicast messages using the message transmission deadlines.

The protocol termination time depends on the message class. The protocol relies on a medium access protocol, which guarantees timely and reliable transmission of hard real-time messages to all destination nodes (Livani, 1998; Livani et al., 1998). Hence, for a hard real-time message the protocol terminates at the transmission deadline and delivers the message to destination objects timely. Under 'normal' load conditions, the protocol delivers soft real-time messages up to one frame transmission time later than their transmission deadlines (cf. rule Time2 in section 3.2). However, in overload situations, the protocol may delay the delivery of soft real-time messages until the end of the overload period plus one frame transmission time.

The mechanism ensures consistent multicast ordering in presence of communication failures leading to inconsistent global view of the CAN bus status among non-faulty sites. Common high-level communication protocols designed for CAN, do not provide consistent multicast delivery order among all application objects in these failure situations.

It was shown, that causal order between events can be globally established by using the deadline-based ordering approach.

An important benefit of this scheme is its efficiency: the algorithm requires no additional communication, like acknowledgements etc. In the Controller Area Network, communication bandwidth is a serious bottle-neck. While the computing power of hardware nodes is increasing rapidly, the bandwidth of CAN will remain limited to a maximum of 1 Mbits/sec. This was the main motivation to develop a multicast scheme which minimizes communication amount for the price of some computational overhead.

6. REFERENCES

Babaoglu, Ö. and R. Drummond (1985). Streets of byzantium: Network architectures for fast reliable broadcasts. *IEEE Tr. Software Eng.* **11**(6), 546–554.

Birman, K.P. and T.A. Joseph (1987). Reliable communication in the presence of failures. *ACM Tr. on Computer Systems.*

BOSCH (1991). CAN specification version 2.0. *Published by Robert BOSCH GmbH.*

Chang, J.M. and N.F. Maxemchuk (1984). Reliable broadcast protocols. *ACM Tr. on Computer Systems* **2**(3), 251–273.

CiA (1993). CAN application layer (CAL) for industrial applications. *CiA Draft Standards 201..207.*

Cristian, F. (1990). Synchronous atomic broadcast for redundant broadcast channels. *The Journal of Real-Time Systems* **2**, 195–212.

Cristian, F. et al. (1985). Atomic broadcast: From simple message diffusion to byzantine agreement. *15th Int. Symposium on Fault Tolerant Computing.*

Crovella, R.M. (1994). SDS: A CAN protocol for plant floor control. *1st Int. CAN Conference.*

Davis, R. (1994). Dual priority scheduling: A means of providing flexibility in hard real-time systems. *Technical Report YCS230, University of York.*

Lamport, L. (1978). Time, clocks, and the ordering of events in a distributed system. *Communications of ACM.*

Livani, M.A. (1998). SHARE: A transparent mechanism for reliable broadcast delivery in CAN. *Informatik Bericht 98-14, University of Ulm.*

Livani, M.A., J. Kaiser and W.J. Jia (1998). Scheduling hard and soft real-time communication in the controller area network (CAN). *23rd IFAC/IFIP Workshop on Real Time Programming.*

Noonen, D., S. Siegel and P. Maloney (1994). Devicenet application protocol. *1st Int. CAN Conference.*

Rufino, J., P. Verissimo, G. Arroz, C. Almeida and L. Rodrigues (1998). Fault-tolerant broadcasts in CAN. *28th Int. Symposium on Fault Tolerant Computing.*

Schneider, F. (1990). Implementing fault-tolerant services using the state machine approach: A tutorial. *ACM Computing Surveys.*

Tindell, K. and A. Burns (1994). Guaranteeing message latencies on controller area network (CAN). *1st Int. CAN Conference.*

Zuberi, K.M. and K.G. Shin (1996). A causal message ordering scheme for distributed embedded real-time systems. *Proc. Symp. on Reliable and Distributed Systems.*

WELL-BEHAVED APPLICATIONS ALLOW
FOR MORE EFFICIENT SCHEDULING

HO. TRUTMANN

*Computer Engineering and Networks Laboratory TIK, Swiss Federal Institute of Technology ETH,
CH-8092 Zurich, Switzerland. E-Mail: trutmann@tik.ee.ethz.ch*

Abstract. Current trends in embedded applications aim towards implementations with sets of inter-operating tasks and elaborate scheduling schemes. The quest for flexibility and universal applicability results in higher resource demands and reduced predictability. If the functional and connection problems are handled separately with the aid of appropriate tools, subsequent code generators can convert design specific issues, such as parallel structures in the functional part and the various periodicities in the connection part to linear and deterministic pieces of code. Such implementations do away with complicated run-time scheduling, make better use of the available resources and can be analyzed exhaustively before installation on a target. *Copyright © 1999 IFAC*

Key Words. Embedded Systems; Specifications; Scheduling; Code Generation

1. INTRODUCTION

Embedded applications used to be tightly interwoven with their run-time environment. Increasing complexity, together with growing requirements and restrictions has favored the tendency to isolate problem concerns in this field, leading to more general results. A collection of singular tasks replaces monolithic programs with the advantage of smaller scale and simpler structures. Tailor-made scheduling for each application has given way to more generic approaches. Modularity and dynamic behavior at the task level should permit system compositions involving new parts as well as legacy code. The drawbacks of these techniques in hard real-time systems are countered with more efficient hardware in the case of speed considerations and testing in the case of unpredictability concerns. In the more cost-sensitive embedded systems segment the claim for additional resources is not well received, because it is mandatory that these systems use the slowest possible processor and a minimum of memory. For these applications progress made in hardware development must result in lower cost or reduced power consumption, rather than compensate implementation deficiencies. Predictable response times, vital in safety-critical systems on the other hand, are hard to achieve without static methods and using components with deterministic properties.

1.1. Current Implementation Techniques

Increasingly, embedded applications that incorporate hard real-time requirements are multi-task implementations scheduled with static or dynamic techniques, the main justification being the possibility to accommodate any number of quasi-parallel tasks on a single processor.

Dynamic scheduling deals with tasks in an arbitrary sequence and unknown start times, where the various tasks operate independently and where interactions are not clearly defined (Liu *et al.*, 1973). Response time requirements are mapped to scheduler priorities, an inadequate measure for systems with non-periodic events. Solving the interaction problem is postponed and these issues are handed down to a later design stage (i.e., run-time). All types of dynamic scheduling are punished with considerable overhead aggravated by conceptual difficulties such as resource depletion and priority inversion; their essential shortcomings with regard to satisfying timing constraints have also been shown (Xu *et al.*, 1993). This is especially true for preemptive methods which create an infinite number of possible schedules for a given problem and introduce additional conflicts concerning resource usage.

Static scheduling on the other hand uses detailed prior knowledge of the tasks' processing requirements to find a feasible schedule (Xu *et al.*, 1993). While this technique can guarantee the correctness of its solutions, it is overly pessimistic regarding asynchronous events which tend to be scheduled too often and therefore waste processing power.

Even combinations of these techniques cannot entirely do away with this disadvantage, as truly asynchronous tasks still have to be scheduled many times only to verify that the associated event has not yet occurred. In both these approaches input and output activities are implemented as tasks and scheduled in the same way as computation. Generally (except for exclusion and precedence relations) no assumptions about the nature of the tasks actually running on a target are made. If such knowledge is available and if the tasks' code confirms to certain limitations, a combination of static and dynamic scheduling techniques is possible that allows for high processor utilization, at the same time guaranteeing low input jitter and predictable behavior. The overhead of a complicated scheduler is avoided, because there are only three basic tasks. No preemption is used and the remaining dynamic decision involves a mere table lookup which hardly introduces any overhead.

1.2. Code Generation

In control system development the most unpleasant problems appear after work on the implementation has begun, when it may be discovered that a design is either wrong, impractical or that alterations to the original intentions need to be made because of implementation constraints. If changes are made at this stage, it is very difficult to keep the design and the implementation synchronized. An obvious solution would be the automatic compilation of a specification into optimized code, thus uniting specification and implementation.

If generated code is to replace manually written implementations altogether, the generators have to meet a number of difficult demands: small code size, fast execution, adaptation to vastly different hardware and operating environments, to just name a few. Adjustments to specific environments should be done automatically from the same specification and tests created in the design phase must remain valid also during the implementation phase. These demands are best met with dedicated models and tools for the two basically unrelated problem concerns in embedded systems:

- correct reaction to events in the environment,
- information transport between this environment and the control system.

If these prospects are exploited, the resulting code can also be made compliant with more restrictive implementation demands. Parallel structures present in the design are implemented by casting them into a single thread with processor loads as uniform as possible and deterministic behavior. With only deterministic components, exhaustive analysis prior to installation becomes possible.

2. PROBLEM DECOMPOSITION

Specifying a system implies creating its abstract model with the purpose of the model defining the abstraction rules. The separation of the three basic activities in control applications, *input*, *processing* and *output* into two main problem areas permits independent development and also sheds some light on the objective of possible models. The *functional problem* comprises all the behavioral aspects of a control system. The interactions with the controlled processes themselves are accomplished via shared phenomena (events and actions in an event driven environment). The appropriate abstraction for solving this problem is to leave out any details not pertaining to the relation between the course of events in the environment and the inner states of a model. The *connection problem* unites input, the extraction of events contained in the new data, the calling of the functional reactive block, and the output. The functionality of the connector that must be built is implementation-oriented with a regular sequential structure, so that a useful abstraction would concentrate on information transport within a generalized form. Based on these precepts problem-oriented development methods with separate solutions for the functionality and the connections to the physical processes can be built.

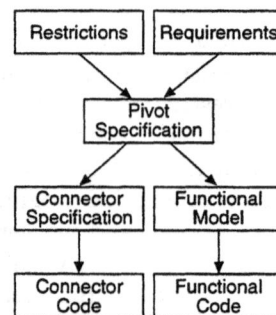

Fig. 1. The design pivot

It is crucial to this approach that the problem areas remain well separated during the entire development. Aside from providing guidelines in the relevant problem areas, architectural restrictions and functional requirements also give rise to an explicit interconnection model, the *pivot specification*. Because it is less prone to frequent changes than the problem areas themselves, development may progress independently behind stable interfaces based on this specification.

2.1. Functional Problem

The functional reactive behavior of a system is modeled with CIP Tool in the form of extended cooperating state machines (Fierz, 1999, and CIP Tool, 1995-99). Control systems in a single- or multiprocessor environment are built with a notation of synchronous processes and asynchronous clusters. Event-based and hybrid systems with integral quasi-continuous processes can both be dealt with. Event messages from the environment cause a transition in the receiving process which activates other processes for his part. Transitions may generate pulses to other processes as well as action messages for the environment, and they can also contain additional code for data manipulation and control functions. A chain reaction initiated by a message cannot be interrupted; its execution depends on the message itself and the internal states of all activated processes.

Consider a simple system with one cluster embodying three processes that model control for an electric motor. It accepts *closeSwitch, openSwitch* (Switch Events), *standstill, sample* (Controller Events) and produces *MotOn, MotOff, MotValue* (Controller Actions), *GreenOn, YellowOn, RedOn, LightOff* (Indicator Actions). The permissible traces in the input stream are determined by the structure and interoperation of the receiving processes; these and other particulars that are of no concern for the generated code will not be considered any further.

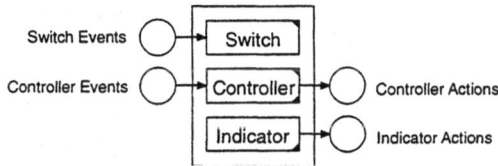

Fig. 2. Motor controller system interface

The transition structure of the controller process (Fig. 3.) shows the motor being either in state *off, running* or *stopping* (the motor runs and stops under continuous control). Its transition triggers are the *start/stop* pulses from the switch process, the *standstill* event and the periodic *sample* event for the controllers from the environment. Its outputs consist of pulses *running, stopping, stopped* to the other processes and actions *MotOn, MotValue, MotOff* to the environment.

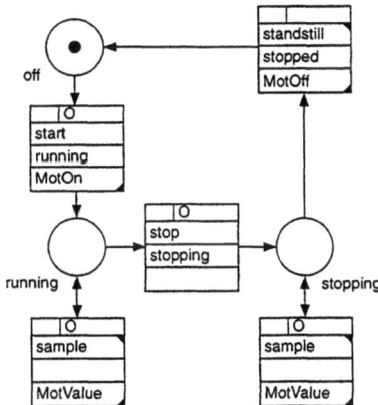

Fig. 3. Controller process

Two different controller algorithms are placed within the transitions triggered by the *sample* event and associated with states *running* and *stopping*, respectively. It follows from the diagram

that only one of these controllers can be active at any time, thus effectively limiting the required processing requirements. The transitions leading to these two states contain the initialization parts for the controller algorithms.

All process interactions are based on a specified net (Fig. 4.) which in comprehensive situations also helps understand system dependencies. It does not suffice, however, as a foundation for the later generation of deterministic code.

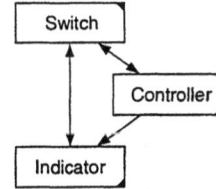

Fig. 4. Motor controller interaction net

To this effect static cascades must be defined that state all possible pulse propagations. In Fig. 5. the cascades for the various processes in the example are shown with the loop *switch-controller-indicator-switch* that is present in the interaction net unraveled. Because the indicator process does not accept events, its cascade is empty.

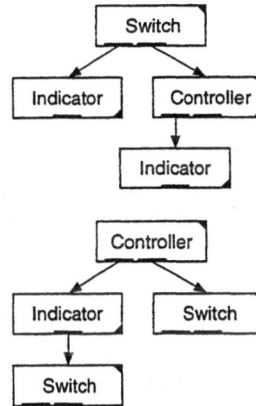

Fig. 5. Motor controller cascades

Code Generation. From a complete model C code can be generated. Along with the modules that implement the functionality, auxiliary code for testing and analysis is provided. One such component is used to verify that and how well an implementation will fit on a specific target. Each specified event sits at the root of possible paths through its pertaining cascade for which a collection of code sequences (minus the action messages) is generated. The worst-case execution path included in this collection must be found by measuring processing times on the target itself. Another generated auxiliary component lists only the sets of action messages belonging to each event with reference to the same possible paths. These will be used to determine upper bounds on the processing needs of the output task.

2.2. Connection Problem

Interfacing functional code to the existent heterogeneous peripheral hardware is a difficult task in itself. Updating input values or the opposite, setting new output values, seem straightforward duties which must be executable within regular structures. On the other hand, input and output operations on some devices are fairly complicated and must be made with multiple accesses, possibly also requiring delay periods in between. In addition, considerable variations with regard to data width, channel number, multiplexer setup and possibly filter and amplifier settings must be accounted for. In present-day systems it is often not possible to exploit all the features of a hardware device, because it would be too tedious to program and impossible to master all potential side-effects. Consider a 16 channel A/D-converter with the possibility to set parameters together with the number of the channel one wishes to convert. Filter and amplifier settings must be issued ahead of the conversion to allow for signal stabilization, i.e., the current setting is influenced by the previous one, introducing dependencies between channels not otherwise related. In practice, this situation is often avoided by using the same settings for all 16 channels and enabling the round-robin conversion option many devices offer.

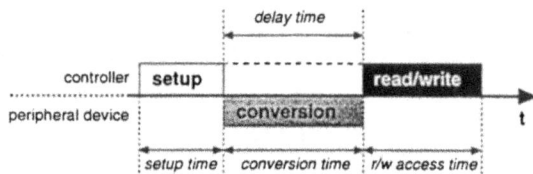

Fig. 6. Device access sequence

Selective specific multi-stage accesses combined with delays as shown in Fig. 6. offer a potential for increasing performance, with the conversion delay unfortunately presenting a dilemma, because the required delay times usually are too short to justify rescheduling at the task level. These delays thus remain periods where the processor idles until the next read or write operation is allowed.

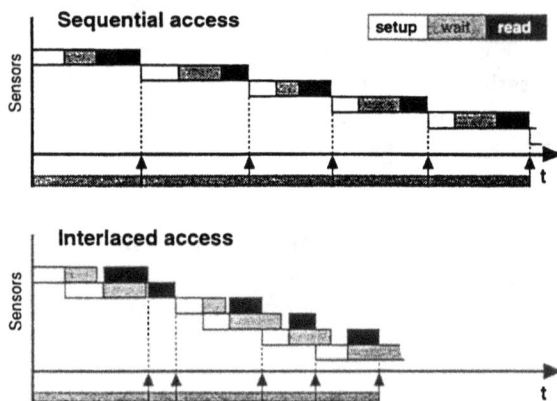

Fig. 7. Optimizing device accesses

Fig. 7. illustrates the benefits of an alternative solution, where the access routines of several devices are interleaved and the processor idle time minimized. This clearly is not a convenient method for manual implementations, but fortunately the affinity to general scheduling problems make it suitable for code generation.

Input. Next to minimizing absolute processor loads the objective of an input system is to distribute work loads evenly among all invocations, a demand even more difficult to fulfill manually. The input task is itself a collection of tasks with precedence and exclusion relations. A feasible schedule is found statically and leads to a single sequential structure. The scheduler algorithm eliminates possible idle times and ensures practically equal processor loads for each invocation.

A textual specification containing the timing requirements for all input signals expressed in terms of the highest timing requirement must be given. Only periodic inputs can be dealt with, and it is desirable to reduce the different timing requirements to just a few frequency classes (e.g. F_{high}, F_{med}, F_{low}). Low level functions (or macros) for setup, access, conversion, verifying and filtering along with their relevant profile data (execution times and required delay periods) that were measured previously on the target are also required. The schedule, consisting of a switch-statement with F_{high}/F_{low} cases, is then generated heuristically. The structure of the specification follows the hardware architecture and therefore also serves as an adequate and up-to-date documentation. In the created schedule the access time of one signal is used to guarantee the necessary delay times for another signal, according to Fig. 7. If no more access routines are available in a slot or where they cannot be used because of exclusion relations (when signals must be acquired by the same input device) delays are inserted into the schedule. Unsatisfactory schedules and wasted processor time reflect deficiencies at the hardware architecture level.

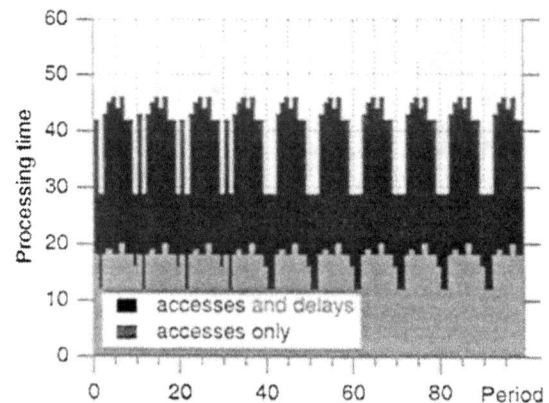

Fig. 8. Processor utilization 42%
(15 input signals, class high: 2, med: 9, low: 4)

In the example in Fig. 8. (with a frequency class relation $F_{high} : F_{med} : F_{low}$ of 1 : 10 : 10) a small collection of input signals with unfavorable exclusion relations causes the accumulation of delay time, resulting in very low processor utilization (42%). A more advantageous setting with five times as many signals including the original ones makes the required processing time rise a mere 30%, practically eliminating the idle delay time. The example also shows that the input task presents a very uniform processor load.

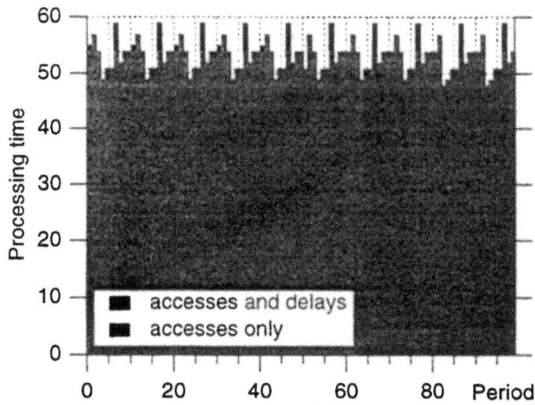

Fig. 9. Processor utilization 99%
(75 input signals, class high: 8, med: 39, low: 28)

Event extraction. Events required to drive the functional reactive block will be extracted at run-time from the newest input data. To generate the tables used by the extractor algorithm another specification must be supplied which contains the conditions for each event. For discrete signals this implies binary patterns, for events buried within continuous signals limit values with an added hysteresis are used. If events shall be extracted according to priorities at run-time, an arbitrary number of priority lists can be supplied in the specification. The currently active scheme will be chosen from within the functional model, where all information regarding system status and desired behavior is available.

Output. Contrary to the input, system output load cannot be distributed statically, as the number of outputs depends on the specific transition structures encountered in a cascade as well as on the occurrence of multi-stage hardware accesses. Stubs for calling the conversion and/or expansion routines from the functional model are generated using the same hardware-oriented specification as for the input schedules, the profile data in this case also including information on access expansions if required. With this information together with the action sets from the functional model (see end of section 2.1.) upper bounds on the processing demands of the output task can be found. A static table with the relevant values is computed to be used by the run-time scheduler.

3. IMPLEMENTATION

The code generators yield code that can be implemented as three distinct tasks either together on a single processor or separately on different processing entities. On a single processor they operate in sequence in every period. The input task with its uniform execution time on each invocation presents a constant load, while the time demands for the processing and the output tasks depend on the occurring events and hence the cascades that must be run through. The performance measurements are used to check if the code can run on a certain target and they will also be needed for run-time decisions.

3.1. Performance analysis

Processing time is present neither in the functional model nor in the specifications for the connector. It is only by examination of the generated code that the performance on a specific target can be evaluated. Because this code is deterministic, the procedure gives the necessary assurance that an implementation does perform according to the requirements. If bottlenecks are detected, a way out can be sought by distributing the work load onto different processors, or by softening the requirements on some events. In this case the measured data together with the event priorities permit quantitative statements regarding the performance degradation.

The time demands for processing the events in the motor controller example in Fig. 10. show that aside from the periodic controllers which are triggered by the *sample* event and must execute first in every period, only a single other event can be handled during the same period, if the worst case paths are taken. Considering the time scale and the improbability of these events arriving at the same time, this may still be acceptable.

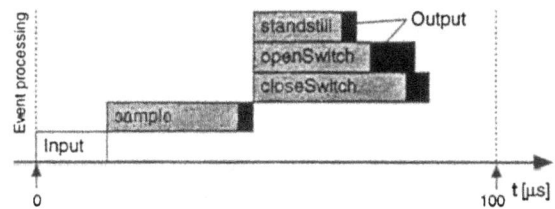

Fig. 10. Motor controller processor time demands

3.2. Run-time scheduling

When the input task yields the processor, the arrival of a new set of input data in the processing task is interpreted as a time event and used by the event extractor to generate the *sample* event for those quasi-continuous processes which run at the highest frequency available in this system. The processing task keeps track of the current time

145

and also records the time demands of all pending outputs, i.e., the ones associated with ongoing event processing. Whenever a new event has been extracted, the expected processing time from the table is compared to the remaining time balance of the current period. If it conforms, the event is fed to the reactive block and the procedure repeated until an event occurs whose execution would not fit any more. It is important to note that in this case the event is withdrawn and not put in a queue, because this particular event may not be present in the input data the next time the extractor runs (this characteristic, combined with sufficiently low sample frequencies can also be used as low-pass filter, e.g., for input key debouncing). Not having a queue for events at the same time effectively prevents processor overloads and is also a prerequisite for switching between different event priority lists.

In the period shown in Fig. 11. processing for the event *standstill* would not fit into the remaining time budget in the current period and is therefore rejected. This type of scheduling is achieved with practically no overhead and it is worthwhile because worst-case execution times are hardly ever exhausted, so that in reality more events can be dealt with than predicted by the static analysis. The slack time nevertheless remaining can be exploited by non real-time tasks, which themselves must be preemptable.

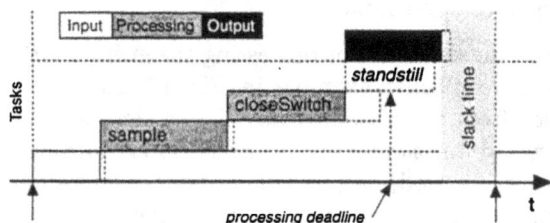

Fig. 11. Task schedule

Output from the processing task is fed to buffers to allow for possible asynchronicities of the output system. The majority of output signals results in a single hardware access, but sometimes expansion to several accesses with time lags in between may be necessary. For this reason two different buffers are provided that may or may not be located on the same processing entity. In the case of a simple output signal, a plain FIFO buffer is used, while for multi-stage accesses an ordered queue with relative delays is necessary. These buffers are read in a round-robin fashion, either until they are empty, a new timer tick occurs or, in case of the second buffer, a first element is encountered with a delay time that has not yet elapsed.

Jitter. Algorithms for quasi-continuous periodic control tasks are sensitive to time variations concerning their input data. This input 'jitter' is usually difficult to control with priority-based or even preemptive schedulers. Because a controller's input data is sampled at precisely the same instant in each invocation, and because the periodic *sample* events are always processed first, jitter is reduced to inaccuracies inherent in the clock hardware (if no other interrupts interfere). Asynchronous events on the other hand may have tight response requirements, but are insensitive to jitter.

4. SUMMARY

Introducing parallel entities into the design of control systems is a well established method when the focus is on specific aspects in a complicated system. While the idea of parallelism is helpful for disentangling complex problems in a conceptual stage, there is hardly any reason for the common practice to implement such a solution using different tasks when they all run on a single processing entity. If the conceptual parallelism cannot be exploited by parallel hardware, it is advantageous to replace it with a single corresponding sequential task. While this may not be easy or even feasible for a set of parallel tasks written in a programming language, it poses no problems for a code generator. Input and output generally use common hardware resources that often impose sequential use. With specifications and code generators efficient solutions for hardware access become possible. Provided that deterministic code results from these efforts, what remains to be done is verifying that the available processing time is sufficient; there is no need for scheduling more than necessary!

The presented approach has been shown to scale well in a number of real-world applications, some with more than 300 I/O signals, sub-ms response requirements and distributed implementations.

5. REFERENCES

CIP Tool - User Manual (1995-99). *CIP System AG.* Solothurn.

Fierz, H. (1999). *The CIP Method: Component- and Model-based Construction of Embedded Systems.* Submitted to European Software Engineering Conference ESEC'99.

Liu, C. L. and J. W. Layland (1973).*Scheduling Algorithms for Multiprogramming in a Hard-Real-Time Environment.* Journal of the ACM, 20:46-61.

Trutmann, HO. (1998). *Generation of Embedded Control Systems.* WRTP'98. Shantou.

Xu, J. and D. L. Parnas (1993). *On Satisfying Timing Constraints in Hard-Real-Time Systems.* IEEE Transactions on Software Engineering 19(1):70-84.

HARD REAL TIME CONTROLLER: A CASE STUDY

Krzysztof Sacha [*,1]

* Warsaw University of Technology, Warsaw, Poland

Abstract: The paper describes hardware and software architecture of a multiple computer control system of a very fast robot. The impact of the requirement for very fast operation on the system architecture is discussed. The control system was designed for research and educational purposes with two goals in mind. First, it had to be flexible and modifiable with respect to control algorithms applied to particular axes. Second, the interfacing of the system to additional devices, such as environmental sensors and other robots, should be possible with only limited effort. The various constraints and considerations imposed by these goals are examined. *Copyright © 1999 IFAC*

Keywords: Hard real time control, robotic controller, multiprocessor system, synchronization, operating systems

1. INTRODUCTION

Real time system is usually defined as a system in which computation is performed in parallel with an external process, in such a way that the results of the computation can be used to control or to respond timely to the process (*IEEE Std 610, Standard Computer Dictionary*, 1990). Two different classes of systems fall into the scope of the above definition. A *soft real time* system should on average respond timely but occasional delays are acceptable. Examples of this type of operation are interactive reservation systems and bank supporting business applications. A *hard real time* system must always respond in time and cannot cease operation even for a moment. This type of operation is characteristic for embedded control applications.

Hard real time constraints imposed on an embedded system typically result from the physical laws governing the technical process being controlled. Any violation of the constraints can disturb or destabilize the process — late computer reactions can be either useless or dangerous. In other words,

the costs for missing deadlines in hard real time environment are very (infinitely) high. Therefore, the overall system architecture must be adapted in order to meet the time requirements. The scope of the adaptation depends on the relation between the speed of the controlled process and the speed of the computer system. If the system is fast in relation to the deadline then the standard software technology of a multitasking operating system can be used. If this is not the case, i.e. the deadlines are short in relation to the required computation time, than specialized hardware and software architectures have to be used.

The goal of this paper is to show the limitations of the current low cost computer technology and the measures which can be applied to adapt the system architecture to hard real time constraints. The "low cost technology" means here popular industrial computer kits offered by many suppliers and based on Motorola 68060/50MHz processors. The technical process under control is the movement of an experimental robot which has been designed at the Warsaw University of Technology.

The mechanical part of the robot consists of a six degrees of freedom robot column (Nazarczuk *et al.*, 1995) driven by three powerful high torque

[1] Partially supported by a Warsaw University of Technology statutory grant 504,036, 1999

direct drive motors and three much smaller conventional alternate current motors. The range of the column movement is unlimited in that the particular axes can move around, making more than a full revolution. The robot movement is very fast and it ranges up to 2 revolutions per second. The resolution of the motor positioning is high and ranges 500 000 distinct points per a single axis revolution.

The robot is designed mostly for research and educational purposes. The users of the device are likely to play with a variety of control algorithms for particular axes. Hence, an additional requirement has been stated, that the structure of the robot control system should be very flexible and a modification of the control algorithm should be possible with only limited effort.

2. SYSTEM ARCHITECTURE

The logical structure of the robot controller (Zielinski, 1997) is layered as shown in Figure 1. The lowest layer tasks, called SERVO, implement control algorithms, such as e.g. PID, of particular axes. Axis control can be based on setting the desired velocity or torque of the motor. Current arm position is fed back to the controller as counts from the motor encoder or resolver. Other status data can also be read from the motor driver. The next layer task, called MASTER, coordinates the robot movement, i.e. it decomposes a movement of the robot column into the individual steps of particular robot axes and performs coordinate transformation. The set-points for the entire robot movement can be issued by the user via a graphical user interface, or by an application.

Typical repetition cycles of the computation within particular layers, documented in a survey paper (Topper, 1991) are about 10ms for the segment generation and about 1ms for servo controllers. In our case, however, an analysis of the robot column dynamics revealed, that much higher frequency of computation is required to take advantage of the possible speed and accuracy of the robot motors. The calculation of SERVO control tasks has to be repeated once per $200\mu s$, while the upper layer segment planning MASTER task should have the repetition cycle of 2ms.

Other tasks can also be performed by the robot control system. In particular, the tasks for communication with advanced environmental sensors, such as video cameras, and for supervisory control of a multirobot workcell. These additional tasks can be executed either in the background of the basic robot control activity in a multitasking environment, or can be shifted to additional computers in a distributed environment. The latter

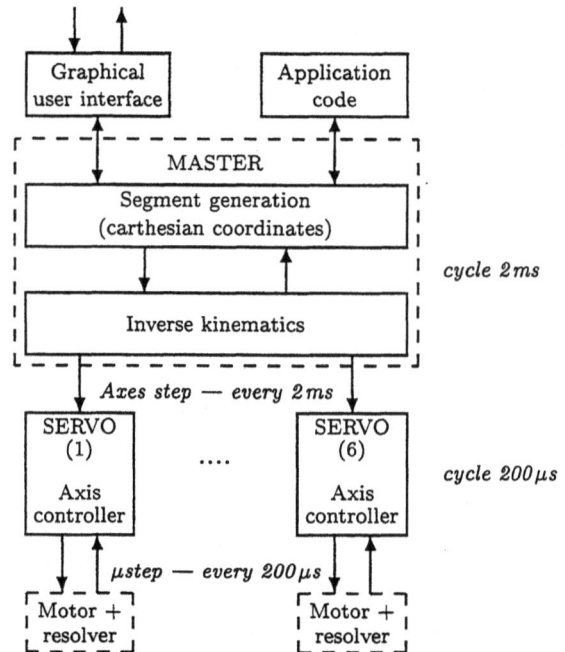

Fig. 1. Control system architecture

solution is more complex and slightly more expensive, but can offer better performance and more modifiable system and application structure.

The requirements on the robot control system, which can be derived from the general objectives of the project stated above, are for:

- high performance, as a very short repetition cycle of the axes controllers is needed,
- high reliability, as the robot can be dangerous for potential users,
- high modifiability, as any part of the software can be subject to changes.

The requirement for high reliability eliminates the use of popular PC-type computers. Economical limitations and the requirement for easy modification eliminate the use of DSP and transputer systems. What remains is an architecture composed of universal microprocessors of a modular industrial computer kit.

The required performance of the control system is very high. The computational characteristics of MASTER and SERVO tasks are different: MASTER computes the kinematic model of the robot and executes a lot of trigonometric functions, while SERVO tasks implement control algorithms and use mostly simple arithmetic operations. This suggests the need for multiple processor architecture with two different types of processors used for running different tasks of the system.

The computer hardware selected for the development of the robot control system has been based on an industrial computer kit from PEP Modular Computers built around VME bus. The hardware is able to work in a multiprocessor configura-

Table 1. Characteristics of the processors

Type of operation	486/66MHz [Mops/s]	68060/50MHz [Mops/s]
Fixed point	3.1	3.6
Floating point	1.6	2.7
Trigonometric	0.3	0.2

Table 2. Characteristics of the operating systems

QNX		OS-9	
Operation	Duration	Operation	Duration
task switch	$6\mu s$	task switch	$5\mu s$
signal	$45\mu s$	signal	$20\mu s$
proxy	$25\mu s$	event	$15\mu s$
semaphore	$15\mu s$	semaphore	$14\mu s$
message	$20\mu s$	interrupt	$18\mu s$

tion composed of Intel 486 and Motorola 68060 processor boards. Basic characteristics of these processors, published in commercial papers, were not sufficient for comparison. Hence, a simple experiment was performed, in which the particular arithmetic operations were executed in long loops. The number of loop cycles was counted and the duration of the loop was measured. The results of the measurements are shown in Table 1. The experiment proved that 486 could better perform in MASTER layer, while 68060 in SERVO layer.

The selected hardware of the robot control system consists of a single Intel 486 processor board and two Motorola 68060 processors boards, connected to VME bus (Figure 2) and mounted within a single rack, together with nearly fifteen interface circuit boards. The processors can communicate with each other through dual-ported ram areas and can exchange interrupt signals. Both 68060 processors act as axes controllers and execute SERVO control tasks, while 486 performs the supervisory MASTER task and provides a graphical user interface. Moreover, the supervisory processor can cooperate through a network with external computers, thus creating a distributed control system with enormous power and flexibility.

To prevent traffic jams on the communication system within the robot controller, a dual bus architecture has been selected in which the interface boards are coupled to local buses of particular processors. VME bus is used as the main system bus for inter-processor communication only. Relatively low actual load on the main bus preserves the potential for future system expansion, as additional robot-related sensors, not assigned to particular axes, can be interfaced to the supervisory processor through VME bus. This is the only possibility for such an expansion, as no local buses for 486 board are offered.

An interface to a motor driver (Kreglewska and Sacha, 1997) consists of an analog torque or speed setpoint signal, quadrature wave from relative resolver and a number of bi-stable status and configuration signals. An external synchronization sensor supplements each axis hardware.

3. SOFTWARE ARCHITECTURE

The operation of Intel 486 is controlled by QNX, a multitasking distributed real-time operating system from QNX Systems Software (*QNX System*

Architecture, 1992). The operation of 68060 is controlled by OS-9, a multitasking real-time operating system from MicroWare (*OS-9 Technical Manual*, 1993). The systems differ from each other in many aspects and cannot cooperate directly using standard tools for intertask communication. The performance characteristics of both systems are, however, similar. The execution times of several system functions, partially measured (Sacha, 1995) and partially found in the literature (Dibble, 1994), are listed in Table 2.

Comparing the times in Table 2 to the requirements of the robotic application one can note that the cycle time of 2ms, required within the layer of the robot coordinator, is not particularly demanding. A 0.5ms real-time clock resolution of QNX is perfectly sufficient. Multitasking capabilities of the operating system can be applied and a number of low priority background tasks, e.g. for network communication, can be executed in addition to the high priority MASTER robot coordinator. This way additional sensors or workcell controllers can easily be interfaced to the robot controller as remote network stations.

Much shorter cycle time of $200\mu s$ within the axes control layer imposes very hard real-time constraints on the controller operation. The requirements exceed, in fact, the real capabilities of the operating system. First, the highest possible resolution of the real-time clock of OS-9 equals 1ms. Hence, no standard timer functions of OS-9 can be used. To overcame the problem the activities of SERVO tasks are driven by an external interrupt signal, generated by an additional timer. The interrupt is requested periodically with the nominal cycle time equal to $200\mu s$. Unfortunately, the latency time introduced by the operating system results in relatively low stability of the interrupt based timing: the deviation of the length of the cycle ranges up to $30\mu s$ and must be compensated by an algorithm within the axis controller.

Timing was not the only problem encountered in designing the software. Axes controllers (SERVO tasks in Figure 1) are basically independent of each other, as the coordination between the axes is ensured by the supervisory MASTER task. Hence, the most flexible and modular structure of the 68060 software could consist of 3 separate tasks, each of which could control a single robot axis. The execution of these tasks could be triggered by

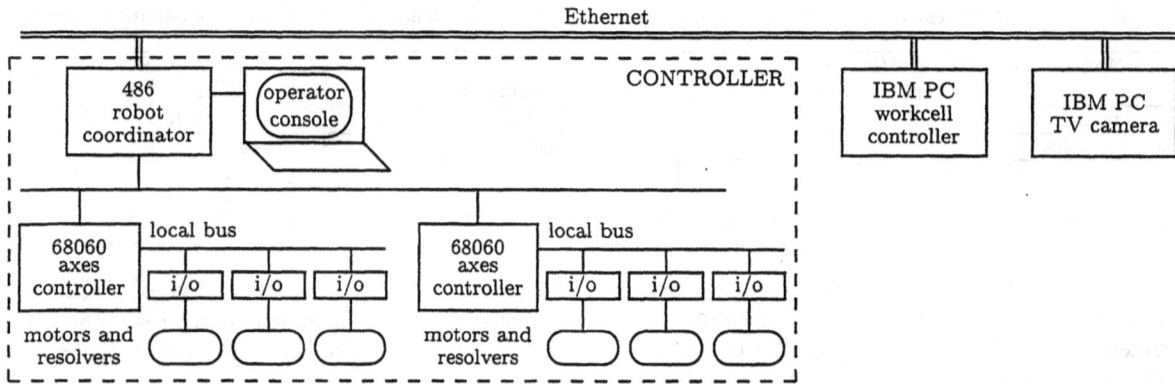

Fig. 2. Hardware structure of the fast robot control system

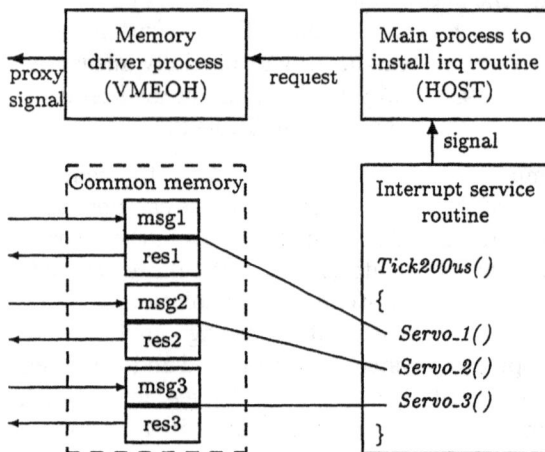

Fig. 3. Software structure of an axis controller

a standard OS-9 signal raised periodically by the timer interrupt service routine (200μs). A quick look to Table 2 shows, however, that housekeeping functions, like: interrupt servicing, generation of 3 signals and 3 task switches would take 93μs, i.e. nearly 50% of the entire repetition cycle!

To solve the problem, a variant of a two-level, foreground-background software architecture has been developed (Figure 3). Control algorithms (SERVOs) of three axes are executed sequentially within an interrupt service routine. Other functions are executed as background tasks at the process level. The execution of the interrupt service routine is triggered cyclically by a timer device. The advantage of this architecture is very low overhead related to a single interrupt servicing (18mus). Disadvantages are: difficult programming and debugging and decreased modularity and modifiability.

Basic form of the communication between MASTER module (executed by 486) and SERVO modules (executed by 68060s) is a periodic polling action executed by MASTER (Figure 3). Each polling cycle conveys a command to selected — in most cases all — SERVO modules, and reads back a status data which describes the current state of the axes controllers. The exchange of data is ac-

complished through designated areas in a common memory. The synchronization of data access uses simple flag-based handshaking principle.

Polling implements regular exchange of messages between the cooperating software modules, but gives no convenient means for mutual synchronization which can be needed in the following cases:

- MASTER process must wait for the end of a command executed by SERVO modules,
- SERVO modules may need to inform MASTER process of exceptional states (e.g. faults),
- MASTER process may need to inform HOST or SERVO modules of some exceptional actions (e.g. end of the system operation).

A set of synchronization functions is implemented by the common memory drivers installed in all processors of the system. The following functions are available: signals passed form SERVO modules to HOST process, signals and proxy triggers passed from HOST to MASTER, and exception commands passed from MASTER to SERVO.

4. IMPLEMENTATION

The support for multiprocessing implemented by the hardware comprises common memory access and inter-processor interrupt capabilities. Common memory subsystem is composed of dual ported RAM areas installed on the processor boards. A particular dual ported RAM area can be accessed by the local processor through the local bus as well as by any other processor of the system through VME bus. Inter-processor interrupt feature is also based on the implementation of the dual ported RAM. Special circuits generate an interrupt signal to the local processor when a specified dual ported RAM location is accessed.

Unfortunately, hardware implementation of the common memory subsystem in this computer is faulty, in that an attempt for simultaneous access of 486 and 68060 processors to the dual-ported

RAM areas of their counterparts can lead to a deadlock. If this occurs, both processors remain blocked until a timeout is raised by a VME controller. The length of the timeout — $85\mu s$ — is too long for the robot controller.

To avoid the possibility of deadlock only dual ported RAM of 486 processor is used. This creates no problems as far as a data exchange is considered. Worse, that also inter-processor interrupt requests to 68060 processors cannot be used. A solution to this problem is, however, simple. Very fast ($200\mu s$) repetition cycle of SERVO modules makes interrupts unnecessary. The requests from MASTER to SERVO modules for exceptional actions can be deposited in the common memory area, and retrieved and executed by SERVO modules in the nearest cycle. Commands to HOST process can be passed this way as well.

Another drawback of the hardware is that no provision for bus or common memory reservation is available. Hence, the mutual exclusion of processors must be implemented entirely by the software.

The instruction list of Motorola processors includes TAS instruction which implements an indivisible Test-and-Set operation. Intel processor has no such an instruction. The implementation of indivisible operations, which has been developed within this project is original. A common resource is granted to a processor on request and on a priority basis.

Basic data structure used for synchronization is a four byte semaphore. Particular bytes correspond to particular processors of the system. A processor requests access by setting the byte to 0x80 and releases by resetting the byte to 0x00. The access is granted when no higher priority request exists. Programs of the semaphore operations PSem (request-and-wait-for-grant) and VSem (release) together with an explanation of data structure are given in Figure 4.

The correctness of the operations PSem and VSem can be proved e.g. using Petri nets. A simple model of the processor synchronization is shown in Figure 5. The initial marking is such that the tokens reside in places: a, u, $s0=0$ and $s1=0$. The synchronization is correct if a marking in which tokens reside in both places e and y is not covered by the net. This can easily be proved using the technique of reachability tree analysis.

The requirements specification of the robot control system includes a requirement for high flexibility and modifiability. The intention of the project is that neither hardware nor software architecture is constant within the whole system lifecycle. At the hardware level this means that the number of processors can vary, and an additional

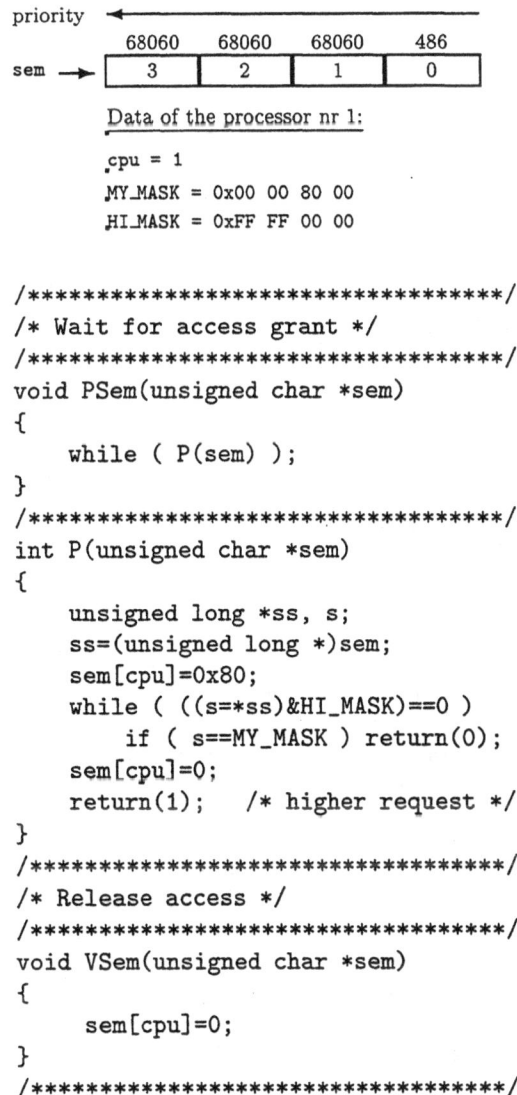

```
priority  ◄─────────────────────────
          68060   68060   68060   486
sem →    │  3  │  2  │  1  │  0  │
         ──────────────────────────
         Data of the processor nr 1:

         cpu = 1
         MY_MASK = 0x00 00 80 00
         HI_MASK = 0xFF FF 00 00
```

```
/********************************/
/* Wait for access grant */
/********************************/
void PSem(unsigned char *sem)
{
     while ( P(sem) );
}
/********************************/
int P(unsigned char *sem)
{
     unsigned long *ss, s;
     ss=(unsigned long *)sem;
     sem[cpu]=0x80;
     while ( ((s=*ss)&HI_MASK)==0 )
          if ( s==MY_MASK ) return(0);
     sem[cpu]=0;
     return(1);    /* higher request */
}
/********************************/
/* Release access */
/********************************/
void VSem(unsigned char *sem)
{
     sem[cpu]=0;
}
/********************************/
```

Fig. 4. Active semaphore implementation

68060 can be attached to give more computational power for the axes controllers. At the software level this means that particular axis controllers (SERVOs in Figure 1) can be shifted and assigned to different processors within the control system — in other words, the code modules which compute axes control algorithms can be shifted from one processor to another one. The contents of those code modules can also be changed.

Any change in the assignment of axes to processors affects the communication between the robot coordinator and the axes control processors. If the low level communication mechanisms (common memory areas and inter-processor interrupts) were used directly, the impact of such changes on the structure of the robot coordinator as well as axes control software could be destructive, in that any change could require a huge number of small modifications to the software code.

The problem of changes identified above has been solved by implementing high level tools for mes-

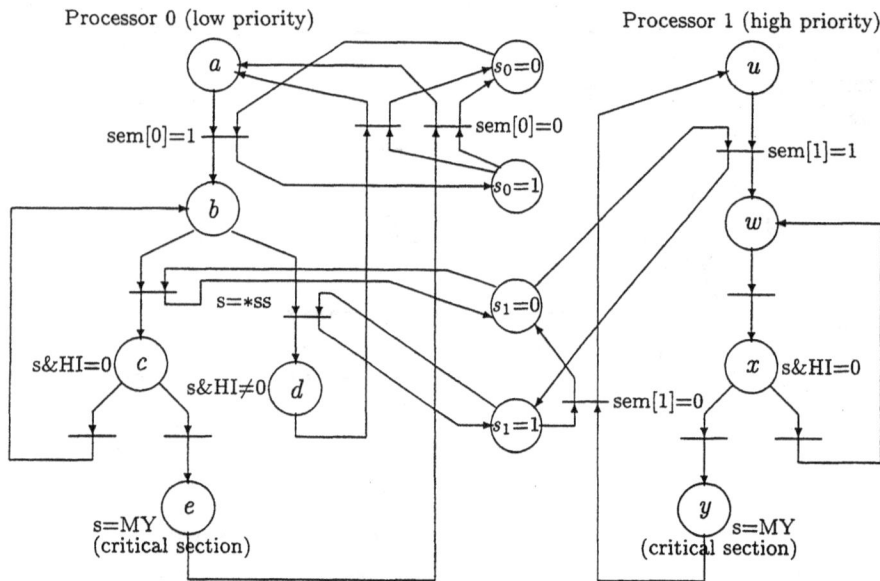

Fig. 5. The cooperation of two processors modeled by a Petri net

sage and signal passing, with the potential for redirection following any reconfiguration of the hardware and software structure. Such a design makes the application software of the robot coordinator (486/QNX) independent of the actual structure of the axes control layer.

The implementation is based on a concept of a port, which can be interpreted as a communication channel between two cooperating software modules. Ports are identified by numbers, and described in a port table stored in the common memory area. A port description consists of a port number, a processor identifier, a task identifier and a poll area address. The port table is filled in during system start-up and the assignment of ports to tasks is constant.

The communication between tasks executed by different processors is implemented by common memory drivers, installed in all processors of the system. A signal, a command or a data message is always sent to the partner through an appropriate port. This way it can reach the destination regardless of the actual placement of the modules within the robot control system.

5. CONCLUSIONS

Controlling a high resolution fast robot is a challenging task which nearly reaches the limits of the current microcomputer technology. Meeting the hard real time constraints requires hardware architecture with multiple processing units and multiple buses for internal data transfer within the control system. Software architecture at the axes control level cannot benefit the advantages of a multitasking operating system, but must be tailored to meet severe real time limitations.

The system described in this paper is currently in the final stage of the development. The computer hardware has been assembled and debugged, interprocessor communication has been implemented, basic software structures and control algorithms have been fixed and prototyped. A complete working robot system is expected to come into the final test phase in summer 1999.

6. REFERENCES

Dibble, P. (1994). *OS-9 Insights, An Advanced Programmers Guide to OS-9*. Microware Systems Corporation. Des Moines.

IEEE Std 610, Standard Computer Dictionary (1990).

Kreglewska, U. and K. Sacha (1997). Versatile control system for a very fast robot. *Proc. 4th International Symposium on Methods and Models in Automation and Robotics "MMAR'97"* pp. 1145–1150.

Nazarczuk, K., K. Mianowski, A. Oledzki and C. Rzymkowski (1995). Experimental investigation of the robots arm with serial-parallel structure. *Proc. IX World Congress on the Theory of Machines and Mechanisms* pp. 2112–2116.

OS-9 Technical Manual (1993). Microware Systems Corporation. Des Moines.

QNX System Architecture (1992). Quantum Software Systems. Kanata.

Sacha, K. (1995). Measuring the real-time operating system performance. *Proc. 7th Euromicro Workshop on Real-Time Systems* pp. 34–40.

Topper, A. (1991). A computing architecture for a multiple robot controller.

Zielinski, C. (1997). Object-oriented robot programming. *Robotica* **15**, 41–48.

Real-time Systems for Mobile Robotic Applications based on a Behavioural Model

F. Buendía, H. Hassan, J. Simó, A. Crespo

Departamento de Informática de Sistemas y Computadores
Universidad Politécnica de Valencia
{fbuendia, husein, jsimo, alfons}@disca.upv.es

Abstract: Behavioural models have been widely used to represent mobile robotic systems which operate in uncertain dynamic environments and combine information from several sensory sources. Previous behaviour-based architectures used deliberative techniques to tackle these questions but they generally ignored the timing analysis of these systems. This paper proposes the use of a real-time task model based on the concept of transactions representing the distributed processing of the behavioural entities. Several methods are studied to check the timing requirements associated with behavioural computations. Techniques from schedulability analysis in distributed systems are applied when the timing requirements are based on the computation of task response time. Alternative methods based on simulation tools are used in the case of more complex requirements, such as those that consider the relationship between transactions executed at different frequencies. *Copyright © 1999 IFAC*

Keywords: Mobile robot, Behavioural model, Real-Time system, Schedulability analysis, Simulation tools.

1 INTRODUCTION

The main goals of this work[*] are firstly; to describe a model of real-time tasks that is powerful enough to represent the functional and temporal features of mobile robotic applications based on a behavioural model, and secondly; to apply techniques for checking their timing requirements.

Behavioural models have been widely used to represent advanced robotic systems operating in uncertain dynamic environments, combining information from several sensory sources, and with different vehicle dynamics. Prior knowledge of the domain may be incomplete, and in a dynamic environment, reasoning must be deliberative and fast enough to respond to unexpected events. Also, the information gained via sensing is incomplete, inaccurate, and uncertain. To function correctly in these kinds of environments, the planning systems must be reactive and take into account information about the current state at regular time intervals. Therefore, mobile robots must combine deliberative goal-oriented planning with reactive sensor-driven operations.

Early mobile robots such as the Stanford Cart (Moravec, 1990) performed *centralised sensor fusion* by gathering all available sensory data, creating a representation of the environment and generating a plan. To overcome the limitations of centralised architectures and to permit higher responsiveness and more accurate plans, *hierarchical architectures* were developed. This type of architecture organises the functions of mobile robots into several levels – the low levels providing reactive operations and the higher levels being based on deliberative planning procedures. This hierarchical structure helps to handle control loops operating at different frequencies and provides an easy way to check timing requirements. Nevertheless, mobile robot applications need to be adaptive and autonomous, and therefore, to perform perceptual activities with more complex timing requirements.

A new generation of *behaviour based architectures* appeared in order to provide more flexibility (Brooks, 1986). The main component of this architecture is the behavioural entity – which is responsible for only a very narrow portion of robot control. Each behavioural entity only receives the information specifically needed for its task, avoiding the need for centralised sensor fusion. A set of behaviour entities, each operating at its own rate, is in charge of controlling the mobile robot. They represent activities such as obstacle avoidance or path following. Some works, such as the proposal of Hughes and Damn (Rosenblatt, 1997), offer the symbiosis of

---

[*] This work has been partially funded by a grant from the Spanish government CICYT TAP94-0511-C02, and the Generalitat Valenciana GV-C-CN-04-059-96

hierarchical and behaviour architectures. However, they do not include techniques to perform the temporal analysis of the mobile robot requirements. Architecture being developed (Benet, 1998) allows the designer to represent the mobile robot system from a behavioural model viewpoint, but it also provides temporal analysis techniques to check timing requirements.

These requirements can be classified according to the hardness of the timing constraints. Thus, there are requirements related to low-level reactions (e.g. detecting and avoiding an obstacle) that must be strictly guaranteed, and others related to high-level deliberative procedures (e.g. exploring the visual references of the environment) which do not require strict guarantees. The reactive level can also handle hierarchical control loops operating at different frequencies (Butaccio, 1996). For example, the frequency of a joint position servo process can be carried out within a period of 1 ms; while a vision process may be executed within a period of 100 ms or more.

Different behavioural and other system entities must be mapped to a task model to check timing requirements. This task model will provide a representation for the execution of the robotic functions on a given computational platform. The most basic model consists of independent periodic and aperiodic tasks which are executed on a single processor. However, current robotic architectures involve the distribution of processing and the use of communication networks (Benet, 1998). This calls for the use of alternative task models. One of these alternatives is based on the concept of *transaction* (Tindell, 1992) – which is defined as a sequence of tasks which can be scheduled and executed on different processors. These component tasks can also be executed at different levels of the robot control hierarchy: ranging from low-level sensory operations to complex tasks based on high-level information coming from lower levels.

Therefore, a mobile robotic application can be implemented as a set of transactions T_i , each of which is composed of tasks. Each task t_{ij} is identified with two subscripts: the first identifies the transaction to which it belongs, and the second shows the position that the task occupies within the transaction. In this way, t_{ij} is the j-th task of transaction T_i. Each task is activated when a relative time (called offset and labelled as o_{ij}) elapses after the occurrence of some event that starts the transaction. Each activation of a task causes the execution of one instance of that task. Depending on the type of event that activates the successive instances of the first task of each transaction, the component tasks can be defined as periodic (period P_i) or aperiodic tasks. Another attribute of a task is

its execution time, called e_{ij}. This can vary in the range $[e_{ij}^-, e_{ij}^+]$, where e_{ij}^- is the best-case execution time, and e_{ij}^+ is the worst-case execution time – and $e_{ij}^- \leq e_{ij}^+$. When a set of such tasks is executed, the response time of a certain task can be defined as the difference between its completion time and the instant at which the associated event starts the transaction to which it belongs. The *worst-case response time* is called R_{ij} a is obtained by computing the response time from the multiple combinations of task phasing times and selecting the higher value. Each task may have an associated global deadline, D_{ij}, which is also relative to the arrival of the transaction event. The most common timing property consists in comparing the value D_{ij} of the final task of a transaction with its R_{ij}. This property can be used to check *end-to-end requirements* – such as those involved in the detection of an obstacle.

This type of requirement is usually checked through techniques of schedulability analysis, originally applied to monoprocessor systems (Lehochky, 1990). (Tindell, 1994) proposed applying the analysis to sets of tasks (transactions) executed in distributed systems. A similar method is used by (Sun, 1995) to compute the response time of these tasks with several synchronisation protocols. However, this solution was restricted to offsets and deadlines smaller than the task periods. A more general technique proposed by (Palencia, 1998) uses the method of time-offsets of (Tindell, 1992) to reduce the pessimism of the original analysis (Tindell, 1994) and extends it to consider dynamic offsets and jitters based on these offsets.

But the end-to-end timing properties based on response time are not the only requirements that must be guaranteed in a mobile robotic system. There are other synchronisation relationships in the multiple processes that collaborate in a given behaviour, each one at its own rate, that must be considered. (Gerber, 1994) proposed some of them: such as *freshness* constraints which bound the time between an output task and the corresponding input task; *correlation* constraints that restrict the difference in sampling time between several inputs used to produce the same output; or *separation* constraints that limit the earliest and latest time between consecutive outputs. These properties are more difficult to analyse with schedulability techniques and alternative methods are applied, such as simulation tools (Storch, 1997), or the bounding of task completion times (Sun, 1996).

The following section describes the behavioural model used to represent the mobile robot system. Section 3 presents the real-time task model that supports the behavioural model. Section 4 introduces an application example to show how a mobile robot system can be represented and how the analysis of its timing requirements is performed. Finally, some conclusions are outlined.

2 BEHAVIOURAL MODEL

The model selected to represent mobile robotic systems is based on the concept of *behaviour*. A behavioural entity defines the knowledge of a specific domain of the robot control. A mobile robotic application is composed of a set of distributed behavioural entities that receive information from the environment through different sensors and send action commands to the controller devices by means of an *arbiter*. A scheme of this model is shown in Figure 1. Each behavioural entity runs completely independently and asynchronously. It uses sensor data, process algorithms, and provides commands to the arbiter at its own rate – and according to its own time constraints. The type of information represented by the commands include speed, trajectory, turn, etc. Each behaviour also has a motivation parameter reflecting importance in performing the control action. The arbiter is then responsible for combining the behaviour commands, taking into account their motivation weights, in order to generate the actions that achieve the objectives.

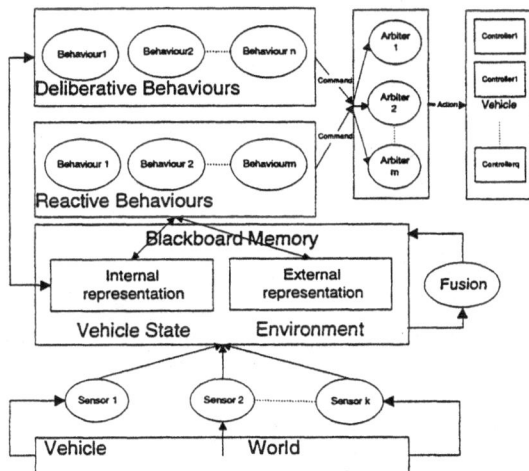

Figure 1.- Components of the behavioural model.

The elements represented in Figure 1 can be classified as passive or active entities. The first are stored in a blackboard memory (Botti, *et al.*, 1995) and they describe the following aspects:

- Vehicle state: The different sensory subsystems collect the odometer information in order to maintain the state of the vehicle. The state of the vehicle includes location, speed and the steering positions of the vehicle. The fusion module could obtain more refined state information.

- Environment: The sonar and the CCD sensory subsystems are responsible for maintaining a coherent representation of the environment in the memory. A map built from the sonar samples reflects the location of the objects around the vehicle. The images captured by the CCD device are processed to extract the visual characteristics of the environment.

The active elements of the model are the following processes:

- Behaviour: Composed of two types, reactive and deliberative. Both processes are related with a specific aspect of the robot. The former are basic algorithms that handle the sensory information and react in a basic, but secure, manner to the environment. Examples include speed control behaviour or obstacle avoidance behaviour. The second type of behavioural entities encapsulate more intelligent algorithms based on rules, neural networks, etc. They generate more precise commands but also take longer. This type of behaviour has the ability, by analysing the state and environment, to reconfigure its motivation.

- Arbiters: They are in charge of receiving commands and motivations from the different behavioural entities and evaluating which are the most important. The selected action is applied to the controller.

- Fusion: These processes are responsible for combining the data generated by the sensory subsystems, and subsequently stored in the blackboard memory, to generate more complex perceptual information.

- Sensors: They represent the driver modules that manage the physical sensor subsystems: the odometer, sonar, and vision using a CCD device.

An example that shows the interaction between the defined elements follows. This example is illustrated by Figure 2 and shows vehicle guidance through an unspecified path where there are some obstacles. The example considers two basic behaviour entities: *obstacle avoidance* and *path following*. Obstacle avoidance acquires the map of the environment and produces a safe trajectory. A fusion process combines the image and odometer information, in order to update the correct situation and trajectory of the vehicle. From the new situation, the path following behavioural entity produces the trajectory to follow. The path arbiter is responsible for deciding which of the two previous trajectories the vehicle must follow. For this decision, the arbiter takes into account the motivation of the behavioural entities. To avoid collision with any obstacle, the obstacle avoidance behaviour motivation is set as the highest.

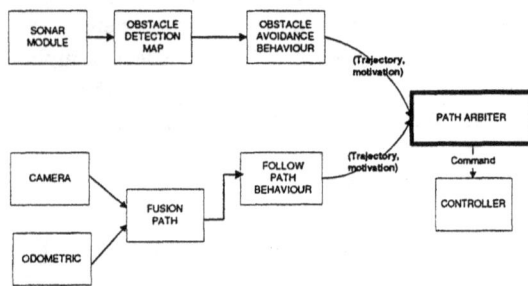

Figure 2.- Example of robotic application based on a behavioural model.

3 REAL-TIME TASK MODEL

Before checking the timing requirements that characterise a mobile robotic system, it is necessary to establish the relationship between the behavioural model entities and the model of real-time tasks defined in the Introduction. Periodic tasks are required when the entity being represented must sample (e.g. a sensor driver); combine information data (e.g. a fusion process); or to decide an action (e.g. an arbiter process). However, other elements, such as behavioural entities, can be modelled either using periodic tasks or alternatively, through aperiodic, or sporadic tasks – depending on the type of event that controls their activation. These non-periodic behavioural entities are generally related to high-level operations based on deliberative procedures. For example, a robot behaviour whose goal is to recognise a certain pattern when an object is detected. Other behaviour entities that are life-critical for the robot, such as obstacle avoidance or path following, will be implemented using periodic tasks.

Another aspect that defines the current robotic model is the distribution of entities over multiple processors linked by communication networks, or control buses. For example, the implementation of the sonar sensor subsystem proposed by (Benet, 1998) is based on three processors: the main controller; a sensor module linked to the main controller through a CAN bus; and a DSP linked via a PCI bus. Another example is the vision subsystem (Buendía, 1999) composed of a set of Pentium processors linked through an Ethernet network. This requires the distribution of tasks and the use of a model of dependent tasks such as the transaction model. The temporal analysis methods based in this task model allows the computation of the worst-case response time of a transaction related with a certain event. An event can be the periodic activation of the task that programs the sensor module in charge of capturing the sonar samples. The response time analysis can measure the time elapsed from the transaction activation to the completion of the

successive tasks. These tasks can include: the reception of information samples though the CAN bus; sending the data to the DSP for deconvolution and filtering; and reception of the processed information. The temporal cost of processing a sonar sample can be fixed at any moment of the considered transaction (equivalent to the worst-case) – and this can determine the period of the transaction and its deadline. Generally, the response time of a transaction can be variable. This depends on the task execution and phasing times and it is essential to compute the worst-case response time using techniques such as that proposed in (Tindell, 1994), (Palencia, 1998).

These techniques can be useful for computing the response time of operations produced by complex situations such as path following. This behavioural entity is based on the processing of visual information obtained through a CCD device and it considers two different transactions, each with their own frequencies. The first transaction controls the short-term trajectory based on areas close to the robot and it applies a low-cost correlation algorithm executed at the image capturing frequency. The second transaction uses global images to plan a long-term trajectory and it executes a sequence of tasks, from pre-processing basic operations to complex recognition algorithms, with variable execution times. The pre-processing operations are executed on the same processor as the first transaction, while a second processor is needed for the recognition algorithms. The computation of worst-case response times can help to determine the maximum time required to detect a change of trajectory.

An additional problem arises when there are several transactions associated with a given behavioural entity and each has its own period. For example, obstacle avoidance uses information from the sonar unit to build two types of obstacle maps, one local and another global. The local map can be determined by means of a periodic task that uses as input the DSP-filtered sonar samples and computes the accumulation of points, indicating the existence of close objects in a grid. The number of sample readings per scan can be variable and this situation can be represented using different period values for the local map task. If the number of samples varies, then the cost of interpolating and computing the local map will also be variable. This scenario presents timing requirements that cannot be evaluated using the response time analysis. Examples of these requirements are the maximum time between two consecutive local maps, or the maximum number of sample readings for computing a local map. These requirements can be evaluated using the end-to-end properties such as separation and correlation constraints.

4 APPLICATION EXAMPLE

The example of mobile robotic application shown in Figure 2, is now implemented using the real-time task model described previously. Table 1 displays the temporal attributes (measured in ms) of the main transactions and their component tasks.

Transact.	Task	Exec. time	Period	Dead-line
Image Acquisition	Image_capturing	2	100	100
	Image_copy	15		
Odometer Processing	Odomtr_process	1	10	10
Sonar Processing	Sonar_sensor	2	50	50
	Sonar_process	8-15		
Fusion	Fusion_process	5	50	50
Window Processing	Window_process	30	100	100
Global Processing	Sobel_process	25	300	300
	Thinn_process	10		
	Hough_process	20-100		
Arbiter Processing	Arbiter_process	5	75	100

Table 1.- Temporal attributes of the system example

Firstly, the behavioural model elements are described:

- Sensor processes are shown by the following tasks: *Image_capturing* (image data), *Odometer_process* (odometer data) and *Sonar_ sensor* (ultrasound data). The *Image_capturing* task is a part of the *Image Acquisition* transaction that completes its work by means of the *Image_copy* task which waits for the recording of the image.
- The Fusion process is represented by the *Fusion* transaction that assigns an estimated location coming from odometer data to each captured image.
- The Path following behaviour is based on image information and it is implemented by two transactions: *Window_process*ing (short-term trajectory) and *Global_Processing* (long-term trajectory).
- The Obstacle avoidance behaviour is implemented by the *Sonar_process* task that uses ultrasound information coming from the *Sonar sensor* task to build the local obstacle map. Both tasks form the *Sonar_ processing* transaction.
- The Arbiter process is represented by the task Arbiter_process that decides the control actions based on the path and obstacle behaviour.

The following shows the results of the evaluation of some timing requirements associated with the application example. All the experiments assume a fixed priority assignment based on a deadline monotonic strategy. Figure 3 shows the response time computation of the tasks belonging to transactions that produce information for the *path following behaviour*. This computation is based on three different techniques: an analytical method proposed in (Tindell, 1994) and whose results are labelled as TIN; (WDO) based on the results from (Palencia, 1998); and finally, a method using a simulation tool called DRTSS (Storch, 1997) to obtain the results labelled as SIM. Although the WDO worst-case response times are lower than the TIN results (especially in lower priority tasks such as those assigned to the Global_Processing transaction), both analytical methods perform worse than the simulation. This difference is due to the specific execution scenario that uses the simulation tool to obtain the response time results.

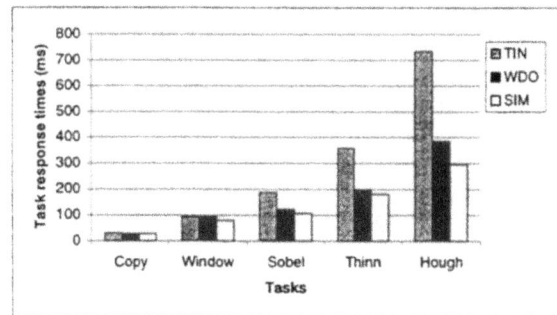

Figure 3.- Comparison between worst-case response time computations

Figure 4 displays a simulated execution of the tasks that form the *obstacle avoidance behaviour* and how these tasks interact with the *path following behaviour* using the *arbiter* process. This simulation has been performed using the DRTSS tool and the obtained events have been represented in a temporal diagram. The vertical axis shows the name of the tasks and the horizontal axis defines the temporal values associated to the execution of each task. The last line in the display, labelled as *Idle_time*, defines the remaining time of the computation.

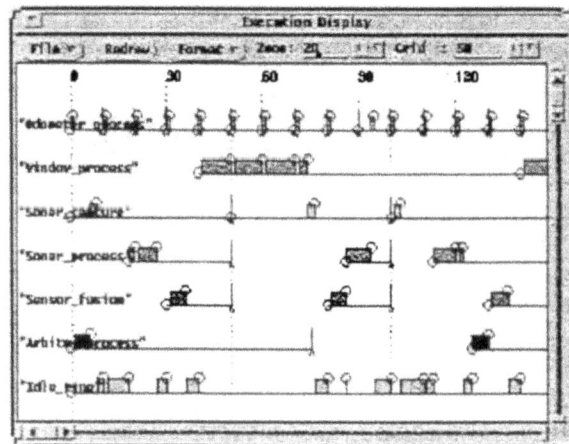

Figure 4.- Example of execution display.

The DRTSS tool also returns the timestamp of the events which have been produced in the system simulation. These temporal values can be used to compute end-to-end properties such as those defined in the Introduction. In this case, a freshness relationship is evaluated between *the Arbiter_ process* task and the *Sonar_processing* transaction. This relationship will indicate the current state of obstacle information causing each arbiter action.

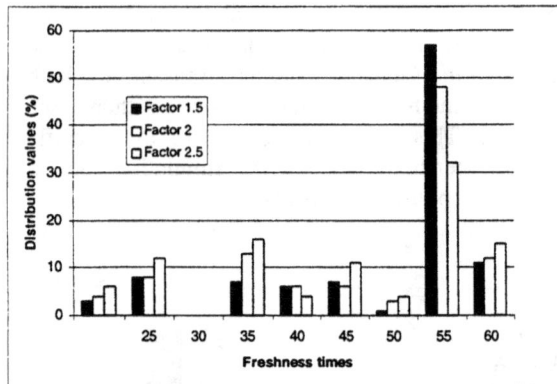

Figure 5.- Freshness evaluation.

Figure 5 shows the simulation results that measure the time between the action generated by the arbiter and the last sonar sensor data that caused this action. These results have been represented through the percentage distribution (Y co-ordinates) associated with each range of freshness times (X co-ordinates); and several experiments have been performed by assigning different period values to the *Arbiter_process* task. These periods are computed as a factor of the period of the sonar transaction (1.5, 2 and 2.5 respectively). The analysis of the results obtained reveals that a higher factor returns a more uniform distribution of freshness times.

5 CONCLUSIONS

A behavioral model for mobile robotic applications is described in this paper. The processes that define this model are mapped to a real-time task model. The application of this task model allows the examination of different temporal analysis techniques such as the worst-case response time computation and simulation tools. An application example is illustrated and its set of tasks is described. Some experiments were performed to check the timing requirements that are usually associated with robot operations. Planned future work includes implementing this real-time model on a real robot application.

6 REFERENCES

Albus, J., McCain, H., Lumia, R. (1987), NASA/NBS Standard Reference Model for Telerobot Control System Architecture (NASREM), NBS Tech. Note 1235, Gaithersburg, MD,.

Benet, G. Blanes, F. Martínez, M. Simó, J. (1998). A Multisensor Robot Distributed Architecture. IFAC Conference INCOM'98. Metz-Nancy. Junio 1998.

Botti, V., Barber, F., Crespo, A., Onaindía, E., García-Fornes, A., Ripoll, I., Gallardo, D., and Hernandez, L. (1995) A Temporal Blackboard for a Multi-Agent Environment. *Data and Knowledge Engineering*. North Holland Elsevier 15 (3).

Buendia, F. (1999) An Architecture for Image Processing in a Real-Time System. Internal Report. Dept. of Computer Eng. Univ. Politécnica de Valencia.

Butaccio, G. (1996). Real-Time Issues in Advanced Robotics Applications. Euromicro Workshop on Real-Time Systems, Jun. 1996

Brooks R. (1986), A Robust Layered Control System for a Mobile Robot, IEEE Journal of Robotics and Automation vol. RA-2, no. 1, pp. 14-23, April.

Gerber, R., Hong,, S., Saksena, M. (1994). Guaranteeing End-to-End Timing Constraints by Calibrating Intermediate Processes. Proceedings of the IEEE Real-Time Systems Symposium, Dec. 1994.

Lehoczky, J.P. (1990). Fixed Priority Scheduling of Periodic Task Sets with Arbitrary Deadlines. Proc. IEEE Real-Time Systems Symposium Dec. 1990.

Moravec, H (1990) The Stanford Cart and the CMU Rover, in Cox, I. and Wilfong, G., Autonomous Robot Vehicles, Springer-Verlag,.

Palencia, J:C, Harbour, M. (1998). Schedulability Analysis for Tasks with Static and Dynamic Offsets. Proceedings of the IEEE Real-Time Systems Symposium, Dec. 1998.

Rosenblatt, J (1997) " DAMN: A Distributed Architecture for Mobile Navigation" PhD. Thesis. Robotic Institute of Carnegie Mellon.

Storch M.F. (1997). A Framework for the Simulation of Complex Real-Time Systems. PhD Thesis, Dept. Computer Science, Univ. of Illinois, 1997.

Sun, J., Liu, J. W. S. (1995) Bounding the End-to-End Response Time in Multiprocessor Real-Time Systems. Proceeding of Workshop on Parallel and Distributed Real-Time Systems, April 1995

Sun, J. W. S. Liu. (1996) Bounding Completion Times of Jobs with Arbitrary Release Times and Variable Execution Times. Proc. Real-Time Systems Symposium, December 1996.

Tindell K. (1992). Using Offset Information to Analyse Static, Pre-emptively Scheduled Task Sets, YCS 182, Department of Computer Sciences, University of York Sep.1992.

Tindell, K., Clark, J. (1994). Holistic Schedulability Analysis for Distributed Hard Real-Time Systems, Microprocessing and MicroProgramming, Vol. 50, n.2-3, pp.117-134.

MODULAR VERIFICATION OF FUNCTION BLOCK BASED INDUSTRIAL CONTROL SYSTEMS

Norbert Völker and Bernd J. Krämer

University of Essex, United Kingdom
FernUniversität, Hagen, Germany

Abstract: IEC 61131-3, the world-wide standard for industrial control programming, is increasingly being used in safety-related control applications. Control loops are built from components taken from domain-specific function block libraries. Code inspection and testing are the two predominant quality assurance techniques. For highly dependable control applications, however, these techniques are not sufficient, in general.

This paper suggests to augment testing with compositional, theorem-prover supported verification. The approach is based on a representation of IEC 61131-3 function blocks in higher-order logic. The verification task is separated into the a priori verification of library components and a separate proof of individual application programs. *Copyright © 1999 IFAC*

Keywords: Safety-critical control systems, dependable software, PLC programming, IEC 61131-3, modular verification, higher order logic theorem proving.

1. INTRODUCTION

Programmable logic controllers (PLCs) form a growing market of special purpose hybrid systems integrating micro-electronic and software components. PLCs are particularly suited to solve application problems in machine logic, process automation and manufacturing. They were developed to replace traditional hard-wired switching networks based on relay or discrete electronic logic.

The rapid development of PLC systems in the 1980's led to a wealth of incompatible vendor-specific PLC programming languages within the process industries impeding the design of more complex, open and distributed control applications. In response, the International Standard IEC 61131-3 for PLC programming (IEC, 1993) was developed. This standard applies to a wide range of programmable controllers and harmonizes the way engineers look at industrial control.

The standard provides a class of five languages that overlap conceptually and share a subset of programming elements. Three languages of the standard, Function Block Diagram (FBD), Ladder Diagram (LD) and Sequential Function Chart (SFC) have a graphical appearance. FBD supports component-based application programming while SFC is mainly used for describing sequential behavior of a control system including alternative and parallel execution sequences.

New capabilities of PLCs, the comfort of the PLC languages, and strong economical demands led to the current situation of increased dependence on PLC-based systems for control and automation functions in safety-related applications. Examples include traffic control, patient monitoring, chemical process automation and emergency shut down systems in power generation.

The growing social awareness of a need to protect the environment, a higher sensitivity to accidents caused by ill-designed technology or processes, and a declining confidence in marketing statements of manufacturers lead to enormous pressure to increase the dependability of safety related applications. In practice however, there is a lack of rigorous proof techniques and robust tools which can be used effectively by practitioners in industry and regulatory authorities. Existing design guidelines and testing practices help to detect design and programming errors but they cannot guarantee the absence of faults because exhaustive testing is limited to rare cases.

The main body of this paper explores function blocks – which represent the engineer's idea of reusable "software ICs" – and sequential function charts in order to develop a modular, theorem prover-based verification framework. By taking components from application-specific libraries of verified standard function blocks, the verification of new applications is simplified considerably because only the correctness of the composition has to be established for each new application.

In the following section, the core concepts of FBD and SFC are introduced. In Section 3, the underlying higher order logic is discussed. The verification approach is based on a semantic embedding of the selected PLC languages in that logic. This embedding is explained in Section 4, while the verification process and the challenges of handling complex continuous systems are sketched in Section 5 and 6. The paper concludes with a brief summary and an outlook on an industrial strength verification tool which can ultimately be used by domain experts with little or no expertise in software verification.

2. FUNCTION BLOCKS AND SEQUENTIAL FUNCTION CHARTS

Function blocks are program organization units with a private state that persists from one invocation to the next. A function block interacts with its environment primarily via input and output variables. The standard also allows global variables but the verification framework does not support these. Besides keeping the semantics simple, this also has the advantage that the execution of function blocks has no side-effects.

From a semantic point of view, function blocks are a special case of deterministic reactive modules (Alur and Henzinger, 1996). Their execution takes place in a sequence of rounds. At the start of each round, the input variables are read. This is followed by an update of the private and output variables. This update is functionally dependent on the current value of the input variables and the previous state of the private and output variables.

The description of a function block can be split into the declaration of its external interface and a specification of the internal implementation. The former is part of the function block signature that specifies the types and names of variables including local instances of function blocks. In

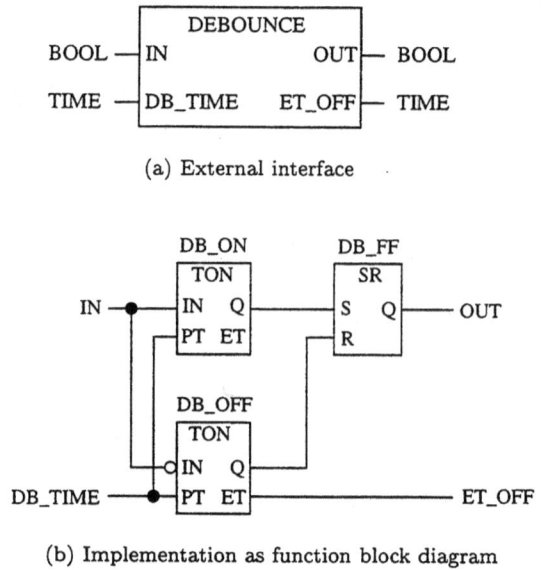

(a) External interface

(b) Implementation as function block diagram

Fig. 1. Function block DEBOUNCE

the context of graphical representations, the input and output variables are also referred to as ports. The internal implementation of a function block body can be carried out in any of the five IEC 61131-3 programming languages or even in another language such as C or Java.

As an example, Figure 1(a) shows a graphical representation of the external interface of the function block DEBOUNCE taken from the IEC 61131-3. DEBOUNCE has two input variables IN and DB_TIME of type BOOL and TIME and two output variables OUT and ET_OFF of the same types.

An implementation of DEBOUNCE as a function block diagram is depicted in Figure 1(b). The function blocks DB_ON and DB_OFF are two separate instances of the timer function block TON while DB_FF is an instance of the standard SR flip-flop. By connecting input and output ports, a diagram is "wired together" from the components. As in the graphical representation of circuits, the open circle at input port IN of DB_OFF indicates signal negation. The named instances of function blocks will usually also be referred to as function blocks. The function block DEBOUNCE is composed from standard function blocks predefined in the norm. Such a composite function block can itself be used in further applications just as if it was one of the standard function blocks. This feature is useful for building an in-house or domain specific collection of function blocks.

The textual IEC 61131-3 language ST is similar in appearance to a structured programming language such as PASCAL. Figure 2 shows an alternative implementation of the body of DEBOUNCE in ST.

```
DB_ON (IN := IN, PT := DB_TIME);
DB_OFF (IN := NOT IN, PT:= DB_TIME)
DB_FF (S := DB_ON.Q, R := DB_OFF.Q);
OUT := DB_FF.Q;
ET_OFF := DB_OFF.ET;
```

Fig. 2. DEBOUNCE in Structured Text

The second graphical language of the standard, SFC, can be regarded as an application of Petri nets. Its language concepts include transitions, steps and actions. They serve to co-ordinate the execution of function blocks that are regarded as asynchronous sequential processes.

The role of SFC is illustrated by a small laboratory plant application. This has been used previously as a benchmark for the tool-aided analysis of discretely controlled continuous systems (Kowalewski et al., 1997). The plant features two cylindrical tanks that are located at different levels (see Fig. 2). The tanks are equipped with three pipes and three valves V_0, V_1 and V_2 which control the flow of liquid between the tanks, at the inlet and the outlet. The liquid level in the second tank is measured by a sensor L. A core safety requirement for this application is to avoid overflow in the coupled tank system.

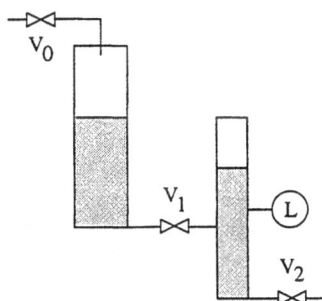

Fig. 3. Laboratory plant

The SFC depicted in Fig 2 controls the behavior of the system. It consists of five steps s0, .., s4. The actions connected with the steps control the state of the valves: the qualifiers S and R denote setting and resetting of an action, respectively. The transitions separating the steps are enabled by Boolean valued expressions representing conditions on the state of the function block.

The encapsulation provided by function blocks together with their flexibility with respect to the internal implementation furthers their reuse in different applications. Hence, it makes sense to develop component libraries. Examples include the collection of standard function blocks of the IEC 61131-3 and a domain specific library of function blocks used by a German manufacturer of chemicals and drugs. This in-house library consists of about 70 function blocks that are suffi-

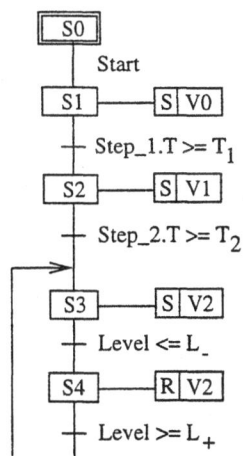

Fig. 4. SFC controller for laboratory plant

cient to specify/program most chemical process automation tasks.

3. HIGHER ORDER LOGIC FOR VERIFICATION

The logic underlying the verification approach is higher order logic (Gordon and Melham, 1993). There are several good reasons for this choice:

(1) The means of abstraction and quantification over functions make this logic very expressive and thus well suited to the concise description of complex theories, see for example the embedding of hardware description languages (Boulton et al., 1992).
(2) HOL is a widely studied and well understood logical system with a remarkably small number of axioms and inference rules. Its expressiveness makes it possible to use definitional extension as the principal method of theory development. Since this method is conservative, logical inconsistencies can be practically ruled out.
(3) Automatic type inference systems for HOL make type annotations to a great extent unnecessary. This shortens formulas and proofs because the information contained in the typing is automatically inferred and propagated.

In comparison to alternatives such as Zermelo-Fränkel set theory, there are also a few disadvantages:

(1) The strict type discipline of HOL leads to a certain loss of expressiveness. This statement is true despite the provision of polymorphism and symbol overloading available in systems such as Isabelle/HOL.
(2) In comparison with first and second order logics, the implementation of the HOL type system is technically more demanding. In particular, the existence of type and function variables complicates unification, the basic

161

method of equation solving. Also, most research in automated theorem proving has been performed in the area of first order theories.

For the purpose of implementing an integrated verification framework, the advantages of HOL outweigh the drawbacks. Its extendibility makes it unnecessary to introduce further, specialized logics. Instead, HOL provides a logical core that can serve as the common semantic basis for both programming languages and specification formalisms.

Furthermore, it is important that HOL is supported by several reliable and efficient mechanical theorem proving assistants. Currently, the verification framework is based on the object logic HOL of the generic theorem proving assistant Isabelle (Paulson, 1994). Noteworthy alternatives include the HOL system and the LISP based PVS system (Rushby and Stringer-Calvert, 1995).

4. THE EMBEDDING OF FUNCTION BLOCKS IN HOL

The foundation of the function block verification framework is a HOL embedding of a subset of Structured Text (ST). The technical details of this embedding can be found in (Völker, 1998). It is a relatively deep embedding, which means that the syntax of function blocks and the assignment of semantics are represented explicitly in HOL. Semantics are defined via evaluation functions for the four different syntactical categories, namely expressions, statements, functions and function blocks. As a result, every function block is associated with a deterministic, but not necessarily finite Mealy automaton in HOL.

Time is treated as an input variable. Like all other input variables, its value stays constant in each round. This agrees with the 0-delay paradigm underlying the synchronous programming languages such as Esterel or Lustre (Benveniste and Berry, 1991). A difference is that function blocks are strictly hierarchical and have a sequential semantics. As a consequence, the issue of instantaneous feedback does not arise. Furthermore, because verification is based predominantly on theorem proving and not on model checking, the efficient compilation of function blocks to finite automata is of less importance than is the case for synchronous programming languages.

The HOL terms that describe the semantics of function blocks are initially cluttered with occurrences of the evaluation functions. In a term rewriting process which resembles a symbolic evaluation, these occurrences are eliminated. This process can be largely automated. It yields HOL terms that resemble simulations of ST function

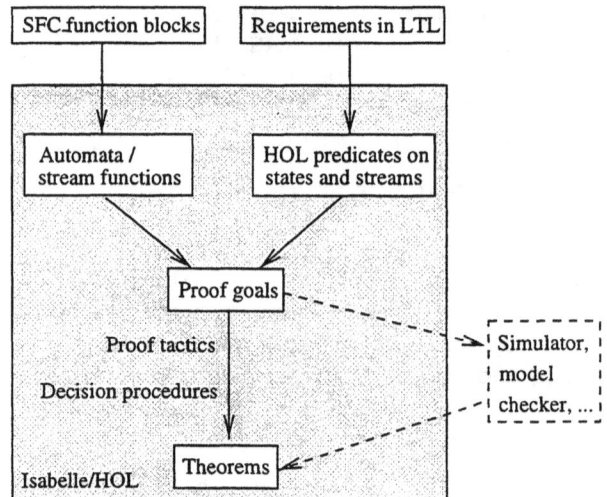

Fig. 5. Function Block Verification Process

blocks viewed as functional programs. In this form, the automata are suitable for verification.

An important aspect of the function block semantics is compositionality. This means that the transition function of the automaton belonging to a composed block is a composition of the transition functions of the automata belonging to the components. Thus proven properties of the components can be reused. Furthermore, by abstracting over component properties, it is possible to prove properties of composed function blocks without reference to the concrete implementation of the components.

In addition to ST, the verification framework also deals with subsets of the two graphical IEC 61131-3 languages SFC and FBD. This is based on interpretations of these two formalisms in ST. The result is in both cases a formal semantics that is sequential and deterministic (Völker, 1998).

5. THE VERIFICATION APPROACH

The deep embedding of PLC programming languages in HOL provides a formal semantics. Furthermore, the semantics given above are operational. Function blocks can thus be evaluated symbolically using a term rewriting tool. Requirements on the behavior of function blocks are translated to HOL predicates and proven formally. Figure 5 shows the verification process for SFC function blocks assuming linear time temporal logic (LTL, (Manna and Pnueli, 1992)) as specification language.

Real-time constraints on the controller behavior are reflected firstly by LTL formulas which depend on the time input variable, and secondly by upper bounds on the controller response time and the time difference between two subsequent controller invocations. Conformance of an implementation to the latter two kind of time bounds is not proven formally but instead established separately using implementation-dependent timing information.

One of the strong points of the HOL based approach is its openness with respect to possible extensions. Adding further programming or specification language constructs is unproblematic as long as it does not affect the model of already embedded language parts. The same holds for the modeling of machine or environment aspects, which might be necessary for the verification of more complex systems. To put it more generally: HOL serves as logical glue that connects different programming and specification formalism and allows their integration and analysis within one framework.

In relatively small examples such as the verification of a liquid container controller presented in (Krämer and Völker, 1997), the standard Isabelle/HOL proof tools are sufficient. Because specifications are mapped to predicates on streams, the basic proof principle is induction over the natural numbers. In the induction step, the validity of a statement in round $(n + 1)$ has to be derived from its validity in round n. Induction is also essential for the proof of auxiliary algebraic equalities and inequalities and the verification of iterated structures such as a generic adder. Other frequently used proof techniques are case distinctions, algebraic simplifications and arithmetic estimations. Isabelle's classical reasoner has been very useful for the automation of these kinds of proofs.

For larger applications, a higher degree of proof automation is essential. This starts off with the automated translation of function blocks into Isabelle theories. Tactics specially adapted to programming or specification language constructs should be tried automatically or offered interactively to the user for selection and parameterization. Relevant automated proof procedures include the symbolic model checking of finite systems (Alur *et al.*, 1996) and algorithms for establishing program invariants (Saïdi, 1997; Halbwachs *et al.*, 1997).

The main focus of the work described above is the correctness of controller implementations with respect to requirements formalized in LTL/HOL. For more complex applications, the formulation and formalization of the controller requirements can itself be a non-trivial task. This suggests an extension of the verification approach so as to model in HOL the whole control loop composed of controller and plant. This could be based on HOL models of plants as non-deterministic timed automata (Alur and Dill, 1994).

6. THE CHALLENGE OF COMPLEX DYNAMICS

Up-to now, the use of theorem prover based tools has been restricted to the verification of systems with relatively simple continuous dynamics. This is partly due to the fact that the treatment of more complex systems would require extensive real/complex analysis libraries. As the pioneering work in (Harrison, 1996) shows, this is a comprehensive task. Even with such libraries, a complete analytic verification of systems such as the two tank laboratory plant sketched in Fig. 4 seems a daunting task.

Besides providing formal models of controllers and abstractions of plant properties, a useful future role for deductive proof tools in this domain is the validation of stability properties. These guarantee that nothing unexpected happens for parameter combinations that have not been explicitly covered during simulation or model checking. This validates intuitive worst-case reasoning and increases the trustworthiness of other verification results.

7. TOWARDS AN INDUSTRIAL STRENGTH TOOL

The main body of this paper has presented a theorem prover based verification technique that supports modular proofs of PLC programs written in FBD, SFC and ST. In general, the development of such proofs with the help of theorem prover assistants requires high skills from quality assurance personnel because the proof assistant relies on sophisticated user guidance. These skills cannot be expected from engineers in the field.

Conversely, people with skills in formal specification and verification techniques normally lack the domain expertise needed to understand functional and safety requirements that are often not made explicit and, if so, are usually presented in an incomplete, ambiguous and informal manner. For example, it took weeks of intensive reading and many hours talking to domain experts before the authors fully understood the requirements in the IEC 61131-3, its German counterpart, and the in-house standard and function block library of a manufacturer in chemical industry.

Hence, in order to make the verification approach viable for automation practice, effective means must be found to carry out the following three tasks:

(1) Comprehensive elicitation of functional, safety, and – if appropriate – timing requirements.

(2) Formalization of these requirements in a suitable logic.

(3) Correctness proof.

The set of standard function blocks that are typically used in a specific control domain ranges from 50 to a few hundred. Furthermore, the complexity of the majority of function blocks in domain libraries is relatively low. In view of the reusability of function blocks, the effort to verify them by computer theoreticians becomes acceptable.

However, for handling individual control applications composed of networks of function blocks, it is necessary to combine an open verification environment with a front-end usable by domain experts. This verification environment may build on a theorem prover as its backbone and comprise other tools such as model checkers, simulators or computer algebra systems. The interface to the front-end must be capable of eliciting enough facts about critical application requirements through interaction with domain experts so that formal requirement statements can be derived. This requires knowledge of terminology of the field, collections of known requirements typical for the domain, and the exploitation of proven properties of function blocks connected to the application interface and the inner "wiring" of the application program. It may also exploit paraphrasing capabilities to verify the adequacy of formalized requirement statements acquired in earlier communications. The work on knowledge intensive software engineering tools conducted by Rich and Waters (Rich and Waters, 1992) might provide prototype solutions for the engineering environment sketched here.

To facilitate the verification task, it is very important to find proof patterns and reusable proof strategies to automate recurring verification steps. In this respect, the integration of automatic model checking procedures such as pioneered by N. Shankar for the PVS system seems particularly promising.

To come up with usable solutions, close cooperation with interested vendors, users and evaluators for PLC controllers in safety critical fields is urgently needed.

8. REFERENCES

Alur, R. and D. L. Dill (1994). A theory of timed automata. *Theoretical Computer Science* **126**(2), 183–235.

Alur, R. and T.A. Henzinger (1996). Reactive modules. In: *Proceedings, 11th Annual IEEE Symposium on Logic in Computer Science*. IEEE Computer Society Press. pp. 207–218.

Alur, R., T.A. Henzinger and P.-H. Ho (1996). Automatic symbolic verification of embedded systems. *IEEE Transactions on Software Engineering* **22**(3), 181–201.

Benveniste, A. and G. Berry (1991). The synchronous approach to reactive and real-time systems. *Proceedings of the IEEE* **79**(9), 1268–1282.

Boulton, R., A. Gordon, M. Gordon, J. Harrison, J. Herbert and J. Van Tassel (1992). Experience with embedding hardware description languages in HOL. In: *Theorem Provers in Circuit Design. Proceedings of the IFIP TC10/WG 10.2 International Conference, Nijmegen, June 1992* (V. Stavridou, T.F. Melham and R.T. Boute, Eds.). North-Holland.

Gordon, M.J.C. and T.F. Melham (1993). *Introduction to HOL*. Cambridge University Press.

Halbwachs, N., Y.E. Proy and P. Roumanoff (1997). Verification of Real-Time Systems using Linear Relation Analysis. *Formal Methods in System Design* **11**(2), 157–185.

Harrison, J. (1996). Theorem Proving with the Real Numbers. PhD thesis. Computer Laboratory, University of Cambridge.

IEC (1993). IEC International Standard 1131-3. *Programmable Controllers. Part 3: Programming Languages*.

Kowalewski, S., M. Fritz, H. Graf, S. Simon J. Preußig, O. Stursberg and H. Treseler (1997). A case study in tool-aided analysis of discretely controlled continuous systems: the two tanks problem. In: *Proceedings of 5th Int. Workshop on Hybrid Systems (HSV), Notre Dame, USA*.

Krämer, B. and N. Völker (1997). A highly dependable computer architecture for safety-critical control applications. *Real-Time Systems Journal* **13**, 237–251.

Manna, Z. and A. Pnueli (1992). *The Temporal Logic of Reactive and Concurrent Systems: Specification*. Springer-Verlag.

Paulson, L.C. (1994). *Isabelle: A Generic Theorem Prover*. Springer. LNCS 828.

Rich, C. and R.C. Waters (1992). Knowledge intensive software engineering tools. *IEEE Transactions on Software Engineering* **4**(5), 424–430.

Rushby, J. and D.W.J. Stringer-Calvert (1995). A less elementary tutorial for the PVS specification and verification system. Technical Report SRI-CSL-95-10. Computer Science Laboratory, SRI International. Menlo Park, CA.

Saïdi, H. (1997). The Invariant Checker: Automated deductive verification of reactive systems. In: *Computer-Aided Verification, CAV '97* (O. Grumberg, Ed.). Vol. 1254 of *Lecture Notes in Computer Science*. Springer-Verlag.

Völker, N. (1998). Ein Rahmen zur Verifikation von SPS-Funktionsbausteinen in HOL. PhD thesis. FernUniversität Hagen, Shaker Verlag.

VERIFICATION OF REAL-TIME SYSTEM REQUIREMENTS: A PETRI NET APPROACH

Marcin Szpyrka

Pedagogical University
Institute of Mathematics
Rzeszow, Poland

Abstract: The aim of the paper is to discuss a Petri net approach to verification of real-time system behaviour requirements, i.e. the elimination of ambiguity, inconsistency, and incompleteness. The main problem is to create a complete description of the system requirements without any negatives. The methodology is based on the coloured Petri nets theory (Jensen, 1996). An algorithm, which produces from a decision table, a coloured Petri net, is formulated in the paper. It allows checking some properties of system requirements and reducing the aforesaid negatives. *Copyright © 1999 IFAC*

Keywords: requirements engineering methods, verification of system behaviour requirements

1. INTRODUCTION

During the specification of requirements phase of the real-time system development, it is necessary to describe the expected external behaviour of the system to be built. It begins with the problem statement generated by clients. The statement in natural language may be incomplete, ambiguous and inconsistent. It can lack essential information, which must be very often obtained from the clients (Rumbaugh, 1991; Braek and Haugen, 1993; Szmuc, 1997). Requirements in natural language are comprehensible for clients but not convenient for automatic analysis.

So the first problem is to find a form of system requirements notations so that it can be comprehensible for all and useful for analysis (Davis, 1988; Macaulay, 1996). In order to do that a kind of decision table is proposed. The second problem is to choose a proper tool for verification of requirements. The approach presented in this paper consists of two steps. First a coloured Petri net is built from the decision table. Second some properties of the net are computed so as to reduce the aforesaid negatives.

The paper is organised as follows. Section 2 presents a natural language specification of a lift driver example and a description of constructing a decision table for that problem. Section 3 describes an algorithm, which produces from a decision table, a coloured Petri net called a *coloured decision Petri net*. This section realises the first step in the verification procedure. In this section basic definitions and notation connected with the coloured decision Petri net are presented too. Section 4 deals with some propositions, which allow reducing the negatives of system requirements. This section realises the second step in the verification procedure. As a result of that, there is the specification of requirements of the external system behaviour, which is complete, unambiguous and consistent.

2. A LIFT DRIVER EXAMPLE

An example of a lift driver will be used to illustrate all new ideas. Consider a problem formulated in the following way. The lift is used in a building with 6 floors. Each floor has its own integer number from 0 to 5, where the ground floor has number zero. For the sake of simplicity it can be assumed that the user calls a lift and chooses the destination floor at the same time. For example somebody on the second floor calls a lift so as to be taken to the fifth floor. Let the ground floor be the start point of the lift. While being called, the lift starts moving up and satisfies all demands connected with this direction. After stopping on the last floor, the lift starts moving down and satisfies all demands connected with this direction. Considering whole such lift control system is not an aim of this paper. Only one object (the lift driver) will be chosen to illustrate the verification procedure. In this example the object *Driver* has only one operation, which will be called *Decision*. The aim is to produce a specification of requirements for the operation *Decision*. While considering all system it is necessary to verify specification of requirements for every operation distinguished in an object model of this system.

A natural language specification of requirements of the operation *Decision* can be presented as a set of implications.

1. If the lift is at the start point and there is a demand, the lift should start moving up.

2. If the lift is moving up to the floor k and there is a demand connected with this floor and consistent with this direction, (somebody wants to get out or get into so as to be taken up) the lift should stop on the floor k.

3. If the lift is moving up to the floor k, k is not the last floor and there is not any demand connected with this floor and consistent with this direction, the lift should keep moving up without stopping on the floor k.

4. If the lift is moving down to the floor k, and there is a demand connected with this floor and consistent with this direction, the lift should stop on the floor k.

5. If the lift is moving down to the floor k, k is not the ground floor and there is not any demand connected with this floor and consistent with this direction, the lift should keep moving down without stopping on the floor k.

6. While serving the demands connected with moving up, the lift stands on the floor k and k is not the last floor, the lift should start moving up.

7. While serving the demands connected with moving down, the lift stands on the floor k and k is not the ground floor, the lift should start moving down.

8. If the lift stands on the last floor then should start moving down.

This set of implications can be treated as the very beginning specification of requirements of the object *Driver* external behaviour. The next step is to construct a decision table. One can avoid a natural language specification and starts from constructing the decision table.

Some attributes of objects, which form considered system, are very important in the lift behaviour. It is possible to distinguish some conditional and decision attributes. In what follows: A_c will denote the set of all conditional attributes, and A_d will denote the set of all decision attributes. For the sake of simplicity it will be assumed that A_c and A_d are nonempty, finite, and ordered sets. For any attribute a the set V_a will denote the set of all possible values (value set) of the attribute a, the set W_a will denote the set of variables with the value set equal to V_a, and L_a will denote the set of all possible expressions composed of elements of sets V_a, W_a, relations operators $\{=, <, >, \leq, \geq\}$ and logical operators $\{not, and, or\}$;

In the considered example there are five conditional attributes:

St - *State* of the lift with the value set $\{up, down, stop\}$;

Po - *Position* of the lift with the value set $\{0, 1, 2, 3, 4, 5\}$;

Se - *Sensor* state on the considered floor with the value set $\{0, 1\}$, (In every moment the state of only one sensor is important, i.e. the sensor of the floor where the lift is.);

Di - *Direction* of the lift moving with the value set $\{dup, ddown\}$;

De - *Demand* with the value set $\{yes, no\}$.

For example:

Situation: $St = up$ and $Po = 2$ means that the lift is approaching the floor 2, while moving up.

Situation: $Se = 0$ and $Po = 2$ means that there is not any demand connected with floor 2 and consistent with this direction of the lift moving.

There are only two decision attributes in considered system. NSt (*New state*) and NDi (*New direction*) with the value set $\{up, down, stop\}$ and $\{dup, ddown\}$ respectively.

After distinguishing all attributes it is possible to construct the decision table. In the considered example this table will include seven columns (number of all attributes) and eight rows (number of all rules - implications). Each cell will include a value of an attribute in the proper rule. There are three possible situations: If there is only one possible value of the attribute a in the rule r, this value is placed in the proper cell of the decision table. If a value of the attribute a is not important

Table 1. Decision table for operation *Decision*

St	Po	Se	Di	De	NSt	NDi
stop	0	s	dup	yes	up	dup
up	p	1	dup	z	stop	dup
up	p < 5	0	dup	z	up	dup
down	p	1	ddown	z	stop	ddown
down	p > 0	0	ddown	z	down	ddown
stop	(p < 5)∧ (p > 0)	s	dup	z	up	dup
stop	p > 0	s	ddown	z	down	ddown
stop	5	s	ddown	z	down	ddown

in the rule r, a variable belonging to the set W_a is placed in the proper cell of the decision table. If a few values (but not all) of the attribute a are possible in the rule r, a logical expression, which restricts a variable belonging to the set W_a is placed in the proper cell of the decision table.

According to this description the requirements can be written as a decision table (see Table 1.).

It easy to see that in some rules, values of some attributes are not important. For example the value of the attribute Di is important only in the rules r_6 and r_7. But in the other rules there is only one possible value of that attribute so it should be placed in decision table. Letters p, s and z denote variables belonging to the proper value sets. In what follows R will denote the set of all rules, and $Tb(a, r)$ will denote the value of the attribute a in the rule r.

3. COLOURED DECISION PETRI NETS

The algorithm of constructing a *coloured decision Petri net* (simply *D-net*) consists of 12 steps (Szpyrka, 1999).

Step 1. Put two places in the D-net. (The first place is connected with the set A_c and will be called a *conditional place* and denoted by M_c, the second one is connected with the set A_d and will be called a *decision place* and denoted by M_d).

Step 2. Set the Cartesian product of the sets V_a, where a belongs to A_c, as the conditional place colour.

Step 3. Set the Cartesian product of the sets V_a, where a belongs to A_d, as the decision place colour.

Step 4. Put a token in the conditional place and leave the decision place empty. (It is important to choose correct entries for that token. For example there should not be taken $St = up$ and $Di = ddown$).

Step 5. Put $card(R)$ transitions in the D-net. (These transitions will be called *decision transitions.*)

Step 6. For every decision transition add an arc from the conditional place to that transition and from that transition to the decision place.

Step 7. For every decision transition set the input arc expression. This should be a vector with $card(A_c)$ entries. Each entry is connected with a proper attribute. For example for the rule r and the attribute a, if $Tb(a, r)$ belongs to V_a or W_a then $Tb(a, r)$ is the proper entry. Otherwise a variable belongs to set W_a which occurs in $Tb(a, r)$ is the entry and the $Tb(a, r)$ is placed in the transition guard.

Step 8. For every decision transition set the output arc expression, which should be constructed in the similar way as mentioned above.

Step 9. Put an extra transition in the D-net. (This transition will be called a *reset transition.*)

Step 10. Add arcs from the decision place to the reset transition and from the reset transition to the conditional place.

Step 11. For the reset transition set the input and output arc expressions as vectors with $card(Ac)$ and $card(Ad)$ entries. Each entry should be a variable of the proper value set.

Step 12. Set the reset transition guards so as to make an appearance of a "bad" token impossible, (for example ($St = up$ and $Po = 0$).

Fig. 1. presents D-net constructed for the Table 1.

4. VERIFICATION OF THE SYSTEM REQUIREMENTS

The D-net is a basis for checking some properties whose definitions will appear in this section.

Definition 1. The set of rules R is called complete if and only if the system can react in every possible situation using the rules, which belong to R.

Let M_0 be an initial marking of the D-net N, consistent with the algorithm of constructing the D-net i.e. there is one "right" token in the conditional place and no tokens in the decision place. $[M_0>$ will denote the set of all reachable markings in the net N. It is easy to check that $[M_0>$ is an union of two disjoint sets C_M and D_M, where C_M is a set of all markings such that there is one token in the conditional place and no tokens in the decision place and D_M is a set of all markings such that there is one token in the decision place and no tokens in the conditional place. Elements of these sets will be called *conditional markings* and *decision markings* respectively.

Proposition 1. The set of rules R is complete iff there is not a death marking in the set $[M_0>$.

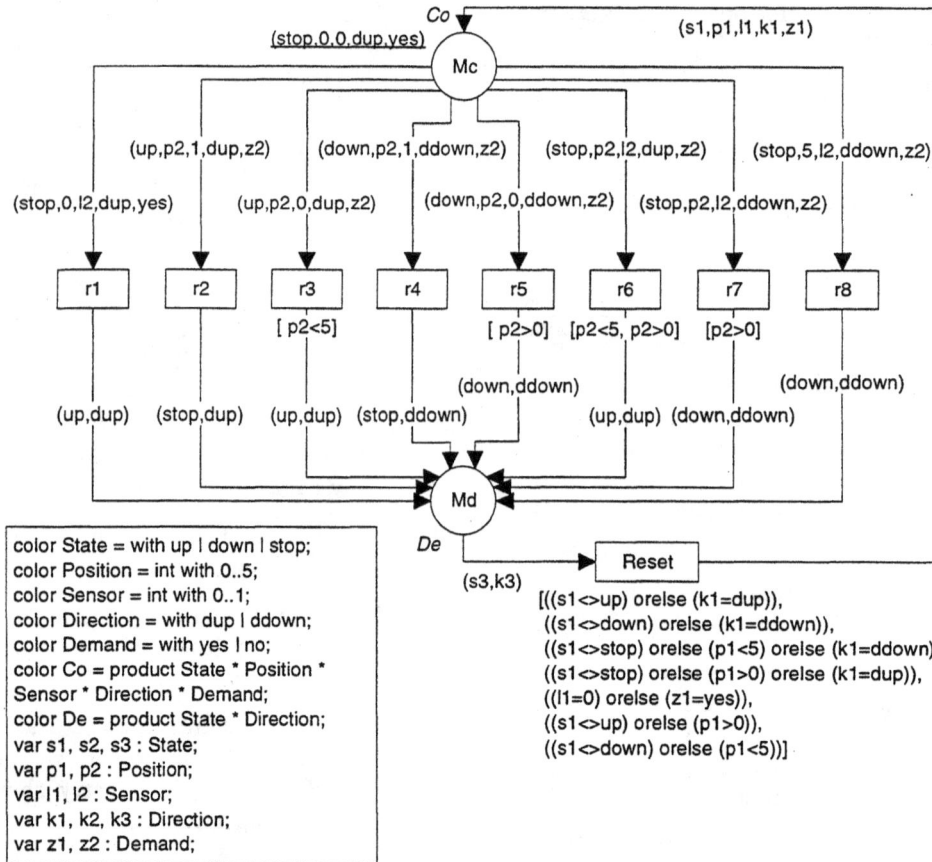

Fig. 1. Coloured decision Petri net built for the operation *Decision* from the decision table presented before

Let M be a marking, which belongs to the set $[M_0>$. If M belongs to the set D_M then the reset transition is M- enabled so M can not be death marking. In order to prove the Proposition 1 it is sufficient to consider only the set C_M. If there is not a death marking in the set C_M then it means that the system can react in every possible situation. Hence according to Definition 1 the set R is complete. On the other hand, if R is complete then the system can react in every possible situation. It means that for every conditional marking M there exists a transition t such that t is M-enable. Hence there isn't a death marking in the set C_M. However from the practical point of view more useful is Proposition 2.

Proposition 2. The set of rules R is complete iff the SCC-graph (graph of strong connected components) of the D-net N is trivial.

Proposition 2 is a direct consequence of the definition of a SCC-graph and Proposition 1.

Remark 1. The set of all terminal nodes of nontrivial SCC-graph corresponds to the set of all situations in which the system can not react.

There are 5 terminal nodes in Fig. 2. It is possible to distinguish three sets of these nodes. The nodes with numbers 1 and 2 correspond to the

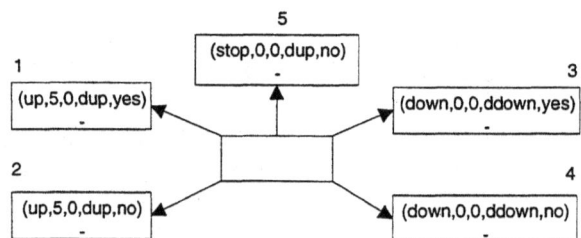

Fig. 2. SCC-graph of the coloured decision Petri net presented in Fig. 1.

situation while the lift is approaching the floor 5. In this situation the lift should stop on the 5 floor. The nodes with numbers 3 and 4 correspond to the situation while the lift is approaching the ground floor. In this situation the lift should stop on the ground floor. The node 5 corresponds to the situation while the lift stops on the ground floor and there is not any demand. In this situation the lift should remain on this floor. So the considered set $R = \{r_1, r_2, r_3, r_4, r_5, r_6, r_7, r_8\}$ is not complete. In Fig. 3 some additional rules are presented which have to be joined in the set R. It is easy to check that the new set $R = \{r_1, r_2, r_3, r_4, r_5, r_6, r_7, r_8, r_9, r_{10}, r_{11}\}$ is complete.

Let r and r' belong to the set R and are M-enable, where M is a conditional marking. The fact that r

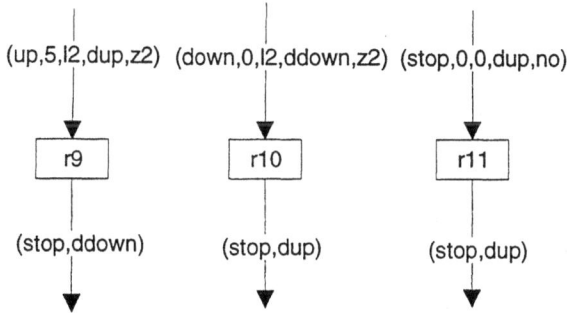

Fig. 3. Additional rules

fires from M to M' will be denoted by $M[r>M'$.
If $M[r>M_1$ and $M[r>M_2$ and $M_1 \neq M_2$ then
the decision transitions r and r' will be called
inconsistent for M. If $M[r>M_1$ and $M[r>M_2$ and
$M_1 = M_2$ then the decision transitions r and r'
will be called *consistent for M*.

Definition 2. The set of rules R is called consistent
if and only if for every decision transitions r and
r' and for every conditional marking M, r and r'
are consistent for M.

It is easy to check that every order pair (r, r_t),
where r belongs to R and r_t is the reset transition,
is a transition invariant. With every transition invariant (r, r_t) is associated a path in the reachability graph of the form $M[r>M'[r_t>M$. This path
will be called a decision cycle. DC will denote the
set of all decision cycles. Let $\sim$ be an equivalence
relation on DC, defined in the following way: For
every d_1 and d_2 belonging to the set DC, if $d_1 = M_1[r_1>M_1'[r_t>M_1$ and $d_2 = M_2[r_2>M_2'[r_t>M_2$
then $d_1 \sim d_2$ iff $M_1 = M_2$.

The equivalence class of a decision cycle d will
be denoted by $[d]$. Let d_1 and d_2 belong to the
same equivalence class of relation $\sim$, and $d_1 \neq d_2$.
There are two possible situations: ($r_1 \neq r_2$ and
$M_1' \neq M_2'$) or ($r_1 \neq r_2$ and $M_1' = M_2'$). It is
easy to see that in the first situation r_1 and r_2
are inconsistent for $M = M_1 = M_2$, and in the
second r_1 and r_2 are consistent for $M = M_1 = M_2$. If d_1 belongs to an equivalence class $[d]$ and
$d_1 = M_1[r_1>M_1'[r_t>M_1$, then the class $[d]$ will
be said to own the decision marking M_1' and the
decision transition r_1. If for an equivalence class
$[d]$, there is $card([d]) = 1$, then the class $[d]$ will be
called trivial. Proposition 3 is a direct consequence
of the aforesaid considerations.

Proposition 3. The set of rules R is consistent iff
for every nontrivial equivalence class $[d]$, $[d]$ owns
only one decision marking.

So from the practical point of view it is necessary
to consider only nontrivial equivalence classes. In
the lift driver example, the set R is not consistent.
It is possible to find two nontrivial equivalence
classes which own two decision markings. These

equivalence classes are presented below. For the
sake of simplicity second part of the decision
cycles will be omitted (the reset transition and
the end conditional marking).

$[d_1] = \{(up, 5, 1, dup, yes)[r_2>(stop, dup)...,$
$(up, 5, 1, dup, yes)[r_9>(stop, ddown)...\}$
$[d_2] = \{(down, 0, 1, ddown, yes)[r_4>(stop, ddown)...,$
$(down, 0, 1, ddown, yes)[r_{10}>(stop, dup)...\}$

To make the set R consistent it is necessary to
change rules r_2 and r_4. (They are not right.) The
conditions $p < 5$ and $p > 0$ should be added to
the guards of transitions r_2 and r_4 respectively.
After this modification the set R is consistent.

Let a rule r belong to a set R, where R is a
consistent set of rules. The rule r will be called
independent iff the set $R - \{r\}$ is inconsistent.
The rule r will be called *dependent*, if r is not
independent.

Definition 3. The set of rules R is called optimal
iff for every rule r belonging to the set R, the rule
r is independent.

The following proposition is based on the aforesaid
considerations.

Proposition 4. The set of rules R is optimal iff
for every rule r belonging to the set R, a trivial
equivalence class $[d]$ exists and $[d]$ owns only the
rule r.

In the lift driver example, the set R is not optimal.
It is not possible to find a trivial equivalence
classes which owns the rule r_8. Some examples of
trivial equivalence classes which own other rules
are presented below.
$[x_1] = \{(stop, 0, 0, dup, yes)[r_1>(up, dup)...\}$
$[x_2] = \{(up, 3, 1, dup, yes)[r_2>(stop, dup)...\}$
$[x_3] = \{(up, 3, 0, dup, yes)[r_3>(up, dup)...\}$

So the rule r_8 can be removed from the set R.
The result of aforesaid verification is the set $R = \{r_1, r_2, r_3, r_4, r_5, r_6, r_7, r_9, r_{10}, r_{11}\}$ which is complete, inconsistent and optimal. In Fig.4 the final
version of the D-net for the lift driver example is
presented.

5. CONCLUSION

The methodology of verification the system requirements presented in this paper is a proposition of
a practical procedure, which allows reducing some
negatives. It is also worth observing that if the set
R is changing, some steps of this procedure may
need repeating. The second important thing is an
order of presented steps. Fig. 5 presents the proper
order of the verification procedure steps.

Presented in this paper coloured decision Petri
nets are very useful not only while preparing

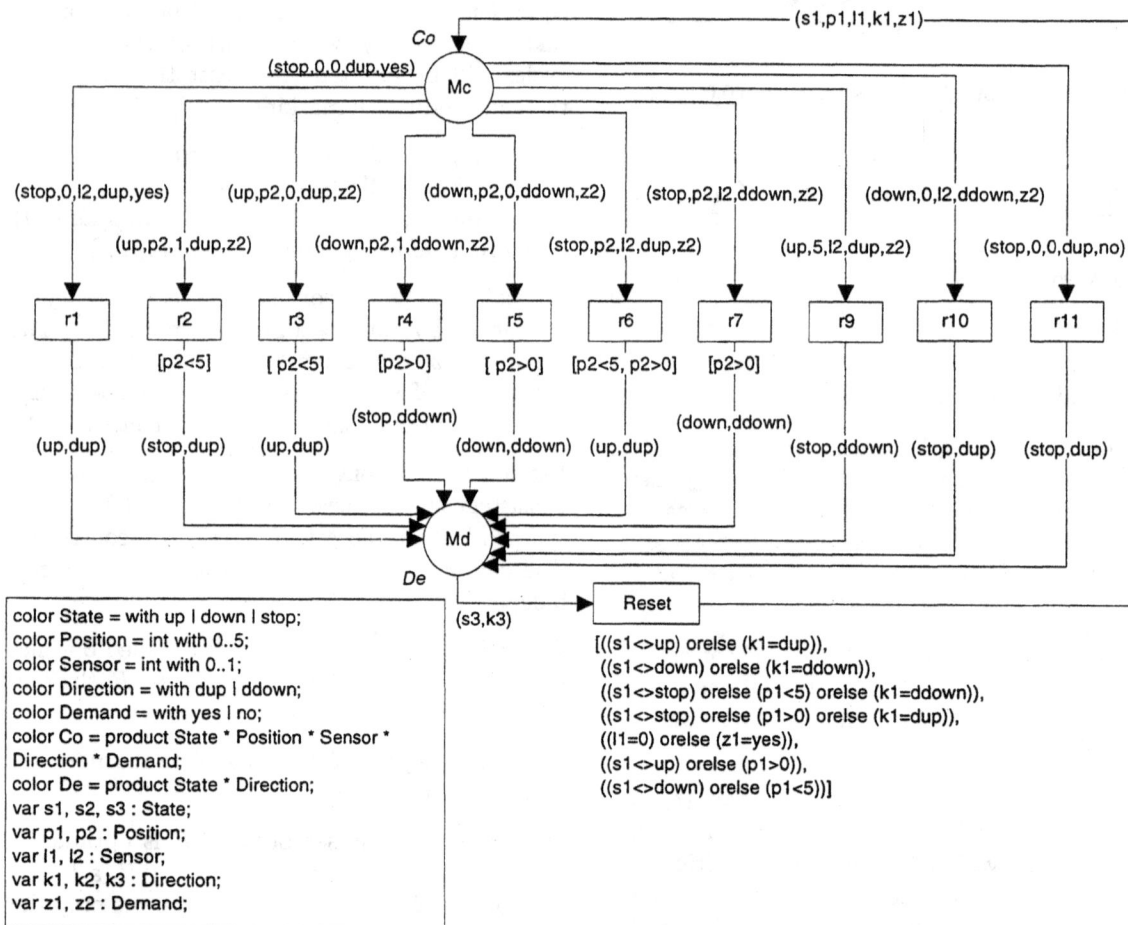

Fig. 4. Coloured decision Petri net built for the operation *Decision* from the decision table presented before

```
color State = with up I down I stop;
color Position = int with 0..5;
color Sensor = int with 0..1;
color Direction = with dup I ddown;
color Demand = with yes I no;
color Co = product State * Position * Sensor *
Direction * Demand;
color De = product State * Direction;
var s1, s2, s3 : State;
var p1, p2 : Position;
var l1, l2 : Sensor;
var k1, k2, k3 : Direction;
var z1, z2 : Demand;
```

Reset
[((s1<>up) orelse (k1=dup)),
((s1<>down) orelse (k1=ddown)),
((s1<>stop) orelse (p1<5) orelse (k1=ddown)),
((s1<>stop) orelse (p1>0) orelse (k1=dup)),
((l1=0) orelse (z1=yes)),
((s1<>up) orelse (p1>0)),
((s1<>down) orelse (p1<5))]

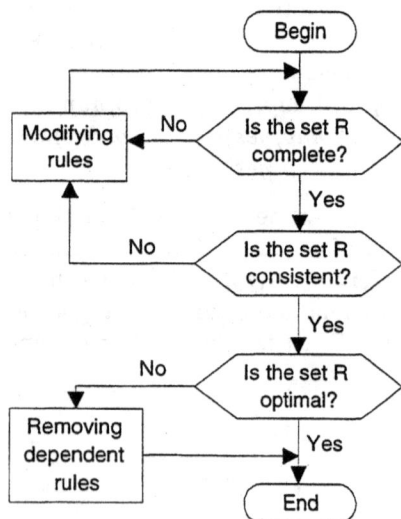

Fig. 5. Scheme block of the verification procedure

specification of requirements. They can be used while designing a dynamic model of the considered system. D-nets are bases for constructing subnets which more precise describe some operations. They can be treated as ready blocks of the dynamic model to be built so the designing phase takes less time.

6. REFERENCES

Braek, R. and O. Haugen (1993). *Engineering Real-Time Systems*. Prentice Hall. New York.

Davis, A.M. (1988). A comparison of techniques for the specification of external system behaviour. *Communication of the ACM* **31**, 1098–1115.

Jensen, K. (1996). *Coloured Petri Nets. Basic Concepts, Analysis Methods and Practical Use*. Vol. 1 and 2 and 3. Springer Verlag.

Macaulay, L.A. (1996). *Requirements Engineering*. Springer Verlag. Berlin.

Rumbaugh, J. (1991). *Object-Oriented Modelling and Design*. Prentice Hall. New York.

Szmuc, T. (1997). Problems of real-time software engineering. In: *Reports of the First National Conference: Computer Systems and Methods in Scientific Researches and Engineering Design*. pp. 1–20.

Szpyrka, M. (1999). Formal verification of real-time systems requirements. *Institute of Computer Sciences Report, Cracow University of Mining and Metallurgy*.(to appear)

AN ANALYZABLE EXECUTION MODEL FOR SDL
FOR EMBEDDED REAL-TIME SYSTEMS

José María Álvarez* Manuel Díaz* Luis Llopis*
Ernesto Pimentel* José María Troya*

* *Dpto. Lenguajes y Ciencias de la Computación.
University of Málaga. SPAIN*
{alvarezp,mdr,luisll,ernesto,troya}@lcc.uma.es

Abstract: The usage of formal description techniques has arisen as a promising way to deal with the increasing complexity of the embedded real-time systems. However, these techniques do not take into account relevant non-functional aspects as real-time constraints and hardware interaction. This is the case of the Specification and Description Language (SDL), one of the most used formal techniques. In this paper, different mechanisms to improve SDL real-time expressiveness and prevent real-time anomalies are presented. In addition, a new predictable execution model for SDL and its associated analysis technique are introduced to make possible the real-time analysis of the systems implemented. *Copyright © 1999 IFAC*

Keywords: formal description techniques, SDL, real-time analysis, embedded systems

1. INTRODUCTION

The requirements of real-time embedded systems are getting more and more complex and they are difficult to manage with the traditional methodologies used by real-time developers. The use of Formal Description Techniques (FDTs) has been proposed as a promising alternative for the development of this kind of systems (Terrier and Barroca, 1997). FDTs provide the basis for an automated design process, allowing simulation, validation and automatic code generation from the specifications. These techniques were originally developed for the design of telecommunication systems, and for this reason, they were designed to cope with specific characteristics like concurrency, reactivity, etc., that are common to embedded real-time systems. One of the most widely extended FDTs is the Specification and Description Language (SDL), which is an ITU standard (Z.100, 1994), and is currently well supported by commercial tools like SDT (Telelogic, 1998) and Object-Geode (Leblanc, 1996). SDL is based on Extended Finite State Communicating Machines

and can be used through all the development cycle, from specification to implementation, although it is better suited for design purposes. Current object-oriented methodologies based on SDL, as SOMT (Telelogic, 1998), use UML (Booch *et al.*, 1997) in the requirements phase, jointly with some other formalisms to express external dynamic behavior, like Message Sequence Charts (MSCs) (Z.120, 1996). MSCs are another ITU standard that can be used in combination with SDL to simulate and verify the system design. However, SDL presents difficulties, common to other message-based models, to express real-time constraints and to prevent real-time anomalies. In addition, SDL does not take into account other important characteristics of real-time systems as hardware interaction.

On the other hand, Rate-Monotonic Analysis (RMA) (Klein *et al.*, 1993) provides a collection of quantitative methods that enable to analyze and predict the timing behavior of real-time systems. This analysis can be helpful to organize processes and resources in the design to make possible the prediction of the timing behavior of the system.

Some works are being developed to try to integrate this kind of analysis in a SDL model. For example, in (Kolloch and Färber, 1998) an earliest deadline semantics is given for SDL, allowing a mapping from a SDL specification to an analyzable task network. However, they do not take into account hardware interaction or real-time anomalies as priority inversion. Another research line in this context is the one of supplementing SDL with load and machine models, as the one described in (Mitschele-Thiel and Müller-Clostermann, 1998), that uses queuing theory to calculate job and message queuing times. These works can be complementary and useful during the first phases of the design, but a final schedulability analysis has to be done for real-time systems and, in order to achieve this, it is necessary to provide SDL with a predictable execution model.

In this paper a new predictable execution model for SDL is presented. The semantics of SDL includes non-determinism, like unpredictable ordering of messages or unpredictable process activation, and cannot be directly used to model real-time behavior. In addition, interactions with the system (including hardware devices) are always asynchronous and are not included in the design model. The new model allows the specification of hard real-time constraints and the interaction with hardware devices is also included as an important model feature. However, it is still consistent with standard SDL semantics, allowing the use of standard simulation and validation tools.

The paper is organized as follows: Section 2 presents some approaches to let SDL deal with real-time requirements. In section 3 a real-time model for SDL is presented. In section 4, a technique for timing analysis is presented. Conclusions and future work are discussed in section 5.

2. A REAL-TIME MODEL FOR SDL

In this section, the SDL features are extended to express correctly the real-time constraints and to avoid the real-time anomalies. The objective is to introduce as few modifications as possible to the language, and to maintain the semantics of timers as close as possible to the standard semantics.

2.1 Expressing real-time constraints

SDL uses two main mechanisms to express real-time constraints:

- A global clock that can be accessed by means of the now function. This function returns a

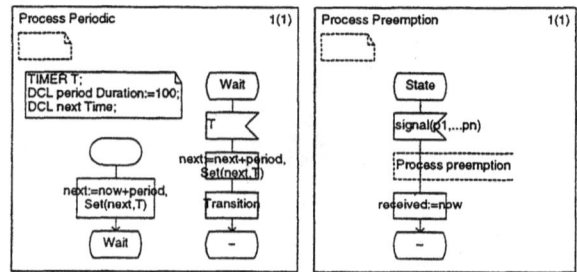

Fig. 1. Real-time requirements in SDL

TIME type value that represents the time in seconds since system initialization.
- Timers: they are special objects that support the SET(time, Timer) and RESET(Timer) operations.

Periodic processes can be specified with these operations (figure 1). The expressiveness of these constructions could be enough, but it depends on the underlying execution model. For example, with the standard FIFO semantics of signal reception, time-out signals are received after all the signals that were in the process queue when the timer expired. With this semantics it is impossible to activate a process exactly when the timer expires and henceforth it is not possible, for example, to specify a periodic process or an exact delay. Another problem with the time semantics of SDL is the difficulty to express real-time constraints involving the sending and reception of signals. There is no way to know when a signal was really sent or received. If in figure 1 the now function cannot be used to know when a signal was received. Since process scheduling can be preemptive, a lot of time could have elapsed between the signal reception and the invocation of the now function.

In order to avoid these difficulties two minimum extensions to SDL have been included:

- Process and transition priorities. Priorities will be assigned to process transitions, depending on the events they are involved and following RMA criteria. Process priorities are calculated taking into account the signals in the process queue and the possible transition on the current state (see section 2.2). Processes can have a default priority that is specified in the process declaration with the keyword with priority.
- Signal Time-stamps. Every signal instance will have two time-stamps recording when the signal was sent and when it was received. These time-stamps can be accessed by means of two predefined functions: sending_time and reception_time, both of them referred to the signal that activated the transition. With these time-stamps it is easy to specify real-time constraints relative to signal sending and reception, as input or output jitter.

2.2 Real-Time anomalies

In addition to real-time expressiveness, SDL presents other real-time anomalies common to other message-based models. For example, data cannot be directly shared among processes and they have to be encapsulated in a server process and, as it is known, this can cause priority inversion. For example, priority inversion can occur when different clients with different priorities try to access to the same server at the same time. In this case, if the standard FIFO semantics is maintained, process with higher priority can be starved. Priority inheritance protocols have to be used in order to avoid these situations.

Precedence constraints can also influence real-time response. In message-based systems, the treatment of any event may involve different related processes in some precedence order. For example, consider two simple external events (Senv1 and Senv2) that are processed by a chain of three processes ProcessH, ProcessM and ProcessL.

The external event Senv1 is initially processed by ProcessH that sends a signal to ProcessM, that, in turn, sends another signal to ProcessL. The other event is processed in a similar way but the precedence relation is reversed. If priorities are assigned to processes in such a way that $priority(ProcessH) > priority(ProcessM) > priority(ProcessL)$, when Senv1 is processed, the behavior of the processes is correct, but when Senv2 is received, its processing can be starved, since ProcessM and ProcessH could also be involved in the processing of other events. This problem occurs because fixed priorities are assigned to processes. If a process can receive signals caused by events with different timing requirements, its priority cannot remain the same to attend the different signals.

Finally, another SDL characteristic that has to be taken into account is hardware interaction. All the interactions with the environment are achieved by means of signals to and from the environment. For this reason hardware drivers have to be active processes independent of the SDL design. In embedded systems, hardware interaction is a great part of the system and must be considered in the design. In order to obtain an analyzable model, hardware timing requirements and driver timing behavior have to be considered.

3. AN ANALYZABLE EXECUTION MODEL

In this section a predictable execution model for SDL is described. This model eliminates non-determinism from the SDL semantics and makes possible the real-time analysis of the implementa-

tions, which are automatically generated directly from a SDL design.

These implementations will run on the final target either with an autonomous SDL run-time library (without operating system support), as the SDT Cmicro library (Telelogic, 1998) or using a real-time operating system to implement the SDL execution model. In both cases, a predictable SDL execution model will be necessary to be able to analyze real-time behavior. When defining an execution model for embedded real-time systems at least the following aspects have to be taken into account:

- Process scheduling. In real-time systems processes have to be scheduled in the appropriate way in order to meet real-time constraints. Preemptive scheduling will be used with fixed priorities that is the base of RMA.
- Process communication mechanisms. SDL uses asynchronous message-passing as the basic mechanism for process communication, although in SDL-92 remote procedure calls have been integrated as an alternative communication mechanism.
- Hardware interaction model. Hardware interaction is not considered explicitly in SDL. All the interactions with the environment have to be achieved by means of asynchronous signal interchange

There are some premises which have to be satisfied in order to be able to define an analyzable execution model: neither dynamic process creation nor implicit signal consumption are considered. The execution model is based on fixed priority preemptive scheduling, however, fixed priorities are not assigned directly to processes but to process transitions. Processes priorities can vary from one state to another depending on the transitions that they can carry out in their current states (taken into account the queued signals). Processes are scheduled according to these dynamic priorities, although the schedulability analysis is based on the transition priorities, which are fixed. Transitions can be preempted by higher priority ready transitions of other processes, but never by a transition of the same process, i.e. if a process transition with higher priority becomes ready while it is executing another transition, this transition is delayed until the current one has finished. This may cause an increment of the response time of events, but this constraint is necessary in order to maintain SDL process execution semantics. Assuming this, processes are preemptively scheduled according to its dynamic priority. In order to show how process priorities are calculated, some sets and functions have to be introduced. Thus, *Process*, *States* and *Signals* denote the sets of system processes, process states and signals, respectively.

The functions *sig: Process × States → ℘ (Signals × N)* and *received:Process → Signals* are used to model, respectively, the signal that a process can receive and the signals actually received by a process. *sig(p,e)* returns sets of pairs *(signal, priority)* indicating the signals that can be received and the priority assigned to the transition associated to that signal. *received(P)* returns the signals that are currently in the queue of process *P*. Initially, a process is assigned its default priority or the priority associated with its START transition. Every time a signal *s* is received in the state *e* by the process *P*, the process priority changes according to the following rule:

$$pri(P) := \begin{cases} max\{pri(P), p\} & if \ (s,p) \in sig(P,e) \\ pri(P) & otherwise \end{cases} \quad (1)$$

The new priority is the maximum between the current priority and the priority of the transition enabled by the signal reception, but only if that signal can be accepted in the current state. If the process was inactive, it can become active and interrupt the current executing process, and if it was active it continues its execution with the new priority. In this case, this priority change can be considered a kind of priority inheritance between the transition currently under execution and the next transition to be achieved by that process. When process *P* changes to state *e* the new priority is:

$$pri(P) := max \left\{ \begin{array}{l} p : (s,p) \in sig(P,e) \\ and \ s \in received(P) \end{array} \right\} \quad (2)$$

Every time a process finishes a transition, its priority is updated for being equal to the priority of the highest priority enabled transition.

Assigning priorities in this way, the transition with higher priority is always executed, except when that transition belongs to the current executing process. In this case, it is necessary to consider the executing transition time as blocking time for the new high priority task. However, this effect can be minimized during design time by trying to avoid having processes attending to possibly concurrent events. This issue is discussed in section 4.

3.1 *Assigning priorities to transitions*

Timing requirements are specified with respect to external events, considering timer expirations as external events too. For each event in the system the sequence of signals exchanged among the SDL processes is considered. This sequence is constructed from the external event by adding to the sequence the signals sent for subsequent processes. If every process only sends a single signal to another process, a sequential string of transitions is obtained, corresponding to actions that have to be carried out on the event occurrence.

The actions achieved for the different processes can be considered equivalent to actions achieved by a single sequential task and, henceforth, all the transitions involved in the same sequential string will have the same priority. This priority will be determined by rate monotonic or deadline monotonic assignment based on the timing requirements of the external event. If a process can send more than one signal in a single transition, each of the signals is considered a new (internal) event, that has to be taken into account in the timing analysis. The priority of the transitions involved in the treatment of this new event will be determined by the ones of the external event. However, when signals are sent to other processes conditionally, the time requirements may be adjusted depending on how often the conditions occur to cause the sending of the message.

Another aspect to take into account is transition sharing. In some systems, transitions may be shared by several events. However, a transition can be executing only on behalf of one event at a time. In this situation, priority inversion can occur if transition priorities are assigned arbitrarily. A possible solution to this synchronization problem is to use the highest locker protocol (González *et al.*, 1991). In this protocol all shared resources have a priority ceiling, that is the highest priority among the processes that can access the resource. Any process trying to access the resource changes its priority to one level higher than the resource priority ceiling. In this way, the process that is accessing the resource will not be preempted by any other process that also wants to use it. This protocol can be directly used to solve the problem of shared transitions, considering the transition itself as a shared resource. In this sense, the transition priority will be one level higher that the maximum priority corresponding to the events in which it is involved, avoiding possible priority inversion problems.

3.2 *Sharing Resources*

As analyzed in section 2.1, resource sharing in SDL can lead to priority inversion situations. This situation can be avoided by using the execution model described above, but data and resource sharing can be very inefficient and difficult to analyze, since it may involve several message exchanges and process context switches. In the model, shared data and resources will be encapsulated into a special kind of processes. These processes are, externally, normal SDL ones, but its behavior is limited in certain ways. (a) They act as passive server processes, i.e. they do not initiate any action by themselves, (b) their only communication mechanism is RPC, i.e. they are always waiting to receive RPCs from other processes, and (c) blocking during transition execution must be bounded.

Each of these processes has a priority ceiling assigned, that is the maximum of the priorities among all the other process transitions where the resource is accessed. In this way, possible priority inversion in data access is avoided and blocking time in shared data access is predictable.

Mutual exclusion is also guaranteed, since all the process transitions will be executed at the higher priority among all the processes that share the resource. Figure 3 shows an example of how a reader and a writer process access shared data using this schema. The priority ceiling of the SharedData process will be 3, since it is accessed from two different transitions with priority 2 and 3, respectively.

In addition, using this schema for modeling data sharing has another important advantage: it can be implemented very efficiently. Although in the SDL model data are encapsulated in a process, this is not actually translated to a real process in the implementation. Each of these processes can be mapped into a set of procedures, one for each transition, inside a module in the target language. These procedures will be called by the processes that share the data after changing its priority to the priority ceiling of the resource.

3.3 Hardware Integration

As discussed in section 2.2, hardware control software must be included in the SDL design. Here, a simple general model for the integration of hardware in an SDL design is provided. Every hardware component is modeled by two different processes:

- A passive process that only executes transitions as a result of a hardware interruption or when its associated driver process calls it. Its priority ceiling will be a hardware level priority and will be system dependent. Only critical hardware operations will be achieved inside this process and with this high hardware priority.
- An active driver process that will interact with the passive one and that will provide the access interface to the rest of the processes in the system. Its priority will be determined by the same rules that the rest of the processes in the system (i.e. depending on the external events it deals with).

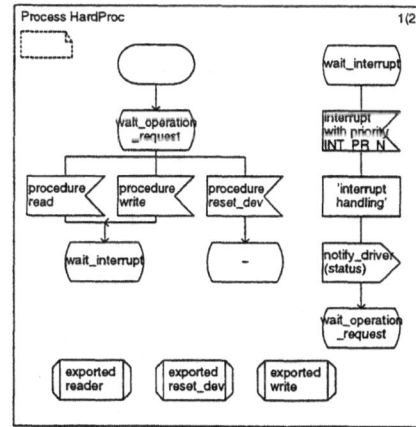

Fig. 2. Hardware Access Process

Hardware registers are described in terms of SDL data types and they will be mapped onto the physical ones during the final implementation phase. These registers are all encapsulated in the passive process.

Hardware interruptions will be modeled by external SDL signals that will be received by the passive processes. The interrupt priority will be indicated in the transition that will attend to that interruption. This transition will be the only one in the state in which the process is waiting for the interruption (in order to prevent the possible interrupt blocking that could occur if the process is attending any call from the driver). Interrupt notifications are sent asynchronously to the driver process (so they can be buffered). With this interrupt handling model, there are three different priority levels, the highest one for attending the interrupt, the one indicated by the passive process ceiling, which indicates hardware operation priorities and the driver priority, that will be determined by the priority of the events it deals with.

In this way the amount of computing carried out at high priority can be minimized, increasing system schedulability. In figure 2 the structure of a passive hardware process and a standard hardware configuration is shown.

4. TIMING ANALYSIS

This section illustrates how the timing behavior analysis of a complete system can be achieved. The analysis is based on the analysis technique presented in (González et al., 1991). With this technique it is possible to calculate the response time for each event in the system. For each transition it is necessary to know the worst case execution time (Ci) and the blocking time (Bi) that can occur in the model under the following circumstances:

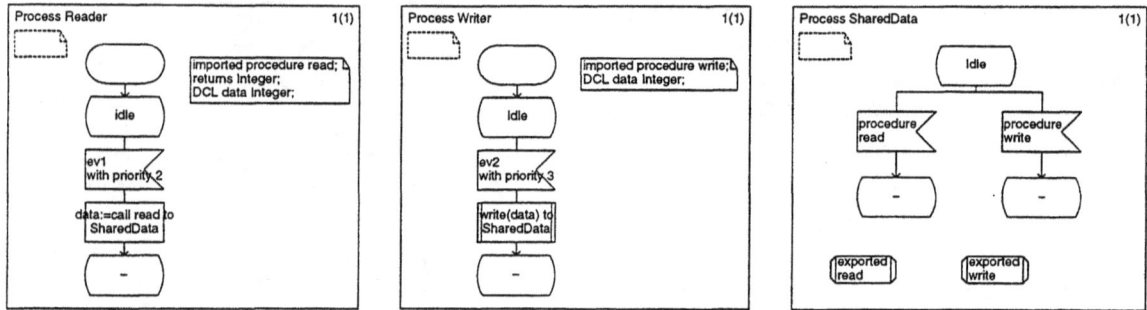

Fig. 3. Resource sharing example

- If a lower priority transition of the same process is under execution. In this case, the worst execution time among all the lower priority transition has to be considered as blocking time.
- If a shared resource is accessed within the transition. The worst execution time among the resource procedures used by lower priority processes has to be considered as blocking time.

In both cases this time is bounded, since immediate priority inheritance is used as described above. SDL atomic actions (timer handling and signal sending and reception) also have to be considered as blocking time. From this data, the response time for each event can be calculated taking into account the transition event string. If all the priorities of the transitions are the same, they can all be considered part of the same sequential process and the response time (Ri) can be calculated with the well-known iterative method given by:

$$R_i = C_i + B_i + \sum_{j \in hp(i)} \left\lceil \frac{R_i}{T_j} \right\rceil \cdot C_j \qquad (3)$$

If there are transitions with different priorities (i.e. the same transition is shared by different events) the method described in (González et al., 1991) has to be used. Other timing characteristics as release jitter, scheduling overheads, etc. can also be easily included in the analysis, but this will be dependent of the SDL run-time system implementation.

5. CONCLUSIONS AND FUTURE WORK

The contributions of this paper could be summarized in two main aspects. On the one hand, Rate Monotonic Analysis is integrated in the SDL development cycle allowing the analysis of the schedulability of the system. On the other hand, a new predictable execution model for SDL has been presented. This model is based on transition priority assignment and priority inheritance, and it considers subjects as precedence constraints, priority inversion and hardware integration.

The conclusions presented at this work are the result of particular experiences of the authors developing embedded real-time software. As future work, an object-oriented methodology including non functional aspects is being developed. It is also interesting the construction of automatic tools to support all the proposed techniques. To do this, there are two possibilities: to extend existing tools, which already support the functional aspects, or construct new ones.

6. REFERENCES

Booch, G., J. Rumbaugh and I. Jacobsson (1997). *Unified Modeling Language. Notation Guide version 1.0*. Rational Software Corporation.

González, M., M. Klein and J. Lehoczky (1991). Fixed priority scheduling of periodic tasks with varying execution priorities. *Real Time Systems Symposium*.

Klein, M., B. Ralya, T. Pollak, R. Obenza and M. Gonzalez (1993). *A Practitioner's Handbook for Real-time Analysis*. Kluwer Academic Publishers.

Kolloch, T. and G. Färber (1998). Mapping an embedded hard real time systems sdl specification to an analyzable task network - a case study. *ACM SIGPLAN Workshop on Languages, Compilers and Tools for Embedded Systems (LCTES'98)*.

Leblanc, P. et al.(1996). *Object-Geode: Method Guidelines*. Rational Software Corporation.

Mitschele-Thiel, A. and B. Müller-Clostermann (1998). Performance engineering of sdl/msc systems. *Workshop on Performance and Time in SDL and MSC*.

Telelogic (1998). Sdt 3.4 manuals.

Terrier, F. and L. Barroca (1997). Object technology and real-time: Problematic and trends. *Object-oriented Technology. ECOOP 97 Workshop Reader* pp. 417–433.

Z.100, ITU (1994). *ITU recommendation Z.100. Specification and Description Language (SDL)*.

Z.120, ITU (1996). *ITU recommendation Z.120. Message Sequence Chart (MSC)*.

H-ASTRAL AND ITS USE IN THE DEVELOPMENT OF REAL-TIME CONTROL SYSTEMS

Klaas Brink* Jan van Katwijk* Hans Toetenel*
Janusz Zalewski**

* Dept. of Technical Informatics, Delft University of Technology
2600 AJ Delft, The Netherlands
** Dept. ECE, University of Central Florida,
Orlando, FL 32816-2450, USA

Abstract: An important issue in the design and implementation of real-time control software is the verification of (time continuous) properties during system development such that even in early stages of the development adequate feedback on design decisions can be given. It is beyond any doubt that formal methods play an important role in this kind of verification. Our research of the past few years focused on applying formalisms in requirements specification and in developing tools for modeling, design and verification of specifications for real-time systems. The development approach builds upon requirements specifications constructed using the formal real-time specification language Astral. In this paper we discuss issues in the use of Astral in development and verification of real-time control systems and give an example of its use. *Copyright © 1999 IFAC*

Keywords: formal specification, real-time software, Astral, inverted pendulum

1. INTRODUCTION

Formal notations provide a concise framework within which software requirements and designs can be expressed unambiguously. The resulting specifications are, at least potentially, suited for analysis and verification. In particular, if the specification is executable, rapid prototyping techniques can be used to develop prototypes.

However, many current formal specification notations do not match the engineering practice existing in application domains. If domain experts are involved in the development of software, formal specification methods are hardly usable, simply because this would require specific skills to be present, that are not.

A similar observation holds in control engineering. Notations and techniques common in the domain are hard to understand for non-domain experts such as average software engineers.

Real-time control systems are a typical example of the class of problems where we have to look for ways to merge expertise from different domains. Indeed, three disciplines are involved: system engineering, control engineering and software engineering.

During the last five years, we have developed an integrated approach for the development of verifiable embedded controllers [14]. In this approach, controllers are modeled and analyzed by *hybrid* notations, i.e., notations that allow the use of both continuous and discrete (time) functions. Central in our approach is the use of Hybrid Astral (H-Astral) [4], an extension of Astral [9] with continuous functions. The notation is supported by a methodology and tools for checking, verification and prototyping.

In this paper, we give an overview of our development approach (Section 2), then briefly introduce the notation, its background and the tool-set supporting the approach (Section 3), discuss a simple case study (Section 4), and derive some conclusions (Section 5).

2. INTEGRATED DEVELOPMENT

In control engineering, system designers focus on modification of system parameters or addition of sub-systems (controllers) to achieve predefined system characteristics [8]. The overall set of desired system characteristics, relating to steady-state behavior, transient response, stability, sensitivity and disturbance rejection, is documented in the system requirements.

During *model creation* a mathematical model of the system to be controlled, including its actuators and sensors, is constructed. Such a model consists of a set of interrelated (partial) differential equations that model nets or circuits of ideal components. Next, a *control strategy* is chosen, based on the physical characteristics of the system to be controlled. The chosen control strategy is used to derive control laws that create the controller structure.

The models of the process together with the mathematical models of the control laws are usually depicted as simple *block diagrams*. To validate a model, analysis and simulation is done using tools such as Matlab, Simulink and alike [16]. By re-iteration, the design is refined into the final model.

The process of software development contains similar phases as in the controller development case. We start with model creation, resulting in a (formal) specification of a model. This model is then analyzed, e.g. by simulation, logic deduction, model checking, etc. The feedback from the analysis is used to re-iterate the model creation.

One of the more intriguing connections between control engineering practice and software engineering practice is the time domain in which the two models are operating [5], [6]. Control models are essentially time-continuous functions, while software models are expressed, by definition, in a discrete time domain. At least one model should be complete, that is, it should specify a closed world, including the interaction between the two components. Therefore, a specification language should be able to handle both discrete and continuous entities.

This need has produced great interest in so called *hybrid* specification notations [11]; hybrid, in the sense that both discrete and continuous entities are present in the notation.

We provide a formal notation that can be used both as a target for a translation from block diagrams and as a source for (formal) analysis and further development. The approach in which the notation is applied is depicted in Figure 1. The control engineer develops a control strategy, which after analysis and simulation is captured in an H-Astral hybrid specification. The software engineer then develops an overall system specification, embeds the controller and analyzes the closed-world specification.

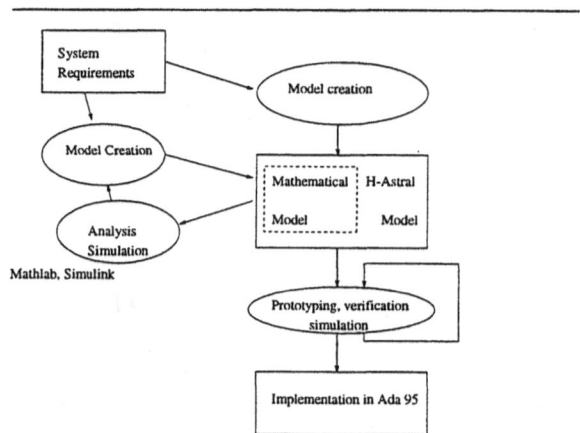

Fig. 1. Integrated approach to embedded controller implementation

The next step in the approach is aimed towards implementation in which the continuous model of the controlled system is replaced by an interface to the real-world in an open-world specification.

3. ASTRAL AND H-ASTRAL

H-Astral builds upon the language Astral, developed by Ghezzi and Kemmerer [10]. An Astral specification can be regarded as a program that runs on a virtual machine, characterized by maximal parallelism, maximal progress and multicast communication. An overview of 'plain' Astral can be found in [7].

In H-Astral [4], we extend Astral in that it allows a specification to contain both continuous and discrete entities. The semantics of H-Astral are based upon hybrid automata [1]. Hybrid automata are a means for formal specification, they lack however, the characteristics to make them useful as end-user vehicle in the specification of systems.

3.1 *H-Astral: An Outline*

An H-Astral system specification consists of a single global specification and a set of process-type specifications. Each process type specification contains:

1. State variable declarations;
2. Transitions (to define state changes);
3. Some (default) process attributes;
4. Specification of the potential communication between different processes;
5. Specification of process properties.

State variables Processes are state-based, that is, their state is defined by the evaluation of its state variables. State variables are declared in a part of the process type specification called the VARIABLE clause

178

Constraints on the initial values of state variables can be specified in the INITIAL clause of the process type. This clause is a predicate over state variables and defines the constraints on the values of the state variables, on startup time of the system. Process instantiation is static, all process instances are defined in the PROCESS clause of the global specification.

Transitions Process behavior is specified operationally, i.e., by defining a set of TRANSITION clauses. Each transition clause consists of a unique transition name, a precondition, a postcondition and a fixed duration.

Relations between state variables The values of the state variables in a continuous process are neither defined nor constrained by transitions. Rather, their value is determined in the RELATION part of the process. This part contains (continuous) time equations or transfer functions, optionally guarded with conditions.

Continuous extensions are based on the transforms used in block diagrams as found in control engineering. The state of a *continuous* process is made up of a set of continuous state variables which are declared in the VARIABLE clause of the continuous process specification.

Exports Communication between a continuous process and other Astral processes is based on a shared data model. In the EXPORT clause of continuous processes only names of continuous state variables can occur, meaning that their value can be read by other processes. Other (discrete) processes can refer to such a variable, for example, in pre-conditions of transition specifications. Transition executions can thus be triggered by predicates becoming TRUE over the state of a continuous process. Communication between continuous processes is disallowed for reasons of simplicity, since from this restriction it follows that interactions between continuous processes are disallowed.

Imports The IMPORT clause of a continuous process specification lists the state variables and constants whose values can be read by the continuous process.

Relations between continuous and discrete entities The values of the state variables of a continuous process are neither defined nor constrained by transitions. Instead the value is defined in a special part of a continuous process specification called the RELATION part. The RELATION part consists of a number of (continuous) time equations or transfer functions optionally guarded with (mutually exclusive) conditions. Each condition defines the situation(s) in which the equations associated with this condition define the values of state variables over time.

3.2 *Tool Support*

Tools are helpful in keeping an overview of the specification (managing complexity) and are useful in pointing out errors or inconsistencies in the specification. In the course of our research project a tool-set for manipulating specifications, in particular for simulation and prototyping of Astral specifications was developed.

To show the feasibility of our approach, we started using the symbolic model checking tools, Hytech [12] and Uppaal [15], and our own model checking tool, Delft PVS [2].

Hytech allows verification of a wide range of properties of real-time and hybrid systems defined as hybrid automata [1] with non-urgent transitions. Uppaal is a tool for verification of timed automata with integer variables. It allows verification of a more restricted set of properties, but is generally more efficient. Delft PVS is a locally developed tool-set for symbolic model checking, using hybrid automata as input.

Most of our tools in the tool-set take H-Astral as input and share a single front end. This front end takes an Astral or H-Astral specification as input and generates an intermediate representation based on a hybrid automaton as output. The other tools in this set take the intermediate output as their input, and are as follows:

1. An Astral to Ada 95 prototyping tool (ast2prot), used to generate a prototype from an Astral specification.

2. A simulator for H-Astral specifications (ast2sim). The translator generates an Ada 95 executable simulating behavior of discrete process specifications from an H-Astral specification. Furthermore, it generates scripts to perform simulations of the behavior of the continuous processes in the specification. The major difference between the simulator and the prototyping tool are the way time is handled. In the simulator, time is simulated, while prototypes are driven by the 'real' time clock.

3. Translators for Astral to some of the more common model checking tools, Hytech and Uppaal, and to the XTG notation, the input language for our own research model checking tool.

Furthermore, we started the development of an Ada to Astral compiler, as a tool to derive suitable abstractions from Ada 95 programs for which formal verification (model checking) is possible.

4. EXAMPLE: THE INVERTED PENDULUM

As a case study we discuss the specification and implementation of a system often called *inverted pendulum*. This problem is typical for the control domain.

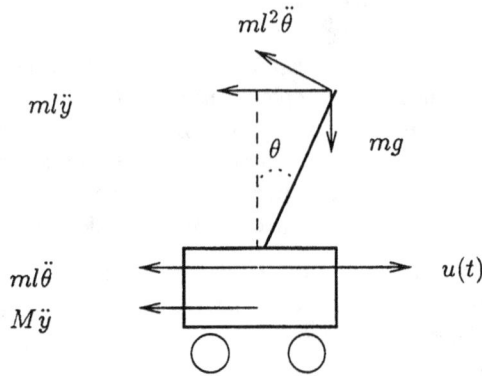

Fig. 2. The inverted pendulum

It is depicted in Figure 2 and its typical behavior is shown in Figure 3.

The pendulum system consists of a cart with a stick mounted on top. A stick (or pendulum) rotates around a mounting point. A force (or acceleration), $u(t)$, can be applied to the cart and the main goal is to control the force, $u(t)$, such that the pendulum is kept in upright position.

4.1 Mathematical Model

In the mathematical model of the system, the following variables are used:

- Angle of rotation θ;
- Mass of the cart M;
- The mass of the stick m;
- Length of the pendulum l.

In case the system is in equilibrium both the sum of the torques (T) and forces (F) equal zero, leading to the following equations:

$$\sum T = ml\cos(\theta)\ddot{y} + ml^2\ddot{\theta} - mgl\sin(\theta) = 0 \quad (1)$$

$$\sum F = M\ddot{y} + ml\cos(\theta)\ddot{\theta} - u(t) = 0 \quad (2)$$

Under the simplifying assumptions that $M \gg m$, $\cos(\phi) \approx 1$, $\sin(\phi) \approx 0$ (when $\phi \approx 0$) and an acceleration u is applied to the cart equations 1 and 2 can be transformed into [8]:

$$u(t) = -l * \dot{\theta} + g * \theta(t) \quad (3)$$

Based on these equations, a state-space formulated system model, representing the controlled system and controller together, (presented in [3]) is defined as follows :

$$\begin{pmatrix} \dot{x}_0 \\ \dot{x}_1 \end{pmatrix} = \begin{pmatrix} 0 & 1 \\ \frac{g}{l} & 0 \end{pmatrix} \begin{pmatrix} x_0 \\ x_1 \end{pmatrix} + \begin{pmatrix} 0 \\ -\frac{K_p}{l}x_0 - \frac{K_d}{l}x_1 \end{pmatrix} u \quad (4)$$

$$\theta = (\, 1 \ 0 \,) \begin{pmatrix} x_0 \\ x_1 \end{pmatrix} \quad (5)$$

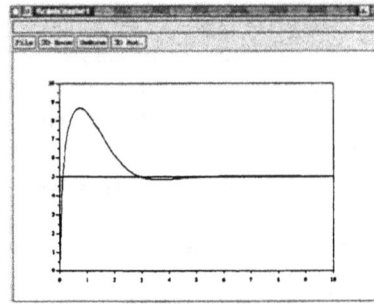

Fig. 3. Example of simulation result

The following state variables are used in modeling the behavior of an inverted pendulum:

x_0	=	Angle of pendulum (θ)
x_1	=	Angular velocity of pendulum
u	=	Acceleration of cart

The elements can be calculated, leading to the following state space equation defining the behavior of the pendulum subject to a controller output u:

$$\begin{pmatrix} \dot{x}_0 \\ \dot{x}_1 \end{pmatrix} = \begin{pmatrix} 0 & 1 \\ 30 & 0 \end{pmatrix} \begin{pmatrix} x_0 \\ x_1 \end{pmatrix} + \begin{pmatrix} 0 \\ -3 \end{pmatrix} u \quad (6)$$

$$\theta = (\, 1 \ 0 \,) \begin{pmatrix} x_0 \\ x_1 \end{pmatrix} \quad (7)$$

From these equations, it becomes obvious that the control value ($u = -h_1 x_0 - h_2 x_1$) is proportional to the current angle (x_0) and the angular velocity (x_1). The values of K_p and K_d are constants and have to satisfy certain conditions to obtain a stable control system [3].

The discrete Proportional Derivative (PD) control strategy applied to control the force applied to the cart, the output of the PD controller is defined by the following equation:

$$u(t) = K_p\theta(t) + K_d\dot{\theta}(t) \quad (8)$$

The relation states that the current value of the control signal $u(t)$ depends on

- the current value of $\theta(t)$ (with weight K_p),
- and the first derivative of $\theta(t)$ (with weight K_d).

Behavior of discrete-time systems is modeled using difference equations rather than differential equations. Discretization (using first-order approximation assuming equidistant observations of the continuous signal) of the equation for the PD controller model (see equation 8) results in the following linear difference equation:

$$u(kT) = K_p * \theta(kT)$$
$$+ K_d * (\theta(kT) - \theta((k-1)T)/T_s) \quad (9)$$

4.2 An H-Astral Specification

It is straightforward to transform the mathematical descriptions given above into a specification. We map the specification of the continuous, external, world into a continuous process, map the PD controller model into a discrete process, and glue the processes together in the global specification part.

4.2.1. The continuous process
The continuous process was described as a relation between $\theta(t)$ and $u(t)$ in the state space notation (equations 6 and 7). This relation is specified in the body of the process, using a notation suitable for processing with tools like Matlab and Scilab.

```
CONTINUOUS SPECIFICATION Inverted_Pendulum
IMPORT u_k;
EXPORT theta_p;
VARIABLE
    theta_p : REAL;
INITIAL
    theta_p=0.0;
RELATION
    CASE TRUE : d[x_0;x_1] =
            [ 0,1; 30,1]*[x_0;x_1]+[0,-3]*u_k;
                theta_p = [1,0]*[x_0;x_1];
END RELATION;
END Inverted_Pendulum;
```

The specification states that the external process (the inverted pendulum) maintains a relation, specified in equations (6) and (7). It allows other processes to inspect the variable $theta_p$, and it imports a value named u_k.

4.2.2. The controller process
The controller process is described by equation 9. A corresponding specification of the controller process then follows:

```
SPECIFICATION PDController
IMPORT Control_Val, theta_p, C_p;
EXPORT u_k;
CONSTANT
  T_c, T_s     : Time;
  K_p, K_i,K_d : REAL;
VARIABLE
    u_k        : Control_Val;
    theta_k1   : REAL;
    prev       : Time;
INITIAL
    u_k=0 & theta_k1_1=0 & prev=0;
TRANSITION Sample            T_c
PRE
    now = prev + T_s
POST
    prev = now &
    theta_k1 = theta_p' &
    u_k = K_p*theta_p +
          K_d*(theta_p-theta_k1')/T_s
END PDController
```

The specification states that

- Three items are made visible that are defined elsewhere: a type $Control\_Val$, value $theta\_p$ that was exported from the continuous process,

and a constant value C_p that is defined in the 'main' program.
- The INITIAL clause ensures that the various entities do have a value;
- The precondition states that the transition is enabled on times $n * T\_s$;
- The postcondition states that whenever the transition is fired, $prev$ is updated, the local entity $theta\_k1$ is updated and u_k is updated. Semantics are freezing, i.e. the value of now taken into account in the first relation of the postcondition is taken to be the one at the time the postcondition completed.

4.2.3. The global specification
The global specification is the part where the individual processes are instantiated and connected. In other words, the global specification relates the processes. The continuous and the discrete process are defined and instantiated immediately. For the inverted pendulum example, it reads as follows:

```
GLOBAL SPECIFICATION I_P_C_S
PROCESSES
    invpend    : Inverted_Pendulum;
    ipcontrol  : PDController;
TYPE
    Control_Val = REAL;
CONSTANT
    C_p =    5;
    Kd  =    3;
    Kp  = 11
END GLOBAL SPECIFICATION
```

4.3 Verification

Since the main objective of using a formalized specification notation is the increased ability for verification, we spent some time in looking for appropriate validation and verification techniques. To validate an H-Astral specification, both simulation and verification techniques are useful.

Our ast2im program (discussed in Section 3.2) is able to generate an Ada95 code that simulates behavior of the discrete processes of the specification and a Scilab [13] script to perform the numerical computations required in the simulation of continuous process behavior. Communication between the Ada95 and Scilab parts of the simulator is implemented such that the simulation can be performed on one machine but also on different machines communicating over a network.

For verification, we rely on model checking techniques. As stated earlier, our tool-set supports the translation of an Astral specification to input languages of three model checkers.

At first, it seemed that the resulting automata were too complex to be handled by the model checkers used.

Based on partitioning techniques, however, a hybrid automaton specification could be constructed. The resulting system specification, giving a conservative approximation of the behavior of the control system specified in equation 5, is useful in verifying the following property with Hytech:

If the system initially starts at

$$0 \leq x_0 \leq 1 \ \wedge \ 0 \leq x_1 < 1 \qquad (10)$$

then the system response stays within

$$-3 \leq x_0 \leq 3 \ \wedge \ -1 \leq x_1 \leq 1. \qquad (11)$$

Using Hytech, it was proved that this property holds for the hybrid automaton specification of the system.

Such verification experiments indicate that possibilities exist for automated verification of simple hybrid automata specifications. They also support the claim that simulation is currently the only feasible means to study behavior of more complex hybrid automata specifications.

5. CONCLUSION

In this paper, we discussed the use of H-Astral as a vehicle to bridge the gap between control engineering and software engineering. On one hand, Astral and H-Astral can be used as target for the translation of block diagrams as they appear in control engineering. On the other hand, Astral and H-Astral can be used as a source for (formal) analysis and further development.

In order to support our approach, we developed a tool-set, including tools that translate a (subset) of Astral into an Ada 95 prototype and tools that facilitate use of several forms of formal specifications. Our experiences with the tool-set and with prototyping are very favorable. In all cases we did, we found mistakes and errors in the original requirements specification.

Our current research addresses some additional problems, in particular the transformation of the basic tool-set from a set of prototype tools into a robust user-friendly tool-set and extending our work on XTG. We found XTG a valuable formalism as a basis for the development of a highly efficient verification tool. Although our experiences with Hytech and Uppaal are favorable, the state explosions limit the usefulness of those tools in real-sized specifications.

6. REFERENCES

[1] R. Alur et al. The Algorithmic Analysis of Hybrid Systems. *Theoretical Computer Science*, 138:3–34, 1995.

[2] M.J.Ammerlaan, R.F. Lutje Spelberg, W.J. Toetenel. XTG: An Engineering Approach to Modeling and Analysis of Real-Time Systems. In *Proceedings of the 10th Euromicro Workshop on Real-Time Systems*, pp. 88-97, IEEE Computer Society Press, 1998

[3] R.H. Bishop R.C. Dorf. *Modern Control Systems. Eighth Edition*. Addison-Wesley, Reading, Mass., 1998.

[4] K. Brink. *Interfacing Control and Software Engineering*. PhD Thesis, Delft University of Technology, 1997.

[5] K. Brink, P.M. Bruijn, G. Frensel, J. van Katwijk, and H. Toetenel. Issues in Real-Time Process Controller Realization. In *Proceedings of the 20th EUROMICRO Conference*, pp. 267–274. IEEE Computer Society Press, 1994.

[6] K. Brink, L. Bun, J. van Katwijk, and W.J. Toetenel. Hybrid Specification of Control Systems. In *Proceedings of the ICECCS'95*, pp. 149–152, IEEE Computer Society Press, 1995.

[7] A. Coen-Porisini, C. Ghezzi, and R.A. Kemmerer. *Specification of Real-Time Systems Using Astral*. Technical Report, University of California at Santa Barbara, September 1996.

[8] R.C. Dorf. *Modern Control Systems*. Addison-Wesley, Reading, Mass., 1992.

[9] C. Ghezzi and R.A. Kemmerer. *Astral: An Assertion Language for Specifying Real-Time Systems*. Technical Report, University of California at Santa Barbara, November 1990.

[10] C. Ghezzi and R.A. Kemmerer. Astral: An Assertion Language for Specifying Real-Time Systems. In *Proceedings of the 8th European Software Conference*, pp. 122–146, Springer-Verlag, Berlin, 1991.

[11] R.L. Grossman, A. Nerode, A.P. Ravn, and H. Rischel (Eds). *Hybrid Systems*, Springer-Verlag, Berlin, 1993.

[12] T.A. Henzinger and P.-H. Ho. A User Guide to Hytech. In *Proceedings of the 1st Workshop on Tools and Algorithms for the Construction and Analysis of Systems*, pp. 41–71. Springer-Verlag, Berlin, 1995.

[13] INRIA. *SCICOS, A Dynamic System Builder and Simulator (User Guide)*, Rocquencourt, France, 1996.

[14] J. van Katwijk, J. Zalewski. Merging Formal Specifications and Engineering Practice in Real-Time Systems Design. In *Proceedings of the 3rd World Conference on Integrated Design and Process Technology*, pp. 57-64, Society for Design and Process Science, Austin, Tex., 1998.

[15] K. G. Larsen, P. Petterson, and W. Yi. Model Checking for Real-Time Systems. In *Proceedings of Fundamentals of Computation Theory*, pp. 62–88. Springer-Verlag, Berlin, 1995.

[16] MathWorks Inc. *SIMULINK, User's Guide*, Natick, Mass., 1992.

DEVELOPING PROVABLY CORRECT SYSTEMS
WITH OBSERV

Shmuel Tyszberowicz* Amiram Yehudai*

* *Computer Science Department, Tel-Aviv University, Israel*

Abstract:
The OBSERV methodology for software development is based on rapid construction
of an executable specification, or prototype, of a system. The specification may
be examined and modified repeatedly to achieve the desired functionality. The
objectives of OBSERV also include facilitating a smooth transition to a target system,
and providing means for reusing specification, design, and coding of systems and
subsystems.
This article describes OBSERV, and demonstrates how the methods used in our approach can be used to develop provably correct real-time reactive systems. Correctness
is checked by means of simulation and formally proved with a model checker.
Copyright © 1999 IFAC

Keywords: OBSERV, real-time reactive systems, simulation, model-checking.

1. INTRODUCTION

The OBSERV methodology for software development is based on rapid construction of an executable specification, or prototype, of a system.
The specification may be examined and modified
repeatedly to achieve the desired functionality.
The objectives of OBSERV also include facilitating a smooth transition to a target system, and
providing means for reusing specification, design,
and coding of systems and subsystems.

This article describes OBSERV, and demonstrates
how the methods used in our approach can be
used to develop provably correct real-time reactive
systems. Correctness is checked by means of simulation and formally proved with a model checker.

In the following section we describe OBSERV
and its environment. In Section 3 we demonstrate
how OBSERV can be used to develop provably
correct systems, by means of both a simulator
(Section 3.1) and a model checker (Section 3.2).
Finally, further work is discussed in (Section 4).

2. OBSERV

2.1 General Description

The OBSERV language combines several paradigms to express the behavior of a system. The
class-based approach [1] provides the basic mechanism for building a system from a collection
of objects, with well-defined interfaces between
them. *Finite state machines* are used to model the
behavior of individual objects. At a lower level,
activities that occur within objects (either upon
entry to a state or in transition between states) are
described with the *logic programming* paradigm
(Prolog), thus allowing a non-procedural description when possible.

In OBSERV, the designer defines a system by
identifying objects and the relations between
them. Whenever one designs an object, one actually defines a type, that can be instantiated
as many times as needed. The concurrency of a

---

[1] According to Wegner's classification (Wegner, 1987),
OBSERV is considered a class-based language, i.e., each
of the objects in OBSERV belong to a class, but there is
no inheritance.

system is expressed by constructing it from independent objects; they communicate by sending messages. Objects are isolated from each other by decoupling the knowledge about the receiver of the message from the sender. Each object declares its own *outgoing interface* in addition to the more conventional *incoming interface*. The outgoing interface is the list of messages the object may send out, with names local to the sending object. The mapping between these and the object(s) to which they are sent is placed elsewhere.

We distinguish between three kinds of objects: *primitive, regular*, and *compound* objects. A primitive object encapsulates a simple value and provides operations that set, retrieve, and modify that value. Primitive objects are pure servers (they have no outgoing messages as they do not require any service from the outside). Primitive types are user-defined, like the other types in OBSERV, and not fixed built-ins. For convenience, however, types like *integer, boolean* and *string* are pre-defined.

The behavior of regular objects is described by means of state machines, which consist of states, transitions, and activities. *State variables*, that are objects of any kind, are used to store state information. Transitions from state to state may be conditional, and they occur either due to an *external event* or as a result of a *spontaneous event*. The former happens when a request is sent by another object. The latter is activated either by the end of a state activity (**end** transition), or if within a certain amount of time following the completion of the state activity, no transition out of the state has occurred (**timer(T)** transition).

Two kinds of activities are concerned with regular objects: (i) *state activity*, performed when entering a state, and (ii) *event activity*, performed when the event is served. Activities cause messages to be sent (outside the object or to state variables), and may change the values of state variables. Activities, as well as conditions, are expressed using Prolog commands for computations, with special forms for message sending.

To reduce complexity, we allow only a limited kind of concurrency: a regular object is allowed to respond to one external event at a time,[2] with the only exception that it can respond to simple, side-effect free inquiries (*immediate events*) concurrently to one non-immediate event. More significant concurrency is achieved by combining several objects to a compound object, which is the main structuring mechanism in OBSERV. A compound object is a collection of objects of any

kind, viewed from the outside as a single object with a well-defined interface. The definition of a compound object includes mapping each service supplied by the object (each entry in its incoming interface) to services (entries in the incoming interfaces) of some of its components. We also need to map each entry in the outgoing interface of each component object to entries in the incoming interface of some other component(s), and/or entries in the compound object's outgoing interface. Control in such an object is distributed among its components, by simply specifying to which components each incoming event is directed. The structuring facility supplied by the compound objects, as well as the ability to endow regular objects with their own state variables, encourages the construction of hierarchical systems.

2.2 *Real-Time Aspects*

The **timer(T)** transition enables the specification of the occurrence of timeouts and the initiation of periodic activities. According to (Dasarathy, 1985), there are two categories of timing constraints for a real-time system:

(1) limits on the response time of a system, and
(2) demands on the rates at which users apply stimuli to the system.

Both the response time and the stimuli rates may be further classified by the following restrictions:

- no more than T time units may elapse between the occurrence of two events, stimuli and/or responses, and
- no less than T time units may elapse between the occurrence of two events,
- an event must occur for T units of time.

The **timer(T)** event enables the designer to specify all these classifications. It is possible to set an unbound parameter[3] T thus enabling the user to dynamically specify values for the time a transition should take place.

At most one **timer** transition may emanate from each state. This is sufficient to enable the description of any event that depends on a clock, whether timeouts or periodic events, possibly at the cost of some additional states. For instance, suppose we are in state S_i and we need to exit that state after T units of time, for a known T, have elapsed. The identity of the target state depends on satisfying various conditions, i.e., the **timer(T)** transitions are conditional. This seems to require more than one **timer(T)** transition emanating from the same state. In order to bypass this obstacle, a new state, say S_j, may be added; it will be entered, unconditionally, after T time

---

[2] These events are called *non-immediate events*; if an object has to respond to more than one of these events, the others are queued.

[3] An argument is unbound if it is not assigned a value.

units in S_i, using the timer(T) transition. There will be only conditional end transitions coming out of state S_j. The conditions associated with these transitions, as well as the target states, are those that were planned for exiting from state S_i, using the conditional timer(T) transitions.

A simple modification enabled us to implement a second version of the simulator which supports a less restrictive semantics. This version enables more than one timer(T) transition to emanate from the same state. This requires that, in all the transitions, T has the same value. The end transition may be interpreted as a timer(0) transition.

In order to ease the specification, we have added the timer(Lower,Upper) construct to the syntax of OBSERV. This is straightforward interpreted: the timer transition is activated within the closed interval $[Lower, Upper]$. During the simulation process, a random value within these boundaries is chosen.

2.3 *Messages and Time*

OBSERV views objects as independent entities acting in parallel; their interaction is achieved by sending messages. In this section we consider message passing between objects and the concept of time as seen in OBSERV.

When an object O_1 sends a message E to another object O_2, several things may occur. If E is not expected to return a value, O_1 may proceed without waiting for E to be acted upon. If E does return a value, then O_1 must wait until it receives the value. This waiting period is the result of the time it takes for the message to reach O_2, the time until O_2 can start serving the message, the time O_2 needs to serve it, possibly including further messages sent to other objects, and the time needed for the result to travel back to O_1.

The three kinds of objects in OBSERV vary in their behavior and have different time notions as well. Compound objects can be seen as a structuring mechanism: they do not initiate message sending, and any message received by them is ultimately mapped to either a regular or a primitive object. Thus, when describing the time concept it suffices to consider regular and primitive objects. These two kinds of objects vary in the amount of time they need to handle incoming messages. We assume that local activities, which are essentially computations, take no time to be completed. If the designer wishes to model a situation in which an object takes a measurable time to perform some computation, this can be achieved by using a state with a timer transition. Activities within a primitive object are considered to be atomic. These activities include only computations which

are performed instantaneously. Therefore, if O_2 is a primitive object then the time the regular object O_1 needs to wait for a result depends solely on the transmit time of both the message and the result. We also assume that it takes no time for a regular object to send a message to a state variable (it is considered a local activity). It also takes no time for a result to be returned by a state variable. Hence, if O_2 is a primitive state variable of O_1, then O_1 has no waiting period, i.e., the result is immediately returned.

A regular object O_2 may receive several messages during the same period of time. Immediate messages will be served immediately and concurrently with other activities. Other messages will be queued and handled in the order they were received. This is a simple model. We have also considered and experimented with a scheme that discards messages that have not begun to be served within a certain period of time.

If the designer wishes to limit waiting time, he may design the interaction between O_1 and O_2 so that returning the value is decoupled from sending E; a second message can be designed to return the value. This message may be from O_2 to O_1 so that O_2 sends the result when it is ready or from O_1 to O_2 so that O_1 may periodically ask whether the result is ready. The designer may also use parameters of messages to pass information that directs O_2 how and when to act. In particular, a parameter may be used to pass a time value, so that O_2 can perform some special activities as specified by the designer if a longer time than expected has elapsed since handling of the message began. Note that this is different from the mechanism mentioned in the previous paragraph. While the former mechanism deals with managing the queue, which is ordinarily beyond the direct control of the object, the latter treats design issues.

2.4 *The Environment*

The OBSERV environment is a collection of tools that enable the construction, browsing, checking, and simulation of OBSERV designs. The tools are all built on top of ProLab (Bäcker *et al.*, 1988), a Prolog programming environment developed at GMD (the German National Research Center for Information Technology).

All the tools share a common user-interface, with windowing and scrolling conventions supplied by ProLab. We have followed the *Model-View-Controller* (MVC) paradigm as suggested in Smalltalk (Krasner and Pope, 1988), a paradigm which is also advocated and supported by ProLab. The MVC paradigm supports highly interactive software.

The tools that can be invoked are a static checker, a browser, and a simulator. Each tool in OBSERV has two main parts, both implemented in Prolog: the application level and the interaction level.

The OBSERV description of a system is mapped to the representation language of OBSERV. This language consists of Prolog facts which are used to define objects and their relationships. The common basis for the OBSERV tools is a representation language, which is essentially a collection of special predefined Prolog predicates. Refer to (Tyszberowicz and Yehudai, 1990) for a full description of the representation language.

2.5 *The Simulator*

The simulator is used to execute the prototype as defined using the OBSERV language. We simulate a single object of a user-chosen type, called the system under simulation. The system being simulated includes the entire hierarchy whose root is the chosen type.

The simulator may detect deadlocks, which occur when there are objects waiting for the result of messages they have sent and no object has any action to perform. The interested reader can find a detailed description of OBSERV in (Tyszberowicz and Yehudai, 1992).

3. BUILDING PROVABLY CORRECT SYSTEMS

3.1 *Using the Simulator to Check Properties*

OBSERV's simulator can be used to check the behavior of a system under development. First, we can examine the behavior of each type separately. Once all the primitive and regular objects seem to be correctly implemented, we can check compound objects, which are actually subsystems at various levels of granularity.

By using the simulator we can check the functional correctness of the system. We can also find desired properties that the developed system does not fulfill. This, however, is no proof that indeed all the properties and constrains are met. For this, a formal method is needed.

Using the trace option of the simulator—tracing the messages sent among objects—we actually create a file that contains a kind of use-cases (Jacobson *et al.*, 1992). Those cases help both in the simulation and during the formal verification.

3.2 *Formal Verification of Properties*

A formal proof that an implementation meets its specification is very important especially in embedded real-time systems. When the program is described as a finite-state machine, which is the case in OBSERV, model checking can be used.

One of the problems with model checkers is the need to write programs as state machines. This is not the usual way programs are written. OBSERV's specifications are written as finite state machines, and this simplifies mechanical translation into an input to SMV, a model checker developed at CMU (McMillan, 1992). As already mentioned, the OBSERV description of a system is mapped to the representation language of OBSERV, which consists of Prolog facts. Each type in this database of facts serves as a module, and is separately translated.

Primitive objects can be seen as state machines with only one state, which is their value. We abstract data, replacing them by messages. Thus, for example, the predefined type boolean has an incoming event *set(N)*, where N is true or false. We replace it by the input signal *set_N*. The model checker will arbitrarily choose the value—true or false—of the input signal. In order to ensure that the model checker will infinitely many times use the value true, we add set_N to the fairness statement of SMV.

The translation of regular objects is almost straightforward. For each incoming event and for each state, we define a boolean variable with the name of the event or state. When the variable associated with the incoming event is true it means that the event occurred at that point in time. When the variable associated with the state is true, it means that we are in that specific state. Only one variable representing a state will be true at any step. We can also prove this property using the model checker.

For a compound object we recursively translate each of its component objects. Then, in order to avoid name clash (the developer can use the same names inside different objects) we use a renaming mechanism.

The mapping mechanism causes some minor activities, such as removing declarations and the fairness statements of variables that are no longer input events, defining the values of internal events, etc.

The detailed translation process is described in (Tyszberowicz, 1998). This article describes how OBSERV was used to implement a provably correct Production Cell system.

The desired properties of an OBSERV implementation are verified in a stepwise manner. First, each regular and primitive object is separately verified, focusing on modular properties. Then, we verify properties of each compound object, the last one being the system itself.

In the Production Cell example we proved safety ("something bad never happens") and liveness ("something good eventually happens") properties. Thus, for example, one of the safety demands was that the belt emits a signal (*ready_to_consume*) which means it is allowed to put a blank on it, only when the belt has no blanks on it (at the state *unloaded*). The CTL formula to verify this property of the belt (i.e., this is checked when the belt is designed) is

$$AG(ready\_to\_consume \rightarrow unloaded\_left).$$

$AG(f)$ means that the formula f holds on every path from the initial state; i.e., f holds globally. In this example we state that the belt emits *ready_to_consume* only when it is in the *unloaded_left* state.

An example of a liveness property that was proved is that once the belt is in its unloaded state, it will eventually arrive its at-edge position (loaded with the blank).

$$AG(unloaded \rightarrow AF\ at\_edge).$$

Both properties have been demonstrated when the belt object was simulated. In order to prove the correctness, we needed the model checker.

4. FURTHER WORK

OBSERV has been used to construct executable specification of systems under development. The properties of the methodology enabled us to specify reactive real-time systems. After running simulations, which demonstrated the behavior of the system, we used a model checker to verify that the system indeed always fulfill some desired properties. The model checker has mainly been used to verify safety and unbounded responsiveness properties. Those properties are important for any system. For real-time systems, however, many of the responses are constrained in time. Failure to fulfill those time constraints means failure of the system. SMV supplies algorithms that compute the minimum and the maximum number of occurrences of a condition on any path between two given events. We are trying now to see how to use them to verify some quantitative temporal properties (like responsiveness within a given number of time units, etc).

MASS (Gafni, 1997) is an activation oriented approach for hierarchical representation of real-time systems that we are currently working on. The language MASS models a system as an act—a collection of tasks, with reactions that prescribe which events should trigger the activation of each task, and what are the real time constraints on the termination of these response tasks. A related real-time logic PLOT for plant modeling and requirements specification was also developed. A deductive proof system is used to verify a MASS design against the PLOT requirements. We plan to use existing model checkers to help the verification process.

5. REFERENCES

Bäcker, A., R. Budde, K. Kulenkamp, A. Meckenstock, K.-H. Sylla and H. Züllighoven (1988). Prolab, a Prolog programming environment user's manual. Technical report. GMD. St. Augustin, Germany.

Dasarathy, B. (1985). Timing constraints of real-time systems: Constructs for expressing them, methods of validating them. *IEEE Transactions on Software Engineering* **11**(1), 80–86.

Gafni, V. (1997). MASS/PLOT, A Specification Framework for Real-Time Systems. PhD thesis.
Tel-Aviv University. ftp://ftp.math.tau.ac.il/pub/amiram/gafni_phd.ps.Z.

Jacobson, I., M. Christerson, P. Jonsson and G. Overgaard (1992). *Object-Oriented Software Engineering – A Use Case Driven Approach*. Addison-Wesley/ACM Press.

Krasner, G.E. and S.T. Pope (1988). A cookbook for using the model-view-controller user interface paradigm in smalltalk-80. *Journal of Object-Oriented Programming* **1**(3), 26–48.

McMillan, K. L. (1992). Symbolic Model Checking: An Approach to the State Explosion Problem. PhD thesis. Carnegie Mellon University.

Tyszberowicz, S. (1998). How to implement a safe real-time system: The OBSERV implementation of the production cell case study. *Real-Time Systems Journal* **15**(1), 61–90.

Tyszberowicz, S. and A. Yehudai (1990). OBSERV–The representation language. Technical Report 169/90. Eskenasy Institute of Computer Science, Tel-Aviv University. Israel.

Tyszberowicz, S. and A. Yehudai (1992). OBSERV–A prototyping language and environment. *TOSEM* **1**(3), 269–309.

Wegner, P. (1987). Dimensions of object-based languages design. *Proceedings of OOPSLA '87, SIGPLAN Notices* **22**(12), 168–182.

FROM TIMED AUTOMATA TO TESTABLE UNTIMED AUTOMATA

Eric Petitjean* Hacène Fouchal*

*RESYCOM
Université de Reims Champagne-Ardenne,
Moulin de la Housse, BP 1039, 51687 Reims Cedex 2, France
Fax : (33) 3 26 05 33 97
e-mail::{Eric.Petitjean, Hacene.Fouchal}@univ-reims.fr

Abstract: This paper presents a method which translates a timed automaton into an untimed one. The purpose of this process is to generate an automaton that we can test by using classical methods for untimed automata. We first present an extended model of Alur and Dill's timed automata in order to take into account more constraints. Then we show the implications of these definitions on the generation of the region graph and its minimization. We give also an overview of the different steps needed for testing these automata. *Copyright © 1999 IFAC*

Keywords: Testing, Timed Automata, Real-Time Systems, Protocol Engineering, Labeled Transition Systems

1. INTRODUCTION

Timing properties take an important place in the development of new technologies as real-time and multimedia systems. Since time-evolving systems dis-functionings may have catastrophic consequences, testing them has become an inevitable, but not really common issue, because of the great difficulties generated by time dynamics. The aim of testing is to ensure that the implementation of a system respects its specification, i.e., the system will operate correctly once it has been implemented. Most of these methods model a system as an automaton where a transition is composed of an input part (a stimulus coming from the environment) and an output part (a reaction of the system). The test generation produces sequences of inputs (actions) from the specification, and the implementation must be able to execute these sequences (called 'test sequences') and produce the correct outputs.

In this paper, we present a technique for testing timed systems, which consists in a translation of the timed automaton into a minimal region graph from which we generate timed test sequences. We apply these sequences on an implementation under test (IUT) using a test architecture which takes into account timing aspects. The paper is organized as follows: Section 2 presents some related works about timed testing. Section 3 gives some theoretical notions about the timed automaton model and the extension we propose. Section 4 details the different steps of the transformation of the automaton. Section 5 presents the different parts needed for testing any system. Section 6 concludes and gives some ideas about future works.

2. RELATED WORK

While untimed testing is well developed yet, timed testing is still a new field. Nonetheless, some studies, from which we have developed our own work, have been undertaken. In (En-nouaary et al., 1997b), the authors present a simple method for timed test sequence generation based on the transition tour of the entire region graph of the control part. A technical report of the university of Amsterdam (Springintveld et al., 1997) gives a general outline of an adaptation of the canonical tester for timed testing. (En-nouaary et al., 1997a), proposes a timed testing methodology based on the transformation of the region graph into an untimed automaton upon which the Wp-method (Fujiwara et al., 1991) is applied. This study has been extended in (Petitjean and Fouchal, 1998b) to test the implementation on each vertex of the clock zones.

3. THEORETICAL BASIS

In this section, we shall recall the definitions of *timed input output automaton* and *region graphs*, which represent the basis of our work.

Timed input output automata have been proposed to model finite-state real-time systems. Each automaton has a finite set of *states* and a finite set of *clocks* which are real-valued variables. All clocks proceed at the same rate and measure the amount of time that has elapsed since they were started or reset. Each transition of the system might reset some of the clocks, and has an associated enabling condition, which is a constraint on the clock values. A transition can be taken only if the current clock values satisfy its enabling condition.

Definition 3.1. **Clock constraints and clock guard**

A *clock constraint* over a set C of clocks is a boolean expression of the form x **oprel** z where $x \in C$, **oprel** is a classical relational operator $(<, \leq, =, \geq, >)$, and z is either an integer constant n, or a clock y, or their sum $y + n$.

A *clock guard* over C is a conjunction of clock constraints over C.

It is important to notice at once that all these constraints can be expressed by a relation of the form $\Theta(x_1, \ldots, x_{|C|})$ **oprel** k where Θ is linear and k is an integer constant.

Definition 3.2. **Timed Input Output Automata**

A timed input output automaton (Alur and Dill, 1994) A is defined as a tuple $(\Sigma_A, L_A, l_A^0, C_A, E_A)$, where :

- Σ_A is a finite alphabet, split in two parts : the input actions, beginning with a "?", and the output actions, beginning with a "!"
- L_A is a finite set of states,
- $l_A^0 \in S$ is the initial state,
- C_A is a finite set of clocks,
- $E_A \subseteq L_A \times L_A \times \Sigma_A \times 2^{C_A} \times \Phi(C_A)$ is the set of transitions.

An edge (l, l', a, λ, G) represents a transition from state l to state l' on input or output symbol a. The subset $\lambda \subseteq C_A$ allows some clocks to be reset with this transition, and G is a clock guard over C_A. $\Phi(C_A)$ is the set of clock guards over C_A

Definition 3.3. **Clock valuation**

A clock valuation over a set of clocks C is a map v which assigns to each clock $x \in C$ a value in $\mathbf{R}^+$ (set of nonnegative reals). We denote the set of clock valuations by $V(C)$.

A clock valuation v satisfies a clock guard G, denoted $v \models G$, if and only if G evaluates to true under v.

For $d \in \mathbf{R}^+$, $v + d$ denotes the clock valuation which assigns the value $v(x) + d$ to each clock x. For $X \subseteq C$, $[X \mapsto d]v$ denotes the clock valuation over C which assigns d to each $x \in X$, and agrees with v over the rest of the clocks.

Definition 3.4. **Clock region**

Let $A = (\Sigma_A, L_A, l_A^0, C_A, E_A)$ be a timed input output automaton.

The equivalence relation $\sim$ is defined over the set $V(C_A)$; $v \sim v'$ iff :

(1) for each linear constraint $\Theta(x_1, \ldots, x_{|C_A|})$ **oprel** k appearing in at least one transition of A, $\Theta(v(x_1), \ldots, v(x_{|C_A|})) \leq k \Leftrightarrow \Theta(v'(x_1), \ldots, v'(x_{|C_A|})) \leq k$

(2) for each tuple $(x_i, x_j, c_i, c_j) \in C_A^2 \times N^2$ such that x_i **oprel** c_i and x_j **oprel** c_j are constraints of A,$((v(x_i) \leq c_i) \wedge (v(x_j) \leq c_j)) \Rightarrow ((v(x_i) - v(x_j) \leq c_i - c_j) \Leftrightarrow (v'(x_i) - v'(x_j) \leq c_i - c_j))$

A clock region for A is an equivalence class of clock valuations induced by $\sim$. Let $[v]$ denote the clock region to which v belongs.

This definition of clock regions differs rather appreciably from the original definition given in (Alur and Dill, 1994). We expose here the main reasons of this modification and its mathematical and logical basis.

The first reason is that Alur and Dill's works deal very well with automata where the clock constraints are only expressed as comparaisons between a clock and a constant value, but do not take into account comparaisons between two clocks, or comparaisons involving the difference between two clocks, which appear in the new definition 3.1, while testing transitions containing such constraints is often a crucial issue.

The second reason is that the definition of clock regions given in (Alur and Dill, 1994) considers comparaisonsbetween clocks and integer constants which do not appear in the transitions of the automaton. This produces almost always artificial and useless separations, and thus create far too many regions, compared to what we obtain when we only take into account the constraints actually expressed in the automaton.

In order to avoid these problems, we have chosen to make the definition of the clock regions much closer to the internal structure of the automaton, and in particular to its clock constraints. We therefore only consider the separations demanded by the automaton, and we express it in the first condition of our definition of the $\sim$ relation.

On the other hand, we always need to know from any moment which one of the constraints will first change its value when time elapses. The second condition of our definition allows us to check it concerning the constraints which are comparaisons between a clock and a constant value. The new constraints x **oprel** y and x **oprel** $y + n$ do not demand this sort of associated c omparaisons because their value (true or false) does not depend on the elapse of time.

Definition 3.5. **Clock zones**
A zone z is a convex polyhedron formed by clock constraints. It consists of a union of clock regions.

Definition 3.6. **Region graph**
Let $A = (\Sigma_A, L_A, l_A^0, C_A, E_A)$ be a timed input

output automaton. A (classical) region graph of A is an automaton $RA = (\Sigma_{RA}, S_{RA}, s_{RA}^0, E_{RA})$ where:

- $\Sigma_{RA} = \Sigma_A \cup \delta$, where δ represents the elapse of time
- $S_{RA} \subseteq \{\langle s, [v] \rangle \mid s \in S_A \land v \in V(C_A)\}$
- $s_{RA}^0 = \langle l_A^0, [v_0] \rangle$ where $v_0(x) = 0$ for all $x \in C_A$
- R_A has a transition, $q \xrightarrow{a}_{RA} q'$, from state $q = \langle s, [v] \rangle$ to state $q' = \langle s', [v'] \rangle$ on action a, iff either
  - $a \neq \delta$, E_A contains a transition (s, s', a, λ, G) and there is $d \in \mathbf{R}^+$ such that $(v+d) \models G$ and $v' = [\lambda \mapsto 0](v + d)$,
  - $a = \delta$, $s = s'$ and there exists $d \in \mathbf{R}^+$ such that $v' = v + d$.

Definition 3.7. **Zone successor**
Let $RA = (\Sigma_{RA}, S_{RA}, s_{RA}^0, E_{RA})$ be a region graph. A zone z' is said to be a zone successor of a zone z for symbol a iff there exists a transition $q \xrightarrow{a}_{RA} q'$ where $q = \langle l, Y \rangle$ and $q' = \langle l', Y' \rangle$ with $z \subset Y$ and $z' \subset Y'$.

4. TRANSFORMATIONS OF THE AUTOMATON

The idea of the translation of a timed automaton, representing a timed system, into an untimed equivalent one, is obviously inherited from (Alur and Dill, 1994) where Alur and Dill introduce the notion of region graph, a labeled transition system where the time constraints are expressed within the states of the automaton, which are now a combination between the original states and a clock zone, i.e. a polyhedron of allowed clock values.

The authors of (En-nouaary *et al.*, 1997b) and (Ennouaary *et al.*, 1997a) have built a methodology around this idea, based on the transformation of the timed input output automata into a flattened region graph. We have adapted their method to give here a simpler way that leads from a timed system to an semantically equivalent untimed one, upon which classical test generation techniques can be applied. We will illustrate the successive steps of this translation by applying them on the example of system represented by the timed input/output automaton of figure 1

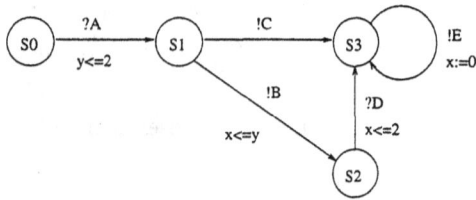

Fig. 1. An example of system

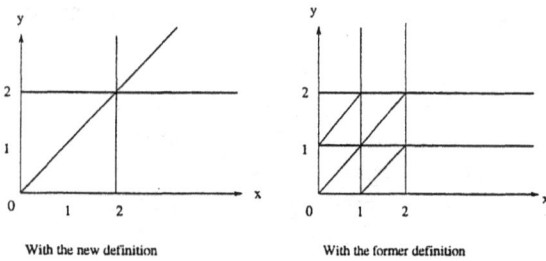

With the new definition   With the former definition

Fig. 2. The regions

Zones	Definition
z_1	$x \leq y \leq 2$
z_2	$x \leq 2$ and $y > 2$
z_3	$2 < x \leq y$
z_4	$y < x$

Fig. 3. List of zones

4.1 Region graph

The first step consists in translating the timed automaton into its associated region graph. The theoretical mechanism of this transformation is well known and given in (Alur and Dill, 1994), and we apply it using the new definition 3.4 of the equivalence relation of the clock values. Using the new relation $\sim$ does not really affect the transformation algorithm since it deals only with the way the regions are computed. However, it leads to more pertinent clock zones, in respect to the temporal behavior of the system, and moreover a great part of the minimization work is already done. The obtained regions are given in the figure 2, where they can also be compared to what they would have been had we used the former definitions.

4.2 Minimization

The second step is the minimization of this region graph. It is often a necessary step since most of the classical test generation methods, on finite state machines as well as on labeled transitions systems,

can only be applied on minimal systems. An algorithm for this minimization is given in (Alur et al., 1992). It is noticeable that the first timed automaton could have been minimized as well using the algorithm of (Yannakakis and Lee, 1993) We obtain, once this step executed, a minimal labeled transition system semantically equivalent to the first times automaton.

Concerning our example, the clock zones obtained in this minimal graph appear in the figure 3.

4.3 Changing labels

The third step is not really a transformation of the LTS and is only due to the way we want to generate test sequences from it, and then execute them on the IUT. Since most test generation methods are based on the transitions, and mostly on their labels, we have to make all the information we need appear on them. Therefore, the idea is to make the time behavior appear again on the transitions, without coming back to a timed automaton. In (En-nouaary et al., 1997a), the authors proposed a method which consisted in splitting again the clock zones into the regions which compose them. The obtained automaton was actually testable but it was also extremely big, because that transformation caused the system to forget its minimization.

In fact, in order to express again the time constraints on the transitions, all we have to do is rename the transition labels as follows : if the transition is labeled by an input symbol, we add the clock zone of the head state of the transition ; if it is labeled by an input symbol or the **tps** symbol (also often denoted as δ) which represent the elapse of time, we add the clock zone of the tail state of the transition. Consequently, a transition $(s_i, z_i) \xrightarrow{?A} (s_j, z_j)$ becomes $(s_i, z_i) \xrightarrow{?Az_i} (s_j, z_j)$.

We finally obtain a minimal labeled transition system upon which we can apply some classical test sequences generation methods based on LTS, or that we can translate into an untimed input/output finite state machine about which a lot of methods have already been developed.

If we start with the system given in figure 1, we finally obtain the labeled transition system of figure 4.

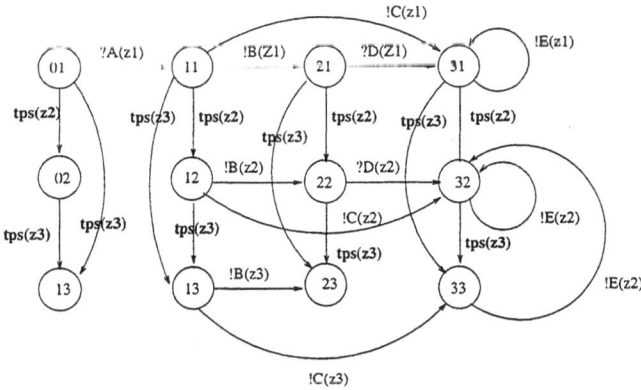

Fig. 4. The obtained labeled transition system

5. TESTING STEPS

5.1 Constraints on TIOA for testing

This study deals with a class of timed input output automata which satisfy the following constraints:

(1) one initial state, i.e. the system can only start from one state with all clocks initialized to zero,

(2) there is no outgoing transition from the initial state labeled with an output action since in testing we apply test sequences to the IUT after bringing it back to its initial state (assuming a reset to initial state exists),

(3) deterministic on the set of alphabet, i.e., from any state, we cannot have two outgoing transitions labeled by the same symbol and whose clock guards are satisfied simultaneously,

(4) the system does not have to choose between an input and an output, i.e. from any state, we cannot have an outgoing transition labeled by an input symbol, and an outgoing transition labeled by an output symbol, whose clock guards are simultaneously satisfied,

(5) each transition in the automaton is executable, i.e. the system is sound where any clock guard can be satisfied by at least one clock valuation.

5.2 Test generation

In the last twenty years some methods have been developed (Chow, 1978), (Sabnani and Dabhura, 1988),(Fujiwara *et al.*, 1991). We have chosen to use the Wp method (Fujiwara *et al.*, 1991) which is one of the most efficient so far. This method

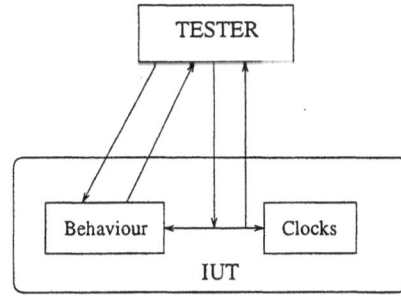

Fig. 5. The test architecture

is based on a characterization of each state by a set of particular sequences of actions. As we deal here with a timed system, we will obtain particular timed test sequences.

Definition 5.1. A timed test sequence is a tuple of test cases $E, Z_1/S, Z_2$ where :

- $E \in I$ where I is the set of input symbols,
- $S \in O$ where O is the set of output symbols,
- Z_1 is the clock zone in which the transition must be fired,
- Z_2 is the clock zone to which the transition must lead.

The elapse of time is represented by the action **tps** and integrated here as an output action. It will be considered as a classic output action during all the test generation, and its temporal nature will be only taken into account during the test execution.

5.3 Test architecture

In order to apply the test cases on the implementation, we use a test architecture inspired by the well-known one defined in (ISO, 1991)and which has been proposed in (En-nouaary *et al.*, 1997b). The IUT (Implementation Under Test) is composed of two parts : the behavior one which expresses the control part of the system and the clock one which contains all the clocks of the system. Both will communicate with the tester via two PCOs (Point of Control Observation), one of which is associated to the behavior part and another to the clock part. Concerning the behavior part, the tester may submit input symbols to it and receive output symbols from it. Concerning the clock part, all the tester can do is asking for the value of one or several clocks, values that it will immediately receive.

5.4 Test execution

During the test execution, the tester submits input signals to the behavior part when the clock valuation belongs to the zone of the test case. When the response is received, it checks whether the symbol received is the same as in the expected one, and if the clock values belong to the clock zone specified in the test case.

6. CONCLUSION AND FUTURE WORK

Our contribution in the study is the extension of the timed automata model of Alur and Dill. We have shown the consequences of this extension on the region graph (which is semantically equivalent to the timed automata) and on its minimization. Then we have presented the important steps required for testing timed systems. We have adapted the classical testing methods for timed automata using a particular test architecture (En-nouaary *et al.*, 1997*b*). For the future, we intend to improve the testing architecture in order to remove the access to the time part of the system to test. We are presently working on a method for testing the region graph without transformation. We are also interested in defining a complete fault model, a first version of which is presented in (Petitjean and Fouchal, 1998*a*). We will focus too on the development of a tool considering the entire methodology, which is currently undertaken in our laboratory. Another interesting issue of this study would be to generate timed test sequences for timed systems described with a high level language as SDL, ESTELLE or ELOTOS. Some ideas have been expressed in (Fouchal *et al.*, 1998).

7. REFERENCES

Alur, R. and D. Dill (1994). A theory of timed automata. *Theoretical Computer Science* **126**, 183–235.

Alur, R., C. Courcoubetis, N. Halbwachs, D. Dill and H. Wong-Toi (1992). Minimization of timed transition systems. In: *Proceedings CONCUR 92*, Stony Brook, NY, USA (R. Cleaveland, Ed.). Vol. 630 of *Lecture Notes in Computer Science*. Springer-Verlag. pp. 340–354.

Chow, T.S. (1978). Testing software design modeled by finite-state machines. *IEEE Transactions on Software Engineering* **SE-4**(3), 178–187.

En-nouaary, A., H. Fouchal, R. Dssouli and A. Elqortobi (1997*a*). Test derivation for timed systems. Report, leri-98-09-01. Département DIRO (Université de Montréal) et LERI-RS (Université de Reims).

En-nouaary, A., R. Dssouli and A. Elqortobi (1997*b*). Génération de tests temporisés. In: *Proceedings of the 6th bi-Annual Colloque Francophone de l'ingénierie des Protocoles*, Lièges, Belgique.

Fouchal, H., M. Defoin-Platel, S.Bloch, P. Moreaux and E. Petitjean (1998). Generation of timed automata from estelle specifications. In: *Proceedings of the International Workshop the Formal description Technique ESTELLE*, Evry, France.

Fujiwara, S., G. von Bochmann, F. Khendek, M. Amalou and A. Ghedamsi (1991). Test selection based on finite-state models. *IEEE Transactions on Software Engineering* **17**(6), 591–603.

ISO (1991). Conformance Testing Methodology and Framework. International Standard 9646. International Organization for Standardization — Information Technology — Open Systems Interconnection. Genève.

Petitjean, E. and H. Fouchal (1998*a*). A Fault Model for Timed Testing. Report: Leri-rs-98-05-02. LERI-RS (Université de Reims).

Petitjean, E. and H. Fouchal (1998*b*). Timed Testing Using Clock Vertices. Report: Leri-rs-98-05-01. LERI-RS (Université de Reims).

Sabnani, K. and A. Dabhura (1988). A protocol test generation procedure. *Computer networks ans ISDN systems* **15**, 285–297.

Springintveld, J., F.W. Vaandrager and P. R. D'Argenio (1997). Timed Testing Automata. Report CS-R9712. CWI. Amsterdam.

Yannakakis, M. and D. Lee (1993). An efficient algorithm for minimizing real-time transition systems (Extended abstract). In: *Proceedings of the 5th International Conference on Computer Aided Verification*, Elounda, Greece (C. Courcoubetis, Ed.). Vol. 697 of *Lecture Notes in Computer Science*. Springer-Verlag. pp. 210–224.

REAL-TIME OPERATING SYSTEMS ON THE TEST-BENCH

M. MÄCHTEL and H. RZEHAK

University of the Federal Armed Forces, Computer Science,
Werner-Heisenberg-Weg-39, D-85577 Neubiberg, Germany

Abstract: This paper deals with the timing behaviour of operating systems, which are mostly relevant for the design of real-time applications. Therefore influences on the run-time performance of an operating system are shown. A measurement method based on monitoring is introduced to determine the system behaviour for the worst case scenario. Finally the paper summarizes the results of measurements of different real-time operating systems using this method. *Copyright © 1999 IFAC*

Keywords: Real-time operating systems, real-time performance metrics, process dispatch latency time, kernel latency time, monitoring, worst case.

1. MOTIVATION

Real-time applications can only work correctly, if time conditions which are given by physical processes are taken into consideration. A computer system, working at its optimal capacity must be able to produce a logically correct result at a specific point in time (deadline).

If a logically correct result is available too early, it can not be used to its full potential. On the other hand if this result is available after the deadline it can result in jeopardizing human life or causing immeasurable costs.

Most of the hardware and software components used in real-time systems are not specifically constructed for these systems. To get the result in time it is necessary to know whether a task or a set of tasks can meet their timing constraints. Due to the high performance rates of standard components, real-time systems can run on a PC or a workstation, making specialized components redundant.

An essential component of a real-time system is the operating system. If this operating system is unable to guarantee timing constraints, therefore real-time capability cannot be fulfilled for the application. In order to assess an operating system it is important to know its influence on the run-time performance beforehand. Manufacturers of operating systems sometimes give specifications like e.g. response times, whereas relevant information on the measurement methods and the measurement environment in use are not published.

The system specifications given by the manufacturer often do not include requirements needed of a real-time system, e.g. upper bound of response times as well as a statement about the determinism. This lack of information by the manufacturer has led to the development of measurement tools to provide the required information. These new tools would allow the user to be no longer solely dependent on the information given by the manufacturer.

This paper is set out as follows: Section 2 discusses the metrics which are used in this paper. Section 3 looks briefly at latency times, more information can be found in (Mächtel and Rzehak, 1995). To measure these latency times the method is presented in Section 4. A summary of the results gained by the method are discussed in Section 5. Finally a conclusion is given in Section 6.

2. PERFORMANCE MEASUREMENTS AND RELATED WORK

The operating systems affects the run-time performance of tasks running at user level. The performance metrics used in the following are explained through the different states of a task (Figure 1). A task can be in the state "Resting", "Waiting',

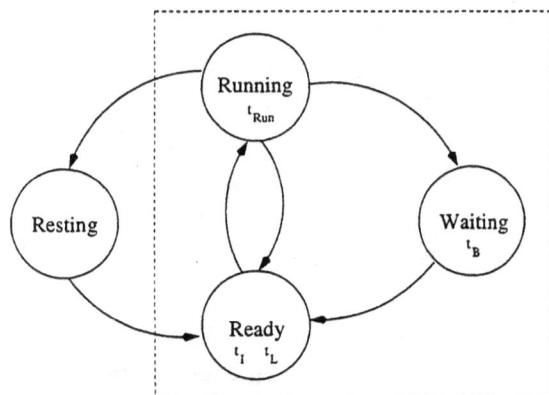

Fig. 1. Task States

"Ready", or "Running". The states are defined as follows:

- "Resting": A task is resting until it is asked to start.
- "Waiting": A task has been started, but must still wait for all requirements for running to be fulfilled, e.g. waiting for input or output or an device.
- "Ready": All operation conditions of the task are completed. The task is ready to run.
- "Running": The task is being processed by the processor.

The response time (t_{Resp}) can be defined as the time that a task has taken to react and complete an event. Often the response time is defined as the time that is required until the beginning of a response. For real-time capabilities this definition does not fulfill the requirements of the system. The response time in real-time systems must be defined until the end of the response where the result is available, not an intermediate response time.

The response time t_{Resp} is the sum of the run-time t_{Run}, the waiting time $t_I + t_B$ and the different latency times t_L:

$$t_{Resp} = t_{Run} + t_I + t_B + \sum_{i=1}^{n} t_L \qquad (1)$$

The run-time of a task (t_{Run}) is the execution time that it takes for the task code to be executed. This execution time has to take into account the run-time on user level as well as the run-time on system level.

The waiting time occurs, when the task is in the state "Ready" or "Waiting" and is further divided into:

(1) The interruption time t_I, which is used for running "tasks" of a higher priority (multi-tasking).
(2) The blocking time t_B, which occurs when waiting for resources which are blocked by other tasks.

The transfer from "Ready" to "Running" must take into account the so-called *latency times*. The internal state of the operating system and changes of this state during the run-time of a task can cause these latency times. Index n specifies the amount of latency times which can occur and is dependent on the system state. Therefore latency times can delay the task.

An obvious sign of latency times is the fact that it is difficult to predict their occurrence. Since latency times can occur at any time within the task it is difficult to estimate the maximum values, since the internal state is not known when an event occurs.

Arising of latency time has been dealt with for the first time in (Doughty et al., 1987) and (Furht et al., 1991). This research concentrates mainly on latency times causing a delay at the start of a response, but not on latency times occurring at a later stage. Latency times occurring at a later stage, e.g. when running kernel functions, were first mentioned in (Rzehak, 1993). Furthermore, the results published in (Furht et al., 1991) and (Furht et al., 1989) can only be seen as average values, not maximum values.

Due to the inconsistency of latency times it is essential for a deterministic system to take account of these latency times. Since latency times rarely occur as peaks one often estimates (with the aid of run-time measurements) an efficient computing system on how the time constraints are fulfilled. Therefore, in order to be more certain, the system is oversized in addition to the estimated value as a safety factor. If taken all these factors into account, run-time measurements can only produce a very general result.

3. LATENCY TIMES

3.1 *Primary response*

The measured latency times are listed briefly in Figure 2. For further information see (Mächtel and Rzehak, 1995) and (Mächtel and Rzehak, 1994).

The reaction of an interrupt starts with the interruption of a running task. Then a branch into an ISR (Interrupt Service Routine), which is re-

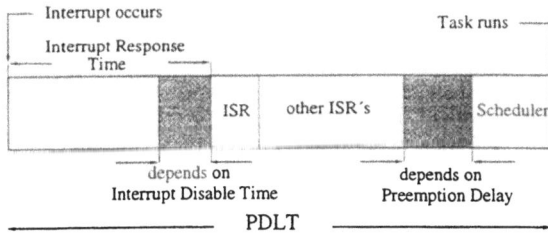

Fig. 2. Process Dispatch Latency Time (*PDLT*)

sponsible for the interrupt follows. This reaction is called primary response. The interrupt response time (see Figure 2) is part of it.

The process dispatch latency time (*PDLT*) can be defined as a time interval. The time interval starts when an interrupt is generated, i.e. the point of time at which the interrupt is caused by an external device. The time interval ends when the first instruction of the task is executed ("Task Runs" in Figure 2). This task is defined as the response task for this interrupt.

The beginning of the response to the interrupt can be delayed by the kernel interrupt latency time (*KIL*). The cause of it is the execution of kernel functions during which an interrupt mask is set for this interrupt ("Interrupt Disable Time").

The interrupt service time (IST) is the time the ISR requires to run the primary response. A part of this response is the notification of the task executing the secondary response, on the condition that this is permitted by the operating system. After the primary response the interrupts of lower priority must be dealt with ("other ISRs").

The call of the scheduler and the context switch (CS) can be delayed by the preemption delay (PD). Due to the finer granulation of the critical region it is possible to use a lower set of the kernel functions in the ISR. The maximum *PDLT* can therefore be calculated as shown in equation 2.

$$PDLT_{Max} = KIL_{Max} + \sum_{i=1}^{n} IST(n) + $$
$$+ PD_{Max} + CS. \qquad (2)$$

3.2 Secondary response

The occurrence of the kernel latency time (*KLT*) during the secondary response is shown in Figure 3. In this example the kernel functions consist of three different regions. Each of these three regions is protected by their own semaphores (1-2-3 and 4-2-5), so that only one task can execute a critical region. Both kernel functions use only one shared object(2). Parts of the kernel function where several competing uses are possible are not shown since no kernel latency time occurs in this case.

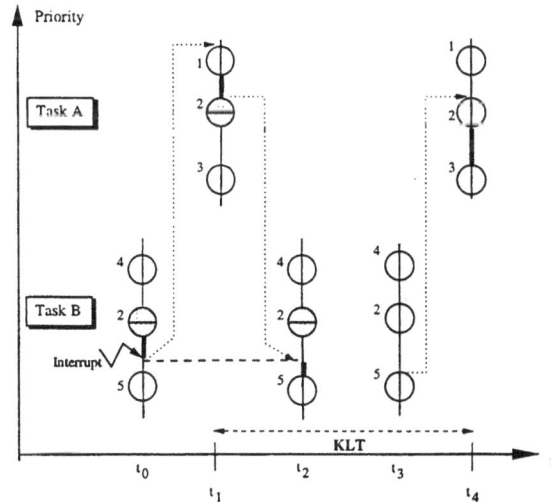

Fig. 3. Kernel Latency Time (*KLT*)

Task A has a higher priority than task B. If an event occurs during a critical region (t_0), the region is protected by the semaphore. A context switch to task A, which is ready to run (i.e. the response to the event) is immediately possible.

At the time t_1 task A is running. Using the same kernel function or a kernel function using elements which have already been locked can cause kernel latency times. This latency time consists of a context switch to task B ($CS : t_2 - t_1$), the ending of the critical region ($CR : t_3 - t_2$) and a context switch back to task A ($t_4 - t_3$).

Therefore the maximum of the *KLT* can be calculated as:

$$KLT_{Max} = 2 * T_{CS} + T_{max.CR} \qquad (3)$$

4. MEASUREMENT METHOD

4.1 States

Figure 2 and 3 show, that latency times can only occur in certain states.

The term *worst-case* describes the situation in which the maximum delay occurs. To find the upper bound of delays in the execution times of the operating-system primitives, as well as the response time of the system, it must be clarified how the worst case arises. Therefore modelling with petri nets can be of great benefit. The petri net in Figure 4 is to be used as an example for the *KIL*. The three states on the left-hand side correspond to the different levels of the kernel function. In case the kernel function runs in level 1, interrupts are masked. In level 2 the kernel function can be interrupted by an event and the primary response (ISR) is run. However, no task scheduling takes place at this level. In level 3 no critical region is executed, this is why dealing with a context switch is permitted. For a general view changing between

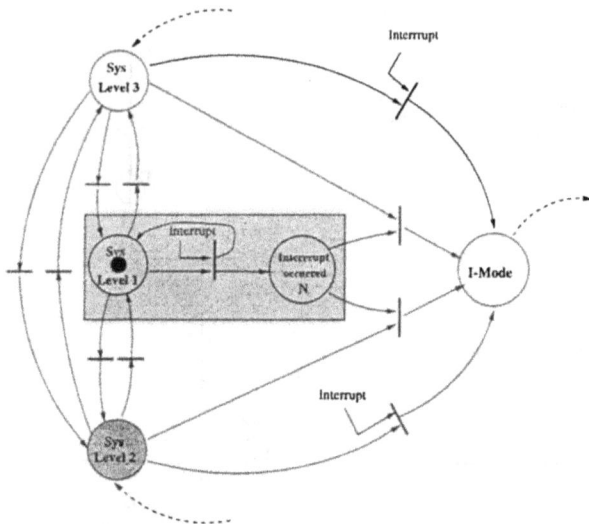

Fig. 4. Petri Net

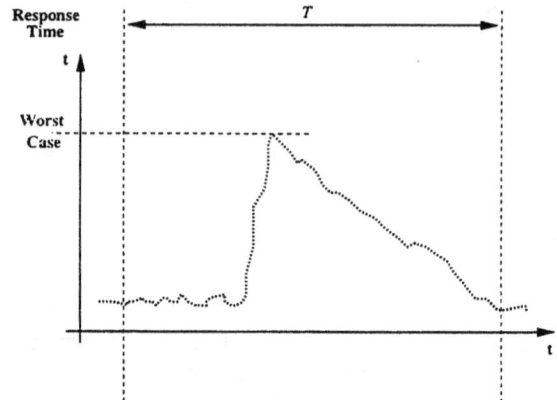

Fig. 5. Principle of Measurement Method

The principle of executing a measurement series is shown in Figure 5, taking into consideration the issues discussed before. The effects on the measurement object can be analysed by the changes of traces. The point of time at which the trace is created, varies during the period of time to be analysed.

all three levels is possible. However, the practical realisation of kernel functions can make certain changes impossible.

If the kernel function runs in level 2 or 3, the ISR immediately runs when an interrupt occurs. If an interrupt occurs in level 1, the place **interrupt occurred** is marked. Since this place is marked, the place "Interrupt Modus" (**I-Mode**) fires, when switching from level 1 to level 2 or level 3.

The maximum *KIL* corresponds to the maximum execution time of the kernel function in level 1, i.e. the longest path in the system code where interrupts are masked.

4.2 *Principle of the measurement method*

In order to assess the respective worst case, information on the system states of the operating system during the measurement is required. Monitors are an excellent measurement method for this purpose.

Information on internal states of the system can be acquired by adding code (event generating) in the operating system code. Among other factors the costs of the operation system code has to be taken into account.

Moreover, analysing a code enables one to analyse critical regions of the kernel functions in order to gain information on the occurrence of the worst case. However, this is extremely time consuming.

Due to the aim of doing measurements in different operating systems, the measurement method should be used on multiple platforms. Therefore, the event generating can not be added into the operating system code.

At the beginning of a period of time T of a measurement series the same start situation is created. The end of the time interval T in which a maximum can occur, can be identified by a specific state.

Analysing the system behaviour during the time period T corresponds to a probing with equidistant steps of time. On the basis of the start state the effects of the following states are recorded by the traces (on the condition that the probing intervals are small enough). According to the duration of the system state, traces following one another can show effects of the same or direct following state.

With the help of recording a sufficient quantity of traces in the respective time interval, the maximum latency time can be analysed.

According to the latency time to be measured, certain start states must be created in order to reach a worst case as a following state. Figure 5 shows the case of a maximum occurring, due to a worst case in the system.

With the help of the models the initial states are determined, which can lead to a worst case. The worst case is produced thus not directly, but occurs as a subsequent state in the executed measurement interval. Therefore the event generation does not have to be implemented in the code of the operating system, since it is not necessary to recognize the worst case by an event.

The time information for the response and processing time can be gained by analysing the time stamps of the respective trace.

The method to measure the *PDLT* and the *KLT* is shown in Figure 6. E_1, E_2 and E_3 are the events, which are recorded (trace).

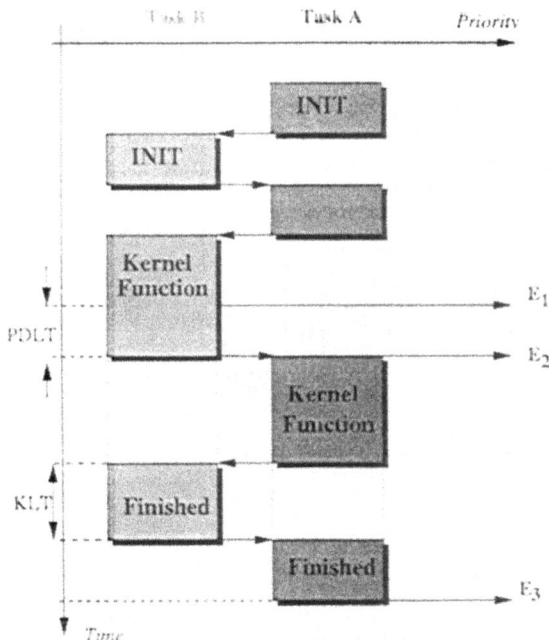

Fig. 6. Measurement Method for *PDLT*

5. MEASUREMENTS RESULTS

To examine the influences of the operating system on the time performance of the total system, the presented measurement method is used. The results for the response time, latency time and execution time of kernel functions of different operating systems are summarized in this chapter. The measurement software has been installed on the systems LynxOS (Singh and Bunnell, 1990), HP-RT (HP, 1992), RealIX (Furht *et al.*, 1991) and QNX (Hildebrand, 1992).

The results of benchmarks are not sufficient to examine the real-time ability of operating systems. With statistical values of benchmarks, no information about the necessary determinism is received. Especially the results of the interaction between different kernel functions is not gathered (shared system data).

Monitoring worked satisfactorily as measurement method. Most measurement errors can be determined by reference measurements. This can be done by recording traces without calling a kernel function. By this the overhead and influence of the monitor can be analyzed. In order to be able to compare results of measurements from different operating systems, the analysis of the reference measurement is inevitable.

By the monitoring likewise conclusions on the implementation of the kernel functions are pos-

sible. Also, predictions about the long-term behaviour of operating systems can be done with this measurement method. Before recording the time stamps, the same initial state is generated. The efficiency of accessing system data structures can be examined during a long period. This influence on the time performance of the operating system can not be neglected.

The measurement method uses as far as possible the respective POSIX (Tec, 1993) interface of the operating system. To a certain extent the results of measurements reflect also how efficient the POSIX interface is integrated. Individual optimizations on different operating systems showed that using specialized kernel functions leads to partial smaller response and latency times.

The examination of the operating system QNX showed unusual latency times during the primary response. The micro kernel of QNX contains smaller non-interruptible critical regions. Within other examined operating systems these regions occurred only when using functions to create tasks e.g. fork().

As an example the *PDLT* of the kernel function access() of QNX is shown in Figure 7, where three critical regions can be seen clearly.

All results presented in this paper have been measured on a INTEL 486 DX2 System running at 66 MHz.

Predictions about real-time abilities depending on the structure of the operating systems cannot be done due to the available measurement results. In addition further investigations are necessary. Advantages of kernel threads (e.g. temporal determinism) are not directly visible in the measurement results.

Depending upon the used kernel function all examined operating systems show more high latencies during the secondary response. This was to be expected, since many of the kernel functions are non-reentrant. A non-reentrant function may lead to unfavorable conditions (interrupting a kernel function) which causes kernel latency times (KLT). Additionally, the influence of the context switch time also affects the maximum of the the KLT_{Max}.

It is particularly remarkable, that the examined kernel functions of QNX showed relatively long execution times. The KLT_{Max} increases as the complexity of the respective function increases.

Particularly problematic is the affect of the *KLT* on systems, which run real-time tasks and standard tasks. If an interrupt for the real-time task occurs, the standard task is interrupted immediately. However, possibly a kernel function is interrupted and thereby, as a consequence the *KLT*

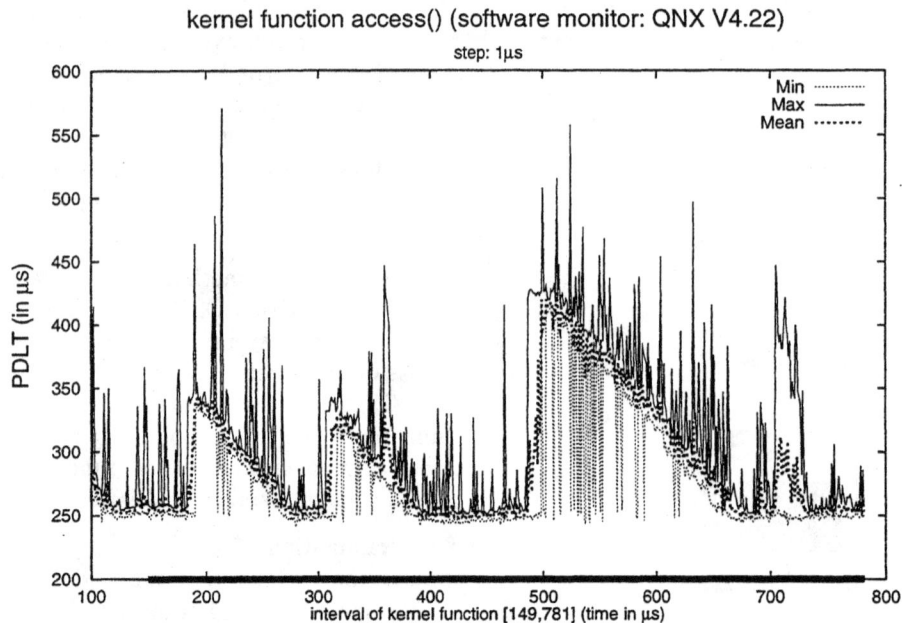

kernel function access() (software monitor: QNX V4.22)

Fig. 7. *PDLT* of access() in QNX

Table 1. Execution Time of Kernel Functions

Kernel Function	LynxOS Version 2.2 in μs	QNX Version 4.22 in μs
open	320	1370
close	35	585
access	330	630
fork	≈ 3000	> 23000

occurs for the real-time task. If different kernel functions access the same system data, this kind of interference may occur. Although the real-time task does not use certain kernel functions due to possible latencies, delays may occur in these systems. A typical example is a system with real-time tasks, on which a standard remote access (e.g. telnet) is running. During an access from the network, the telnet Daemon creates a task by using fork().

In Figure 8 the *KLT* of open() of QNX is shown. KLT_{Max} is about 800 *us*. The function open() is executed by standard tasks at various intervals. For more results see also (Mächtel and Rzehak, 1995).

Taking into account the ever faster becoming processors, the response time and the latency time are reduced. However, extensions of functions are to be implemented in such a way, that the real-time characteristics are not becoming worse.

6. CONCLUSION

The following points result for the construction of real-time operation systems:

- critical regions which increase the *PDLT* should be kept as short as possible. This

is implemented within the most examined operating systems.

- the kernel functions have to be reentrant. The *KLT* can be reduced by a finer granulation of the critical regions, which manipulate system data. This was already be detected in some of the measurement results.

- a clear separation of the system data for real-time tasks as well as standard tasks: So there exist an own kernel for real-time tasks and for standard tasks. Therefore, effects by standard software on real-time tasks can be prevented. But, beyond the complexity of the system and the difficulties of the data exchange between the two classes of tasks, latency times occur when shared hardware resources are used. This can be prevented, when implementing the standard operating system as virtual machine, since no direct access to hardware resources takes place. That is, however, connected with a considerable expenditure of the implementation.

- Multi-level Scheduling: Executing kernel functions of standard tasks is permitted only if in the period, which is necessary for handling the function, no activities for real-time tasks take place. But this presupposes the knowledge of the temporal consequence of the events for real-time tasks.

- Running time-critical tasks on their own platform: The is a dedicated system only running real-time tasks. The functions, which lead to high latencies (e.g. fork()) are called only once, when the system is initialized. When using kernel functions during operation the maximum latency thereby possible (*KLT*) must be determined and taken into

200

kernel function open() (software monitor: QNX V4.22)

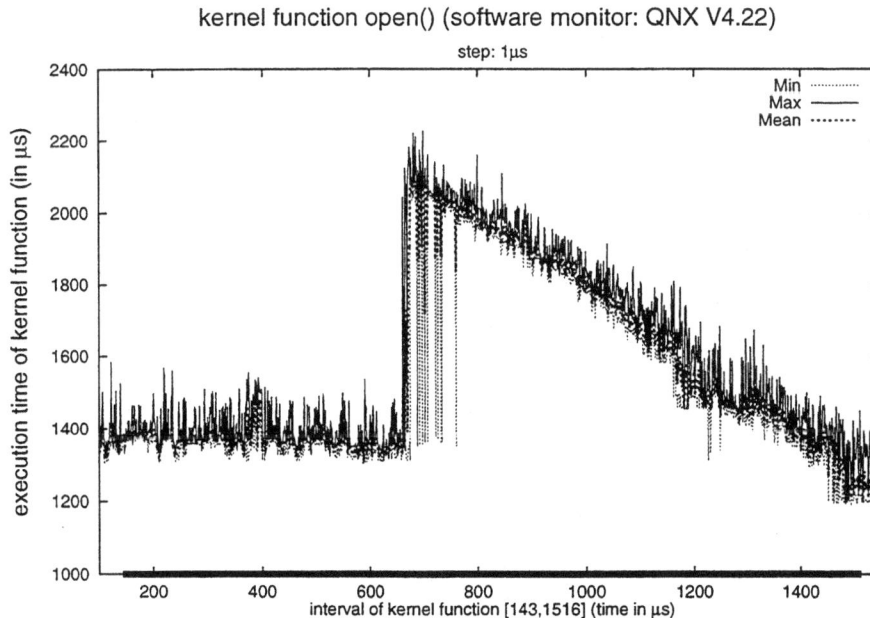

Fig. 8. KLT of open() in QNX

account when the system is designed. Using standard software and the associated advantages are omitted.

Partly these results led to code modifications in the operating system by the manufacturer. To that extent the results of measurements can be referred to problems, which were not considered with the design of the operating system.

Due to the unavailability of the source code all measurement records could not be interpreted to the most detailed level during the analysis. In order to be able to determine the causes of all factors an analysis of the source code is required. It is obvious the measurement method must be executed on an operating system that source code is available.

The operating system LINUX and the extension RT-Linux (Barabanov and Yodaiken, 1997) are good basis for examination since the source code is available freely. Therefore the hypotheses for the causes of the results of measurement can be verified within the code. On the other hand the effects of code modifications of the operating system can be studied by the measurement records.

7. REFERENCES

Barabanov, Michael and Victor Yodaiken (1997). Introducing Real-Time Linux. *LINUX Journal.*

Doughty, S. M., S. F. Karry, S. R. Kusmer and D. V. Larson (1987). UNIX for Real Time. In: *Proceedings of the UniForum 1987.* Washington D.C.. pp. 219–230.

Furht, B., J. Parker, D. Grostick, H. Ohel, K. Kapish, T. Zuccarelli and O. Perdomo (1989). Performance of REAL/IX - A Fully Preeemptive Real-Time UNIX. *ACM Operating Systems Review* 23(4), 45–52.

Furht, Borko, Dan Grostick, David Gluch, Guy Rabbat, John Parker and Meg McRoberts (1991). *Real-Time Unix Systems.* Kluwer Academic Publishers.

Hildebrand, Dan (1992). An architectural overview of QNX. In: *USENIX Workshop on Microkernels and other Kernel Architectures.* USENIX. Seattle, WA. pp. 113–126.

HP (1992). *Application Programming in the HP-RT Environment.* HP Part No. B3127-90001 First Edition.

Mächtel, Michael and Helmut Rzehak (1994). On Realtime Operating Systems: How to compare Performance. In: *IFAC/IFIP Workshop on Real Time Programming.* pp. 139–144.

Mächtel, Michael and Helmut Rzehak (1995). Measuring the Influence if Real-Time Operating Systems on Performance and Determinism. In: *IFAC/IFIP Workshop on Real Time Programming.* Ft. Lauderdale, FL.

Rzehak, Helmut (1993). REAL-TIME UNIX: WHAT PERFORMANCE CAN WE EXPECT. *Control Eng. Practice* 1(1), 65–70.

Singh, Inder M. and Mitchell Bunnell (1990). LynxOS: Unix Rewritten for Real-Time. In: *Seventh IEEE Workshop on Real-Time Operating Systems and Software.*

Tec (1993). *Portable Operating System Interface for Computer Enviroments (POSIX)— Part 1: System Application Program Interface (API)—Amendment 1: Realtime Extension [C Language].* P1003.1b-1993.

TIMING ANALYSIS OF PL PROGRAMS

Man Lin

*Department of Computer and Information Science, Linköping
University, S-581 83 Linköping, Sweden*
linma@ida.liu.se

Abstract: Estimation of Worst-Case Execution Time (WCET) of code is of crucial importance since it can be used to determine the needed computation resources. This paper studies the WCET of PL programs which specify cyclic computing applications. The structure of a PL program differs very much from that of a sequential program. A PL program contains declarative information of the data to be operated on and declarative information of the periodic processes. The WCET of a PL program is defined as WCET for each period of the cyclic application. The processes of a cyclic application may run in different execution modes depending on the context. Not every combination of modes is feasible. Two methods are provided to calculate the WCET of a PL program while taking the infeasibility constraints into account. One method uses Integer Linear Programming technique and the other uses heuristic search-based technique. Timing analysis for multiple-period PL programs is also studied in the paper. The calculated WCET can be used to validate the timing constraints of the system or to help to decide the sampling rates of the system. *Copyright © 1999 IFAC*

Keywords: timing analysis, WCET, cyclic application, execution mode, constraint.

1. INTRODUCTION

Cyclic computing (Lawson, 1992) paradigm is widely used in embedded systems, especially in control-oriented applications. The triggering mechanism of this paradigm is time-based. That is, the processes of a system running under such paradigm are triggered at pre-defined time instances. A simple cyclic controller runs with one sampling rate. At each cycle, the controller reads the sensor input, performs computation and then sends output to the actuators. A complex controller can run with multiple rates, where different processes can have different periods. For more complex controllers, the periodic processes can run at different *modes* depending on the context.

The *Process Layer Executive* (PLX) (Morin *et al.*, 1992) developed at Linköping University is to provide a generic framework to run periodic processes. The processes to be run can have multiple

periods and multiple modes. A language called *Process Layer Configuration Language* (PLCL) was developed to specify the periodic processes and data in operation including the input, output and state data. The programs written in PLCL are called PL programs. PL denotes the Process Layer, which is the lowest layer of a generic layered architecture. The generic architecture (please refer to (Morin *et al.*, 1992) for details) provides a framework to develop autonomous real-time systems.

One requirement of real-time software is to meet timing requirements. Estimation of Worst-Case Execution Time (WCET) of code is of crucial importance since it can be used to determine the needed computation resources. This paper concerns with the WCET of PL programs. As it will be seen later, the structure of PL programs differs very much from that of sequential programs. A PL program contains declarative information of the data to be operated on and of the periodic processes including the running period and the interface of the procedures to be called at different modes. The processes of a PL program will be activated periodically. We define **WCET** for a PL program as WCET for each period of the cyclic application running under PLX. Note that there could be several WCET for a PL program since there might exist multiple periods for a cyclic application.

Intuitively, the WCET of a period for a PL program can be calculated by summing the WCET for all the processes running within this period together with the WCET for PLX needed in this cycle (Morin, 1993). However, the estimation is too pessimistic since the modes of processes are not taken into account. Each process may execute in different *modes* in different situations and it is unlikely that all the processes will take their maximal time to execute in the same cycle. In this paper, we provide a model to express modes of processes and constraints among them. We then describe two methods to automatically derive the WCET for a set of processes while taking the infeasibility constraints into account. The first method uses Linear Programming technique. The second one is a search-based method. Pruning, looking-ahead and back-jumping are applied during the search procedure.

The paper is organized as follows. First, we provide a brief description of the PLX and PL programs in section 2. Then the model of single-period PL programs is provided in section 3. After that, two methods are provided to calculate the WCET for single-period PL programs. ILP-based solution can be found in section 4 and search-based solution can be found in section 5. WCET for Multiple periods is studied in section 6. Finally, the related work and the conclusion are given.

2. PLX AND PL PROGRAMS

The Process Layer Executive (PLX) is an executive running cyclic applications. It maintains the tick counter, the data in operation and the execution of the periodic processes. The data include input, state, and output variables. The data are organized as vectors where the flipping period of the data in the same vector is the same.

The PLX algorithm is defined as follows:

```
INIT vectors;
LOOP
    WAIT FOR tick;
    FLIP;
    EXECUTE appropriate processes;
    DUMP;
END
```

The *flip* operation copies the current state to the previous state and acquires a new sample for the input variables. The *dump* operation outputs output variables. Processes are activated and executed according to their period. When a process is activated, the module functions associated with its active mode will be executed.

The vectors and the processes are specified in a PL program written in PLCL. A vector consists of the period of the vector and the frames(variables) which represent input, state or output. There is one process structure associated with each vector. A process structure contains the descriptions of different *modes* that the process may operate in. Mode switching takes place at run time. But this should not affect the triggering frequency. A process is always triggered with a fixed rate. Each mode is realized by a sequence of transformations called *modules*. The code of a module is obtained from a library (supplied externally). The WCET time of each module can be derived using timing analysis for sequential code (Park, 1993; Puschner and Koza, 1989; Puschner, 1998). The WCET of a mode of a process is thus the sum of the WCET of all the modules defined for this mode.

A simple PL program is shown below.

```
VECTOR gateV is
INTERVAL 1;
SELECT gateM:tSwitch;    INIT registerSelector[];
                         GET  receiveFrame [];
FRAME g:int; VALUE 0; INIT registerInt [];
                      PUT  sendFrameg [];
FRAME dg:int; VALUE 0; INIT registerInt [];
                       GET  receiveFrame [];
END -- vector: gateV

PROCESS gateP@gateV IS
MODE lowering is
  MODULE m(IN *gateV#dg:int, OUT gateV#g:int)
          USE gate_lowering_func [];
END        OR
------- other modes are omitted
MODE open is
  MODULE m(IN *gateV#dg:int, OUT gateV#g:int)
          USE gate_open_func [];
END        OR
ACTIVE open;
END -- process gateP
```

The vector `gateV` has three variables where `gateM` is the mode selector, g represents the angle of the gate. and `dg` represents the change rate of the angle. The `lowering` and open modes are two modes for the process `gateP`:

The time needed for a PL program in one cycle is the time needed by PLX for the management in one cycle and time needed for all the periodic processes running within the cycle. The WCET for PLX has been studied by Morin (Morin, 1993). In this paper, we derive WCET for all the processes of a PL program while considering the constraints among the modes of processes. The model of the processes of a PL program is given in the next section.

3. A MODEL FOR PROCESSES

We first provide the model to express modes of processes which run with a single period and the constraints among the modes.

A system consists of a set of processes which are triggered at every cycle. Each process has a finite number of execution modes. Each mode of a process has a worst case execution time (*wcet*). At each cycle, one and only one mode of each process is executed. Not all the combinations of modes can be executed together. The restrictions come from the semantic of the application and are expressed by a set of constraints. Therefore, a system can be modeled by the fourtuple:

$$(N, N_m, E, C)$$

where

- N is the number of processes.
- N_m is a vector representing the number of modes for each process. $N_m(i)$ denotes the number of modes for the ith process. M is defined as the maximal number of modes for a process, that is: $M = MAX_{i=1}^{N}(N_m(i))$.
- E is a $N \times M$ matrix storing the *wcet* for each mode. $E(i,j)$ represents the *wcet* of the jth mode of the ith process. $E(i,j)$ is set to 0 if $j > N_m(i)$. We assume that the modes of each process are ordered according to their *wcet*, that is, for a process i, $E(i,j) \leq E(i,k)$ iff $j \geq k$.
- C is a set of constraints. A constraint is expressed as a set of pairs indicating the process number and the mode number. The constraint $\{\langle 1,1 \rangle, \langle 2,3 \rangle, \langle 4,5 \rangle\}$ indicates that the 1st mode of process 1, the 3rd mode of process 2 and the 5th mode of process 4 cannot be executed together. The number of pairs in one constraint, called the length of the constraint, can vary from 2 to N.

By selecting one mode from each process, we get a mode combination. The WCET for the set of processes is defined as the sum of the *wcet* of all the selected modes.

A mode combination *violates* a constraint c if and only if it selects all the modes appearing in c. A mode combination is infeasible if it violates any constraint within C.

Given a system (N, N_m, E, C), the goal is to find the maximal execution time for one cycle for any possible feasible mode combination of the processes. This amounts to finding a mode combination $S = \langle m_1, m_2, \ldots, m_N \rangle$, where m_i is the mode selected for the ith process, such that the following holds.

- S is a feasible mode combination with respect to C.
- No other feasible mode combination can have larger WCET than S.

The problem is a *combinatorial optimization problem* (Papadimitriou and Steiglitz, 1982). Next, we provide two methods to solve the problem.

4. INTEGER LINEAR PROGRAMMING TECHNIQUE

The first method to solve the problem is the Integer Linear Programming (ILP) technique. What needs to be done is formulating the problem as an ILP problem.

First, we need to define the variables of the system under consideration. Then we shall express the goal and the constraints over the defined variables.

For each mode of a process, we define a variable Y_{ij} where i represents the process number ($1 \leq i \leq N$) and j represents the mode number ($(1 \leq i \leq N_m(i))$. Y_{ij} being either 1 or 0 indicates that the jth mode of ith is selected or not selected.

The goal function can be expressed as

$$max\Sigma_{i=1}^{N}(\Sigma_{j=1}^{N_m(i)}(Y_{ij} \times E(i,j))).$$

There are two sets of constraints:

- First, one and only one mode of a process is active. Therefore, for each process i, we have one constraint:

$$\Sigma_{j=1}^{N_m(i)}Y_{ij} = 1.$$

- Second, we should translate the constraints in C into equations. Each constraint c corresponds to one equation e. Suppose c has q pairs. Let $\langle c_i^p, c_i^m \rangle$ denote the ith pair. The corresponding equation e is:

$$\Sigma_{i=1}^{q}(Y_{c_i^p c_i^m}) < q - 1.$$

The meaning is: not all the $\langle process, mode \rangle$ pairs can be selected at the same time.

5. USING SEARCH TECHNIQUE

The second method is a search-based method. Next we formulate the problem as a search problem on a graph and later we will use search technique to solve the problem. The algorithm is efficient since some heuristic is applied to reduce the search space.

5.1 Search problem formulation

We construct a graph out of a model (N, N_m, E, C). The graph contains N layers. Each layer contains several nodes. A *node* is a pair $\langle p, m \rangle$ where p indicates the process number and m indicates the mode number. Node $\langle i, j \rangle$ resides at the jth position of the ith layer.

A *path* $P = m_1, m_2 \ldots m_l, (l \leq N)$ on the graph is a sequence of nodes chosen from each layer in order where l is the length of the path. A path can be *complete* or *incomplete*. A complete graph contains one node from each layer while an incomplete graph contains less nodes than N.

The constraints expressed in the original model can be seen as constraints over nodes in the graph. A constraint over a graph is a set of nodes which cannot be selected at the same time.

A path $P = m_1, m_2 \ldots m_l$ *passes* a node $\langle i, j \rangle$ iff $m_i = j \wedge i \leq l$. A path P *violates* a constraint c iff P passes all the nodes in c. That is:

$$\forall n (n \in c \rightarrow P \text{ passes } n).$$

Given a set of constraints C, a path P is *infeasible* if and only if P violates any constraint in C. Otherwise, P is said to be *feasible* with respect to C.

Given E, the *WCET* of a path $P = m_1, \ldots, m_l$, denoted by WCET(P), is defined as:

$$\text{WCET}(P) = \sum_{i=1}^{l} (E(i, m_i)).$$

A complete path $P = m_1, m_2, \ldots, m_N$ corresponds to a mode combination $S = \langle m_1, m_2, \ldots, m_N \rangle$ in the original problem. The aim for the search problem is to find a feasible complete path with maximal WCET.

5.2 Problem solution

To solve the problem, we use depth-first search (Dean *et al.*, 1995) with backtracking. We have developed two kinds of heuristics to guide the search procedure and cut branches. The two types of pruning factors are the following.

- Violation of infeasibility constraints: some branches can be pruned because the incomplete path tested is already infeasible. We called such pruning *Infeasibility-Pruning*.
- The impossibility to exceed the current WCET bound: some branches can be pruned since its WCET can not exceed that of the feasible path found before. We called such pruning *WCET-Pruning*.

5.2.1. *Infeasibility-Pruning*

The search is realized by expanding nodes and backtracking. It wastes time in continuing expansion when a path is found to be infeasible. This is because any path with an infeasible path as prefix is still an infeasible path. Therefore, the paths expanded from an infeasible path can be pruned. Let's consider an example with 4 processes (Fig. 1). Each process has 3 modes. Suppose the constraint set is $\{\{\langle 2, 2 \rangle, \langle 3, 1 \rangle\}\}$. Suppose the current path is $P = \langle 1, 1 \rangle, \langle 2, 2 \rangle, \langle 3, 1 \rangle$ (see Fig. 1 (b)). Since $\langle 2, 2 \rangle$ and $\langle 3, 1 \rangle$ can not execute at the same time, path P is not a feasible path. We should not extend this path any further. That is, the paths shown in Fig. 1 (e), (f), and (g) will be pruned. The immediately subsequent path is shown in Fig. 1 (c).

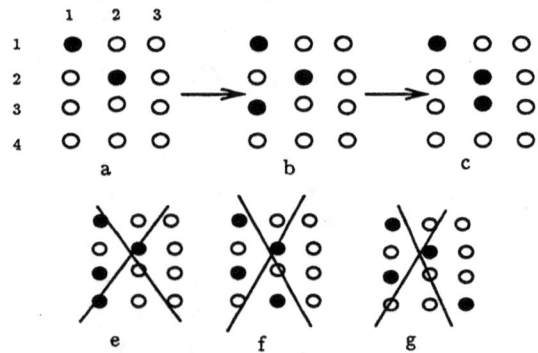

Fig. 1. Infeasibility-Pruning

5.2.2. *WCET-Pruning*

When a complete feasible path is found, the search procedure can not stop since the path found does not necessarily have the largest WCET. However not all its subsequent paths are of interest since some of them obviously have smaller WCET than that of the current path. It is easy to prove (please remember that the modes of each process are ordered by their WCET) that path $P' = m'_1, \ldots, m'_N$ has smaller WCET than $P = m_1, \ldots, m_N$ if

$$\forall i ((1 \leq i \leq N) \rightarrow (m_i \leq m'_i)).$$

P' has the following property w.r.t. P: at any level, the node selected for P' is on the right hand side of that selected for P. However, paths having such property are not continuous meaning that the immediately subsequent path of a path having such relation with P may not still have such relation with P. Therefore, we can not eliminate all these paths from the search space by one pruning. What we can do is to eliminate all the continuous paths with such property at one time. The pruning adopts look-ahead and back-jumping mechanisms.

Instead of backtracking only to the parent node, the search may jump a few levels up. The level to jump to is derived based on the following two observations.

- An increment at the final level of a path won't lead to larger WCET than the original path. Therefore, the level is at most $N - 1$.
- If the node of a path P at any level from L to N is in the leftmost position, then increasing P at any level between $L - 1$ and N won't give larger WCET than the original path P. Therefore, the backtracking level can be set to $L - 2$.

The levels to jump to are $3, 1, 0$, respectively, for the paths shown in Fig. 2 $(a), (b), (c)$. The immediately subsequent path is shown in Fig. 2 $(a'), (b'), (c')$. Note that there is no subsequent path for the path shown in Fig. 2 (c) meaning that the search can terminate.

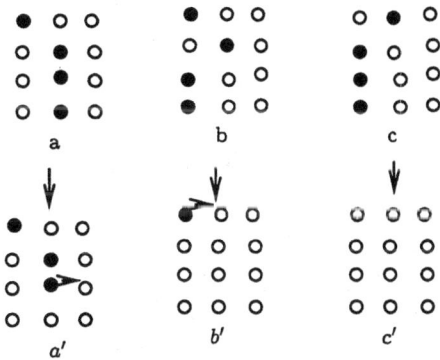

Fig. 2. WCET-Pruning.

5.3 Factors to speedup the search

The fact that the modes of each process are ordered makes WCET-Pruning possible and helps to speedup the search. There are other measures to increase the speed of search. The general idea is to reorder the processes.

- One measure is to put those processes which appear in many constraints, especially in short constraints on the top of the graph. This allows the search to conduct infeasibility pruning as early as possible.
- Another measure is to try to avoid the leftmost nodes of the final levels appearing in any constraints. This helps to jump to smaller levels when conducting WCET-Pruning.

6. EXTENDED MODEL: MULTIPLE PERIODS

So far, the process model concerns only with unique period for all processes. This is usually not the case for most cyclic applications. Different processes may have different sampling rates and computation periods. Therefore, there is a need to extend the original model and compute WCET for multiple periods for the new model. Below, we provide the new model.

The model is a quintuple: (N, N_m, I, E, C), where I is a $N - ary$ vector. I_i represents the factor (an integer) between the period of ith process and the smallest period. Note that any period is assumed to be a multiple of the smallest period. Therefore, I is a vector of integers. The other symbols N, N_m, E, and C have the same meaning as before except that any constraint in C only constrains those modes whose corresponding processes have the same period.

The problem is to compute the WCET for all the periods. The problem is solved in two steps. We first partition the processes of a model into clusters according to their periods. The processes in the same cluster have the same period. The clusters are ordered according to their associated period. The kth cluster can be modeled as

$$cl^k = (N^k, N_m^k, p^k, E^k, C^k)$$

where

- N^k, N_m^k, E^k and C^k restricted to the processes in cluster k have the same meaning as that of N, N_m, E and C in the single-period model. The assumption that C^k only contains constraints among processes in the same cluster (with the same period) makes the partition possible.
- p^k is a number representing the factor between the period of kth cluster and the smallest period of the system.

Then we can apply either method developed in the previous two sections to compute W^k which represents the maximal time needed for executing all the processes in the kth cluster. Note that W^k is just the WCET for running the processes with exactly period $p^k \times T_1$, where T_1 is the smallest period.

The time required for running all the processes within period T_k is defined as $WCET^k$.

$$WCET^k = \sum_{j=1}^{k} (\lceil p^k / p^j \rceil \times W^j),$$

where $\lceil p^k / p^j \rceil$ takes the closest upper integer of p^k / p^j.

To validate whether the timing constraint of the single-period cyclic application is fulfilled, one important criterion is to compare the WCET of the unique period (corresponds to the worst-case reaction time of the controller) and the required minimal sampling time. The WCET should be smaller than the required minimal sampling time. Otherwise, the environment may have already changed during the reaction. For multiple-periods cyclic application, the above criterion can be relaxed since not all the computations should be finished within a single period. To check whether an arbitrary period T_k can hold all the computations needed within such period, we can verify the following resource adequacy condition: $WCET^k \leq p^k \times T_1$ called the *relaxed resource adequacy condition*.

The relaxed resource adequacy condition helps to validate the timing constraints of a cyclic application if the sampling time has been decided. If the sampling time has not been decided, then the relaxed resource adequacy condition can help to decide the sampling time. The basic (smallest) cycle T_1 can be selected as the smallest number which can satisfy the relaxed adequacy condition for any period. After the minimal cycle is decided, all other periods can be decided since the factors are known.

7. RELATED WORK

The focus of this work is to study the WCET of PL programs which can be used to validate whether the cyclic systems can react to continuous changes in time or to decide the sampling time of the system. As opposed to common cyclic applications which only consider independent tasks (Son, 1995), we take into account constraints among the processes of PL programs.

High-level timing analysis is not a new topic in real-time programming community. The major difference between this paper and the studies in the literature are the language constructs. The language constructs studied in the literature are sequential statements (Puschner and Koza, 1989; Shaw, 1989; Park, 1993) such as assignments, functional calls, conditional or loop structure, etc., or reactive rules (Wang and Mok, 1993; Lin and Malec, 1998) while the structure of PL programs contains declarative information of the data to be operated on and the periodic processes. The nature of PL programs is the periodic execution of processes. This differs from sequential code which executes only once. The definition of WCET for a PL program is therefore different from sequential programs. It is defined over periods of the system.

8. CONCLUSION

In this paper, we have defined and studied the WCET for PL programs used to specify generic

cyclic applications. Two methods have been developed to calculate the WCET of one period of a PL program while taking into account the infeasibility constraints of the processes of the PL program. The first method uses ILP technique. The second one uses heuristic search-based technique. It is efficient since pruning and back-jumping are used to reduce the search space. Discussions on how to speedup the search have also been provided. WCET for PL programs with multiple periods can be derived by clustering the process set and then using the two methods developed previously.

9. REFERENCES

Dean, T., J. Allen and Y. Aloimonos (1995). *Artificial Intelligence.* Chap. 4. Addison-Wesley Publishing Company.

Lawson, H. W. (1992). An approach to the engineering of resource adequate cyclic real-time systems. *Real-Time Systems* 4, 55–83.

Lin, M. and J. Malec (1998). Timing analysis of reactive rule-based programs. *Control Engineering Practice* 6, 403–408.

Morin, M. (1993). Predictable cyclic computations in autonomous systems: A computational model and implementation. Licenciate thesis 352. Department of Computer and Information Sciences, Linköping University.

Morin, M., S. Nadjm-Tehrani, P. Österling and E. Sandewall (1992). Real-time hierarchical control. *IEEE Software* 9(5), 51–57.

Papadimitriou, C. D. and K. Steiglitz (1982). *Combinatorial Optimization: Algorithms and Complexity.* Prentice-Hall. Englewood Cliffs, NJ.

Park, C. (1993). Predicting program execution time by analyzing static and dynamic program paths. *The Journal of Real-Time Systems* 5, 31–62.

Puschner, P. (1998). A tool for high-level analysis of worst-case execution times. In: *Proceedings of 10th Euromicro Workshop on Real-time Systems.* Berlin, Germany. pp. 130–137.

Puschner, P. and C. Koza (1989). Calculating the maximum execution time of real-time programs. *The Journal of Real-Time Systems* 1(2), 159–176.

Shaw, A. (1989). Reasoning about time in higher-level language software. *IEEE Transactions on Software Engineering* 15(7), 875–889.

Son, Sang, Ed.) (1995). *Advances in Real-Time Systems.* Chap. 9. Prentice-Hall.

Wang, C. K. and A. K. Mok (1993). Timing analysis of MRL: a real time rule-based system. *Journal of Real Time Systems.*

USING TIMED AUTOMATA FOR RESPONSE TIME ANALYSIS OF DISTRIBUTED REAL-TIME SYSTEMS

Steven Bradley * William Henderson ** David Kendall **

* Department of Computer Science, Durham University, South
Road, Durham, DH1 3LE, UK
** Department of Computing and Mathematics, University of
Northumbria at Newcastle, Ellison Place, Newcastle upon Tyne,
NE1 8ST, UK

Abstract: Rate Monotonic Analysis (RMA) is a well-established technique for assessing schedulability of periodic and sporadic tasks which share a processor resource using fixed priority scheduling. Adaptations of this technique have been made to perform Response Time Analysis (RTA), accounting for jitter, blocking, distributed systems and end-to end timing constraints. However, the nature of the analysis means that, while good bounds can be given for uni-processor systems with relatively little interdependency, the response times calculated for more complex systems can be very conservative.

An alternative approach to analysing such systems is to build a model which represents the behaviour of the system more dynamically, taking into account the dependency between the tasks. To do this, we introduce a simple language for describing the tasks which comprise a system and the precedence relationships between them. From this a timed hybrid automaton is generated which can be analysed automatically to predict end-to-end response times.

Applying this technique in practice yields promising results, with response times lower than those calculated with RTA. However, there is a trade-off to be made between the complexity of the hybrid automaton analysis (which suffers from the state explosion problem) and the conservatism of the more standard RTA approach.
Copyright © 1999 IFAC

Keywords: Scheduling algorithms, Timing analysis, Real-time tasks, Real-time operating systems.

1. INTRODUCTION

Since rate-monotonic scheduling (RMS) and rate-monotonic analysis (RMA) were first proposed (Liu and Layland, 1973), work has been carried out to extend the basic model of computation from a set of independent periodic tasks with fixed execution times sharing a single processor, and to extend the analysis from simple schedulability to system-wide end-to-end response times (Audsley et al., 1993). Audsley et al. (1995) report on the development of the theory supporting fixed prior-

ity pre-emptive scheduling, including extensions to account for interdependence of tasks (through blocking) and the analysis of distributed systems. Analyses of end-to-end response times in distributed systems have been carried out using TDMA (Tindell and Clark, 1994) and more recently with CAN (Henderson, 1998) as the communication mechanism.

Although more recent work has taken account of extra delays incurred through interdependence of tasks, ensuring that the analysis remains conservative, one area which has not received so much attention is to take advantage of the restrictions on possible execution paths brought about by interdependence. This has meant that predicted response times can be overly pessimistic. In this paper we aim to address the problem of pessimism by explicitly modelling precedence (or ordering) relationships between tasks, and performing an exhaustive analysis of all possible execution paths through a system. The analysis is carried out be performing reachability analysis on a timed hybrid automaton (Alur *et al.*, 1995).

The motivation for this work is demonstrated in section 2 through an example which exhibits pessimism under a standard static response-time analysis (RTA) (Audsley *et al.*, 1993). In section 3 we introduce our system model, and describe a simple language (PG) in which such systems can be expressed. We then define a translation from PG into timed hybrid automata in section 4. Sections 5 and 6 explain how response time analysis is carried out, and evaluate the results of the analysis. Finally, our conclusions and relationships with other work are presented in section 7.

2. EXAMPLE OF PESSIMISM

Our simple example consists of a distributed system with two processors connected by a single CAN-style bus, with non-destructive priority-based arbitration. On processor 1 there are two similar tasks, Sender A and Sender B. These tasks are periodically triggered, and they each read a sensor, the values of which need to be sent to corresponding receiver tasks, Receiver A and Receiver B, on processor 2. This is achieved by sending messages Message A and Message B via the bus. We shall assume that each of the tasks and messages has a fixed resource requirement of 200, so we have essentially two clone sub-systems, system A and system B, each of which has a periodically triggered sender task, which triggers a message upon completion, which in turn triggers a receiver task. The tasks and message of system A have higher priority than those of system B.

In a control situation where the receiver task acts upon the information provided by the sender task to provide feedback into the controlled system, a crucial factor in determining whether the system can be controlled in a stable way is the response time from the reading being taken to the control being applied. Therefore, as well as examining schedulability issues on the processors and bus, it is important to be able to predict the end-to-end response time from the start of the sender task to the end of the receiver task.

The analysis of the end-to-end response time for subsystem A is reasonably straightforward,

but subsystem B is more interesting. A standard response-time analysis adds the response times for each section, with each section having contributions from interference by higher priority tasks, blocking by lower priority tasks, and execution time. All sections (Sender, Message and Receiver) of subsystem B are subject to interference of up to 200 time units from the corresponding section of subsystem A, giving a total response time of 600(interference)+600(execution) = 1200. However, taking into account the precedence constraints, we can try to construct the actual worst case for the response time of subsystem B. This occurs when Sender B is released and just fails to complete before Sender A is released, which then pre-empts Sender B. The sequence of events which follows is shown in figure 1. Because of the precedence relationships, it is impossible for competition to occur between subsystems A and B for all three sections if both systems use their maximum resource requirement (200) in all cases. This gives an actual worst-case end-to-end response time of 1000; this kind of pessimism has been noted experimentally (Henderson, 1998).

Such an argument is difficult to construct convincingly by hand, even for a relatively simple example, but the consideration of the many different cases that may occur seems a good candidate for automation. To automate the analysis, however, we first need to declare our assumptions about the systems we analyse, and define the language we are going to use to describe the systems.

3. PG: A SIMPLE LANGUAGE FOR DESCRIBING DISTRIBUTED SYSTEMS

In a simple response time analysis, the precedence relationships between tasks are not made explicit, but the period of a task which is triggered by the completion of some other task is inherited from the triggering task. Our model of computation is very similar to that used in rate-monotonic style analyses, except that we explicitly model these precedence relationships. The most basic components in our model are *tasks*, which can be in one of three states: waiting (to be released), ready (to run, but not running) and running. Each task has a simple life cycle: it is released at some *trigger* event, after which it competes for some *resource* (e.g. access to processor) until completion. The time taken to complete will depend on its *bounded resource requirement* (e.g. processing time) and upon the competition for the resource. Competition is managed using fixed priority preemptive scheduling, although tasks may be declared as *nonpreemptible*, in which case they will retain access to the resource from their first access until their completion. A trigger event can either

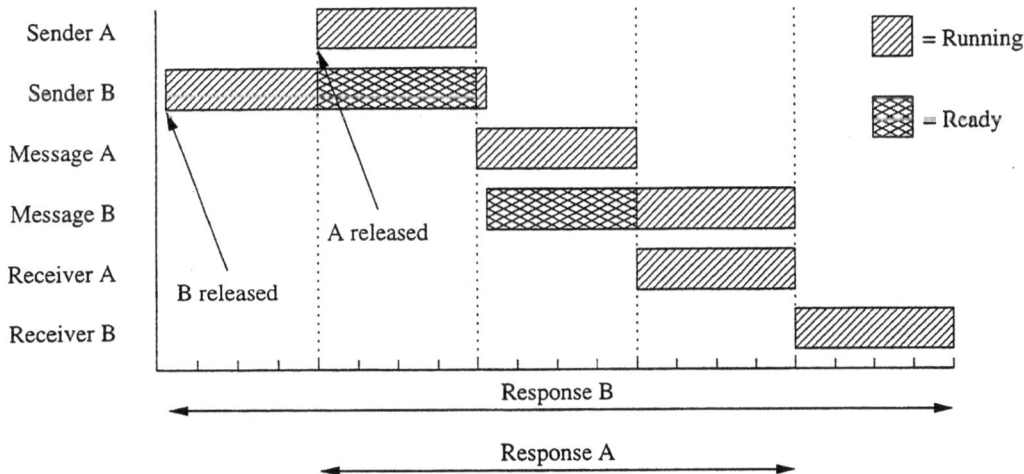

Fig. 1. Example of pessimism

be the elapsing of a fixed period, or the completion of some other task (or message). Communication is assumed to take place via a CAN-style bus with non-destructive priority-based arbitration. In this model, messages have the same properties as nonpreemtible tasks. Critical sections can also be modelled as nonpreemtible tasks. The following assumptions are made:

- each task/message uses exactly one resource
- each task/message has exactly one trigger
- no task/message can be triggered by a task/message that it directly or indirectly triggers (there are no 'loops')
- bounded resource requirements account for any overheads (such as kernel activity, memory management etc)

Important differences between this model and a more standard response time analysis model are that

- Precedence (i.e. triggering or ordering) relationships are made explicit, rather than assuming that all tasks/messages are periodic, with period inherited from their trigger.
- Jitter (i.e. possible delay of release after period has elapsed) is not included, as this is caused by the inheriting of periods from triggers: if an 'upstream' task or message does not complete in constant time, then its completion will not be purely periodic.

In summary, each task or message has a name, an associated resource, a bounded resource requirement, a priority, and a trigger. These are expressed using a very simple syntax: the example of section 2 can be expressed in PG as

```
task senderA on processor1 needs [100,200]
    at priority 0
    triggered by period 1300
task senderB on processor1 needs [100,200]
    at priority 1
    triggered by period 1400
```

```
message messageA on can needs [100,200]
    at priority 0 nonpreemptible
    triggered by senderA on processor1
message messageB on can needs [100,200]
    at priority 1 nonpreemptible
    triggered by senderB on processor1
task receiverA on processor2 needs [100,200]
    at priority 0
    triggered by messageA on can
task receiverB on processor2 needs [100,200]
    at priority 1
    triggered by messageB on can
```

This can also be presented graphically, a more complex example is illustrated in figure 3 where the precedence relationships are shown with arrows (periodic triggers are shown as squares labelled with the period). It is this presentation, as an acyclic directed graph where the nodes are tasks or messages, and the edges are precedence relationships, which gives rise to the name PG: precedence graph.

4. MODELLING DISTRIBUTED SYSTEMS WITH HYBRID AUTOMATA

Before explaining how we translate system descriptions in PG into hybrid automata (section 4.2), we first briefly review the definition of hybrid automata (section 4.1). For a more detailed description including a more formal definition see Alur et al. (1995).

4.1 Hybrid Automata

Timed hybrid automata are an extension of timed automata (Alur and Dill, 1990). The basic notion is that of a finite state machine, extended with real-valued variables. These variables can be used to form enabling conditions for transitions (by comparing with a constant value), and to form invariant conditions for states. Variables can be reset when transitions take place, but cannot be assigned a value other than 0, and cannot be compared with each other. In a standard timed automaton all variables increase their value in line with the increase in global time, and are known as clocks. In a hybrid automaton, the rate at which a variable changes may vary with the state, but this rate will always be a natural number (an integer $n \geq 0$). For the purposes of this paper, we will only need to consider rates of 0 or 1. A hybrid automaton is defined by

- a set of states S
- a set of transitions between states $T \subset S \times S$
- a set of variables V

A *variable valuation* is a function $v : V \to Real$ which assigns a real value to each variable.

Associated with each state in S is

- an invariant function which takes a variable valuation and specifies whether it is possible to remain in the state with that valuation
- a rate of change for each variable. We will use the convention that *clocks*, with variable names based on C (e.g. $C1, C_{period}$) always have a rate of 1, so do not need to be given with the state information. *Integrators*, with names based on I will have a rate of 0 or 1, depending on the state.

Associated with each transition in T is

- an enabling function which takes a variable valuation and specifies whether the transition is allowed with that valuation
- a reset function which specifies which variables are to be reset to 0 when the transition is taken.

4.2 *From PG to hybrid Automata*

In order to translate from a distributed system described using PG to a hybrid automaton, we first of all need to construct the set of states. Each task or message can be in one of three conditions: waiting, ready, or running. (In the following, read 'task or message' for 'task'). The whole state space consists of all possible combinations of each of these conditions for each of the tasks, plus state(s) corresponding to failure. Transitions between states correspond to the triggering or completion of tasks. If a task is released in a state where it is not already runnable, then a transition is made to the state where that task is runnable. The condition of each of the other tasks is not changed, unless the task that has been released has a higher priority than a preemptible running task which shares the resource. In this case, the preemptible task has its condition changed from

running to ready. If a task is released in a state where it is already runnable, then this indicates a problem with schedulability, so a transition is made to a fail state. Invariants are added to states to ensure that some transition is made (either trigerring or failure) once the period has elapsed.

A clock variable is required for each periodically triggered task. This clock will be reset at the start of the period, and will be used to evaluate the enabling of transitions corresponding to the release of the task. An integrator variable will also be required for each task, to keep track of how much resource time has been used by that task. When a task is released, the integrator for that task is reset. In a state where a task is running, the corresponding integrator will have rate 1; otherwise the integrator will be 0. Transitions corresponding to task completions are enabled by the integrator for that task being greater than the minimum resource requirement. Invariants are added to ensure that completion transitions are taken at or before the maximum resource requirement has been used. The destination state of a completion transition takes into account the change of condition of the completing task to waiting, and also the release of any tasks triggered by the completion. As before, if a triggered task is already in a runnable condition then the destination state will be a failure state. Figure 2 shows the timed automaton corresponding to the PG definition given as

```
task A on processor needs [100,150]
   at priority 0 triggered by period 250
task B on processor needs [50,75]
   at priority 1 triggered by period 300
```

This example does not demonstrate all features of building the graph, as there is is only one resource, there are no non-preemptible tasks, and there are no completion triggers (as opposed to periodic triggers), as any example which covers any or all of these yields too large an automaton to present easily. See section 6 for a discussion of the size of the resulting automaton.

5. RESPONSE TIME ANALYSIS

Having built an automaton which models the behaviour of the system we wish to analyse, we now discuss how to perform the analysis upon the model. If we are only interested in schedulability then we need to check whether there is any allowable sequence of events which leads to a failure state. However, we are interested in more than schedulability: as we stated earlier, end-to-end response times are of interest, particularly for distributed systems. We extend our language slightly to define any end-to-end properties we are interested in, such as

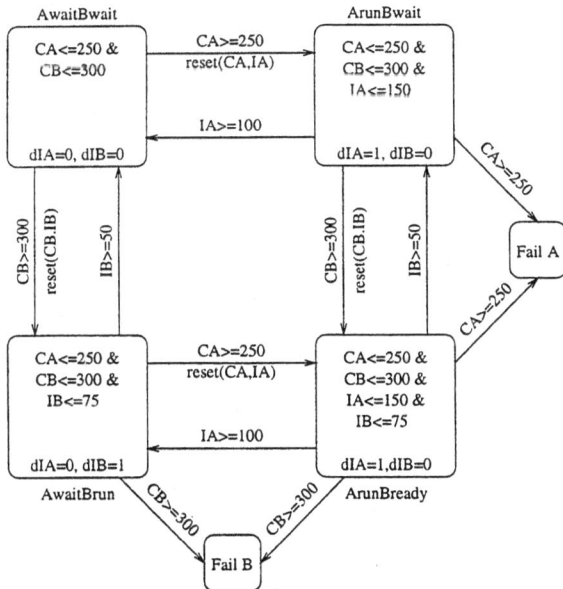

Fig. 2. Automaton for two periodic tasks

Size	RTA		PG		T_{calc}
	A	B	A	B	
5	1600		1000		1.3
3 + 3	800	1200	800	1000	24
4 + 4	1400	2400	1200	1400	108
5 + 3	1800	1600	1200	1000	87
4 + 5	1400	3000	1200	1600	271
5 + 4	1800	2600	1400	1400	280
5 + 5	1800	3400	1400	1600	1083

Table 1. Results for reachability analysis

```
property loop_A
  from start senderA on processor1
  to end receiverA on processor1
```

for the example shown in figure 3. Each property adds one further state and its associated transitions to the automaton. The state is to be used during reachability analysis, and is entered only when the property is violated. A new clock variable is added, and this clock is reset at the start of the first named task. All states which are on the trajectory of this end-to-end response have a transition to the fail state added (the assumption that each task has only one trigger is needed to unambiguously identify whether a state is on a given trajectory). Specific claims about end-to-end response times (e.g. that the response will occur within 1000 units) can be checked, by adding a within clause to the end of the property. In this case, the enabling condition of the fail transitions will be that the property clock has not exceeded the deadline set.

A more sophisticated analysis can be carried out by using the parameterisation mechanism within HyTech (Henzinger et al., 1995). Instead of giving a literal constant, a declared parameter can be used instead. The constraint-solving engine which underlies the model-checker will then calculate the conditions on the parameter(s) under which reachability can occur. In our case, the set whose reachability we are interested in is the failure set, so by fixing the enabling condition on failure transitions to be bounded by a parameter, the model-checker will calculate the least value of the parameter for which failure will occur; in other words, the end-to-end response time. Note that because the property clock is reset at the release of the first named task, the response time recorded

will be at most the minimum inter-release time (the period) for this task.

6. EVALUATION OF RESULTS

There are two main attributes of our technique which we wish to evaluate:

(1) the values for response times given
(2) the time taken to calculate the response times

We wish to compare the response times with those achieved experimentally (we always want to be conservative), and with those achieved using static analyses (we want to be less pessimistic). The calculation times stand by themselves, as comparison with static analyses will almost always be unfavourable: standard RTA is computationally cheap. The question to be asked here is whether the reachability problem is tractable, and whether any gains made in predicting better response times are worth the computational effort. To make the comparison, a range of systems were chosen, with varying sizes (number of tasks and messages) and varying amounts of inter-dependence (length and number of chains of precedence). Initially, the assumption that all messages and tasks had the same bounds on resource requirement was kept. The example of section 2 can be seen as two cloned precedence chains, each of length three. This example was extended to chains of length four and five, by adding first another message, and then a further task. The largest example considered had two chains of length five; the precedence graph for this example is shown in figure 3. Periods were chosen to have a GCD of 100, to ensure that the results were not sensitive to slight changes in period. Table 1 shows the results obtained using on a 233MHz Pentium II under Linux with 64MB RAM and 256M swap space. Calculation times shown are reported user time (in seconds) to perform the reachability analysis. Time taken to build the automaton is not always small, but is insignificant compared to the reachability analysis.

The examples may appear contrived, as conflicts are forced by having two essentially identical systems operating. However, this example is one of

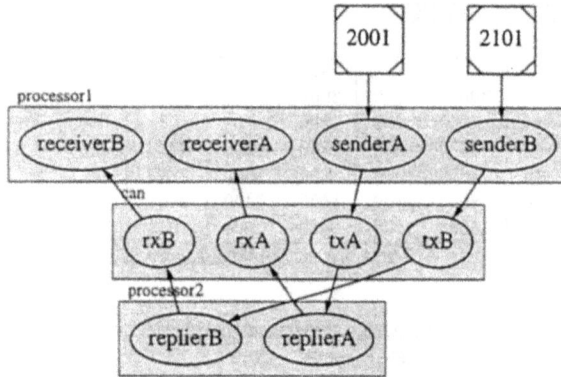

Fig. 3. Two precedence chains of length five

the simplest that demonstrates the kind of dependency that we can usefully identify. Typically, and unsurprisingly, the more dependency that can be identified in a system, the more can be gained from a detailed analysis of those dependencies. Most benefit is gained when chains of dependency are comprised of equal-length segments; in other words, the advantages of considering global properties are maximised when the response time is not dominated by local properties. Also, the results show that most benefit is gained on lower-priority chains of precedence.

7. CONCLUSIONS

We have argued that existing analysis techniques for response time can yield overly pessimistic results for distributed real-time systems with precedence constraints, and have presented a language (PG) for describing such systems. The translation of system definitions written in this language into hybrid automaton models has been discussed, along with the use of parameterised reachability analysis to derive end-to-end response times from the models. This technique has been applied to a range of examples with some positive results.

Related work has been carried out in two main areas:

- The adaptation of standard Response-Time Analysis (RTA) techniques to account for *offsets* which can be used to make the analysis less pessimistic (Tindell, 1994). However, Audsley and Burns (1998) have shown that when task periods are co-prime, this technique fails.
- Corbett (1996) has used timed automaton modelling of Ada tasking programs to carry out scheduling analysis. His work is related to ours, but is based on an explicit model of the scheduler, yielding less tractable (larger) models, and is only applied to uni-processor systems.

Further work is needed in the area of comparing the results we have obtained with those achieved in practice, and also with those obtained by standard and adapted RTA. There is also the possibility of adapting the approach to include more complex scheduling behaviour, such as that given by the priority ceiling protocol or with the use of dynamically assigned priorities (e.g. earliest-deadline first).

8. REFERENCES

Alur, R and D Dill (1990). Automata for modeling real-time systems. In: *17th International Colloquium on Automata, Languages and Programming (ICALP 90)*. number 443 In: *Lecture Notes in Computer Science*. Springer.

Alur, R, C Courcebetis, N Halbwachs, T A Henzinger, P-H Ho, X Nicollin, A Olivero and J Sifakis nad S Yovine (1995). The algorithmic andalysis of hybrid systems. *Theoretical Computer Science* **138**, 3–34.

Audsley, N, A Burns, M Richardson, K Tindell and A J Wellings (1993). Applying new scheduling theory to static priority pre-emptive scheduling. *Software Engineering Journal* **8**(5), 284–292.

Audsley, N C, A Burns, R I Davis, K W Tindell and A J Wellings (1995). Fixed priority pre-emptive scheduling: An historical perspective. *Real-Time Systems* **8**(2/3), 173–198.

Audsley, N C and A Burns (1998). On fixed priority scheduling, offsets and co-prime task periods. *Information Processing Letters* **67**(2), 65–70.

Corbett, J C (1996). Timing analysis of Ada tasking programs. *IEEE Transactions on Software Engineering* **22**(7), 461–483.

Henderson, W D (1998). An holistic approach to performance prediction of distributed real-time can systems. In: *5th International CAN Conference, San Jose*. To Appear.

Henzinger, T A, P-H Ho and H Wong-Toi (1995). A user guide to HYTECH. In: *Tool and Algorithms for the Construction and Analysis of Systems: (TACAS 95)* (E Brinksma, W R Cleaveland, K G Larsen, T Margaria and B Steffen, Eds.). Vol. 1019 of *Lecture Notes in Computer Science*. Springer. pp. 41–71.

Liu, C L and J W Layland (1973). Scheduling algorithms for multiprogramming in a hard real-time environment. *JACM* **20**(1), 40–61.

Tindell, K (1994). Adding time-offsets to schedulability analysis. Technical Report YCS 221. Department of Computer Science, University of York.

Tindell, K and J Clark (1994). Holistic schedulability analysis for distributed hard real-time systems. *Microprocessing and Microprogramming* **40**(2-3), 117–134.

DYNAMIC CPU SCHEDULING WITH
IMPRECISE KNOWLEDGE OF COMPUTATION-TIME

Saud A. Aldarmi, Alan Burns

Department of Computer Science
University of York, U.K.
(saud,burns@cs.york.ac.uk)

Abstract: The majority of the studies conducted in scheduling real-time transactions mostly concentrate on concurrency control protocols, while overlooking the CPU as being the primary resource. Consequently, there are various techniques for scheduling the CPU in conventional time-critical systems; meanwhile, there does not seem to be any technique that is adequately designed for scheduling such a resource in Real-Time Database (RTDB) systems. In this paper, we construct an efficient CPU scheduling scheme that minimizes the preemption rate in order to reduce the frequency by which synchronization protocols must be invoked, along with their inherited performance degradation. In addition, we also introduce a new timing model upon which the newly introduced scheduler is incorporated in order to utilize the system's imprecise knowledge of computation time estimates.
Copyright © 1999 IFAC

Keywords: CPU Scheduling, lowering-preemption, timeliness-functions, and imprecise computation estimates.

1. INTRODUCTION

An *RTDB* system's scheduler is a mechanism for ordering the execution of the outstanding transactions on the system's resources according to some predefined criteria. Thus, an *RTDB* scheduler consists of two separate modules; i.e., a *CPU* scheduler and a Concurrency-Control mechanism. While there are several views on the roll of priority cognizance in resolving contention and conflicts over the database, priority ordering is necessary for resolving contention over the *CPU*.

CPU scheduling algorithms used in current real-time *task* models assume a priori *accurate* knowledge of the tasks' resource requirement; i.e., worst-case computation-time. Such an assumption instigated various scheduling techniques for the task model. Meanwhile, it is nearly impossible to *accurately* determine the *actual* computation-time of a transaction, since the runtime behavior of a transaction is dynamic in the sense that it depends upon the values it finds in the database *at run-time*. Therefore, a worst-case computation time may be very *pessimistic* in an environment such as an *RTDB*. Thus, *CPU* scheduling schemes that relay heavily on *accurate* computation-time may not perform as well when operating with *imprecise* estimates.

In this paper we construct a *CPU* scheduling scheme that does not rely on accurate knowledge of computation time, while simultaneously minimizing preemption and its undesirable consequences, and is thus, most suitable to operate under an *RTDB* environment.

This paper focuses on scheduling "Soft-Deadline" transactions. That is, the transactions that complete their execution before their deadlines are considered successful and impart a full value, *V*, to the system. Whereas transactions that complete after their deadlines are known to be *tardy* and only impart a portion of their net value that is proportional to their *tardiness*; i.e., lateness. The rest of this paper is based on the following assumptions. All transactions are *aperiodic*, and scheduling is *preemptive*. The system's scheduler learns of the following set of attributes *only* at (*A*), the transaction's arrival time.

- *I* – an *importance* level, reflecting the transaction's *significance* to the overall functionality of the system. Note that $V = I$ at arrival time.
- D^a – *absolute* deadline.
- D^r – *relative* deadline, such that $D^r = D^a - A$. Note that whenever we use *D* without a superscript, we are referring to D^r.

The remainder of this paper is organized as follows: Section (2) presents previous work and the motivation of this study. Section (3) introduces a new timing model upon which our newly introduced scheduling scheme is incorporated. Section (4) introduces a new scheduling scheme called *Aging Timeliness Cumulative Computation (ATCC)*. Section (5) presents our simulation model along with its assumptions and the performance metrics used in this paper. Section (6) presents a comparative study exhibiting the effectiveness of the newly introduced scheme. Section (7) concludes this research.

2. MOTIVATION AND PREVIOUS WORK

In the task model, *Value-Density* (*VD*) (Jensen *et al.*, 1985, and Lock 1986), i.e., V/C, is a value-based scheduling scheme that is known to outperform many other scheduling algorithms under overload situations. The manner, in which *VD* scales the task's value by its execution time, causes scheduling priority to be derived on the task level. On the other hand, *Dynamic Value Density (DVD)* (Aldarmi and Burns, 1999a), i.e., $V/\overline{C}^2$ (where $\overline{C}$ represents the remaining computation time), enhances the performance of *VD* by assigning an extremely low priority to newly arriving tasks. Such small priority increases in correspondence with the amount of time that a task executes. Consequently, the scheduling priority is not derived *statically* on the task level; rather, it is derived *dynamically* for the individual execution unit(s). The consequences of such behavior are that preemption is significantly lowered, and execution resumption after preemption is also increased. *DVD* was shown in (Aldarmi and Burns, 1999-a) to better utilize the *CPU* as being the system's primary resource.

(Huang *et al.*, 1989) introduced a technique known as *"Scheduling by Criticalness and Deadline First"* *(CDF)*, that was intended to operate in an *RTDB* environment. The technique is stated in the following function:

$$Priority\ (P) = \frac{Relative\ Deadline}{Criticalness} \equiv \frac{D}{I} \quad (2.1)$$

Function (2.1) uses the transactions' *significance* to the system and *relative deadline*, both of which are not concerned with the computation time of the corresponding transaction(s). However, function (2.1) relies on two *static* attributes; and thus, the derived priorities are *static* and *fixed* from the instant that they are derived until the transactions' termination, where termination means commit or abort. Therefore, *CDF* is subject to similar weaknesses and limitations as *VD*. Since transactions dynamically undergo several changes as they migrate through the system, monitoring and utilizing the transactions' dynamic attributes can provide the means of a better *CPU* scheduling scheme; hence, better overall performance. In this paper we wish to alter *CDF* and construct a scheduling technique that is capable of *mimicking* the effective behavior of *DVD*, in addition to minimizing the preemption rate in order to suite *RTDB* environments.

3. IMPRECISE-TIMELINESS-FUNCTIONS

Before we introduce our scheduling scheme, we wish in this section to construct the timing-model; i.e., *Imprecise-Timeliness-Function(s) (ITF)*, upon which our scheduler will be incorporated.

If a transaction does not start executing before $t + \overline{C} = D^a$ (for t being the current time), then the transaction is guaranteed to become tardy. Thus, if a *waiting* transaction is subject to becoming tardy due to having an *infeasible* deadline, then it should not receive a relatively high scheduling priority in order not to delay the scheduling of other transactions that have a better chance of meeting their deadlines. *Timeliness-functions* (Aldarmi and Burns, 1998) address such an issue, and therefore, they start reducing the system's internal representation of value at $(D^a - \overline{C})$ in order to project, at the scheduling instant, the final value to be gained at the finishing moment.

Abstractly define Δ to be the diminishing-rate by which a transaction's value is computed when the transaction becomes tardy, and $\psi_{(t)}$ being a transaction's tardiness at time t. For V and $\overline{T}$ respectively being a transaction's *value* and *timeliness*, timeliness-functions introduced in propose shifting V to the left by $\overline{C}$ to produce $\overline{T}$, both of which are illustrated in figure (1), and respectively computed in functions (3.1) and (3.2).

Figure (1)

$$\Psi_{(t)} = max\ (0,\ t - D^a)$$
$$V_{(t)} = I - [\ \Delta \times \Psi_{(t)}\ ] \quad (3.1)$$
$$\overline{T}_{(t)} = I - [\ \Delta \times \Psi_{(t+\overline{C})}\ ] \quad (3.2)$$

The proposed technique works well if the transaction's computation-time estimate is accurate. However, if the actual computation-time is much less than the worst-case estimate, then a transaction might be erroneously aborted; i.e., *aborted prematurely*, meanwhile, the transaction could have finished with a significant value. Hence, since transactions in *RTDB* systems do not generally have accurate computation-time estimates, function (3.2) may be more applicable for conventional real-time systems (the task model) than *RTDB* systems (the transaction model). In the next subsection, we address the different types of transactions that might exist in an *RTDB* system.

3.1. *Transactions' Characteristics*

In general, a transaction's computation-time either depends on the nature and functionality of the transaction, and/or on the environment and operating conditions. Several classes of transactions have been identified in previous studies; e.g., (Kim and Son, 1996), each of which has different characteristics and runtime requirements. If a transaction's body were closely inspected, it would either contain a single path with a fixed functionality; e.g., *canned transactions*, which require a specific runtime, and thus, a *precise* computation-time estimate, or it might contain multiple paths. A compiler can certainly approximate the maximum/minimum-required computation-time, which can be used to generate an expected error bound; i.e., a degree of imprecision, within the transaction's worst-case estimate. Consequently, we classify transactions into two categories, those with *precise* estimates, and those with *imprecise* estimates.

Case (1) – Precise Computation-Time

For all transactions with *precise* computation-time, it is reasonable to start diminishing a transaction's timeliness at $(D^a - \overline{C})$, since $\overline{C}$ accurately reflects the remaining computation-time.

Case (2) – Imprecise Computation-Time

For a transaction with an imprecise computation-time estimate, timeliness must not use $\overline{C}$ as stated above, but rather use an estimate of $\overline{C}$ that corresponds to the degree of imprecision. In order to do so, define $\tilde{C}$ to be the total amount of time that a transaction has computed so far. In addition, define α be the degree of imprecision in the transaction's computation-time estimate; i.e., error bound, which implies that the minimum computation-time for the corresponding transaction is "$C \times (1-\alpha)$", which we refer to as C_{min}. That is, for a transaction with a computation-time estimate that is imprecise by *25%*, then $\alpha = 0.25$. For example, for a transaction with $C = 100$, and $\alpha = 0.25$, then the *minimum* computation-time $\equiv C_{min} = 75$. Hence, for a transaction with an accurate estimate $\alpha = 0$. Consequently, a transaction's timeliness can be computed at "$D^a - C_{min} - \tilde{C}$". Hence, a transaction will not be aborted unless its *minimum remaining* required computation-time is infeasible. Thus, timeliness may be computed as given in function (3.3). For further analysis of function (3.3) the reader is referred to (Aldarmi and Burns, 1999b).

$$\overline{T}_{(t)} = I - [\Delta \times \Psi_{(t + C_{min} - \tilde{C})}] \qquad (3.3)$$

4. AGING-TIMELINESS-CUMULATIVE-COMPUTATION (ATCC)

Imprecise-Timeliness-Functions *ITF* described in the previous section; i.e., function (3.3), can be utilized to account for many of the transaction's attributes and thus, *ITF* allow the scheduler's decisions to be made *deadline-cognizant*. Therefore, *CDF* (described in section 2) would be a better scheme if it utilized the transaction's time-variant $\overline{T}$ instead of I. In addition, since we wish preference to be given in an *increasing* order of the priorities; i.e., opposite to the

order of (Huang *et al.*, 1989), we alter function (2.1) above as shown in function (4.1) below.

$$P = \frac{\overline{T}}{D} \qquad (4.1)$$

However, the scheduling priority is still derived on the transaction level. Recall that *DVD* achieves its performance significance (superiority) over the traditional *VD* by increasing the scheduling priority by an amount that corresponds to V for every execution unit. Therefore, function (4.1) must also increase by (at least) a corresponding amount if it is to mimic the effectiveness of *DVD*. Adding $(\overline{T} \times \tilde{C})$ to the numerator of function (4.1) as shown in function (4.2) allows the scheduler to weight each execution unit according to the corresponding transaction's timeliness. Therefore, the scheduler not only becomes aware of various valued execution units, but also is able to credit each transaction for all the units it executed with respect to the transaction's own timeliness.

$$P = \frac{\overline{T} + \overline{T} \times \tilde{C}}{D} = \frac{\overline{T} \times [1 + \tilde{C}]}{D} \qquad (4.2)$$

Furthermore, transactions in an *RTDB* environment have to block for synchronization, I/O, commit protocols, buffer management, etc. Therefore, we should account for such delays in the scheduling decision. Thus, define ω to be the total amount of time that a transaction has been waiting in the system since its arrival, excluding any $\tilde{C}$. That is, $\omega_{(t)} = t - A - \tilde{C}_{(t)}$, which is a *dynamic* attribute corresponding to *aging*. In order to account for a database inherited delays, we need to add such an attribute to function (4.2). In addition, we must also treat it in the same manner that we treated $\tilde{C}$; i.e., multiply it by $\overline{T}$, in order to distinguish between a waiting highly significant transaction vs. a waiting less-significant transaction. However, if the priority of a waiting transaction increases in a manner similar to that of an executing transaction, then the waiting transaction might interfere with an executing transaction, and *thrashing* would be expected due to the corresponding high preemption rate. Therefore, we need to scale down ω first, then multiply it by $\overline{T}$.

Note that if a transaction is short then it is not going to require a substantial amount of system's resources and it should be given scheduling preference. We argue that in order to increase the system's throughput, then we must speedup the execution of all short transactions. Based on such an observation we propose scaling ω down by C. However, since C might involve a certain degree of imprecision, then such imprecision must be utilized when considering C. Thus, define $\Omega_{(t)}$ to be a weighted aging factor at time t such that:

$$\Omega_{(t)} = \frac{\omega_{(t)}}{C_{min}}$$

With Ω defined as such, function (4.2) is mapped into function (4.3) below, which may be called *Aging Timeliness Cumulative Computation* (*ATCC*), owing its name to all the parameters involved in it. Note that function (4.3) utilizes the underlying timing-model by using $\overline{T}$ in weighting $\tilde{C}$ and Ω.

$$ATCC = \frac{\overline{T} \times [1 + \tilde{C} + \Omega]}{D} \qquad (4.3)$$

Function (4.3) clearly shows the feasibility of utilizing information about runtime behavior; e.g., *computation-time*, in priority assignment schemes for the transaction model. The function involves only simple arithmetic, and utilizes the transactions' timeliness (thus value with respect to deadline), in addition to the amount of computation-time that has already been consumed and remains to be acquired. Furthermore, the function also accounts for the amount of time that transactions wait in the system (whether for I/O or concurrency control). Many details have been omitted due to space limitation, all of which can be found in (Aldarmi and Burns, 1999b).

5. SIMULATION MODEL

The simulator we use in this paper is based on *CSIM*; C-based process oriented language (Schwetman, 1994), and has the following parameters and assumptions.

- The levels of importance are randomly assigned to transactions from a *uniform* distribution from (1.0, 5.0), which may be viewed as {low, mid-low, mid, mid-high, high}.
- The worst-case computation-time, C, is randomly assigned to transactions from a *uniform* distribution from (100.0, 200.0). However, the simulated load has the following characteristics: 1/3 of the transactions have *precise* computation-time; and thus, $\alpha = 0$. The 2nd 1/3 of the transactions have *imprecise* computation-time such that $0.5C \leq \hat{C} \leq C$, for $\hat{C}$ being the actual computation time; thus, $\alpha = 0.5$. The 3rd 1/3 of the transactions have *imprecise* computation-time such that $0.1C \leq \hat{C} \leq C$; thus, $\alpha = 0.9$.
- The time it takes to process each data granule is 10 time units, and the number of data granules requested by any transaction corresponds to the transaction's actual length; e.g., a transaction that is 100 units long requires 10 locks. However, since we are assuming memory resident database, there is no I/O blocking delays; rather, there is only concurrency control blocking delay.

- All required data granules are requested from the database manager, which employs *Priority-Abort* (*PA*) (Abbott and Garcia-Molina, 1988) for conflict resolution.
- When a transaction is submitted to the system, it is assigned its *deadline* (D^a), such that:

$$D^a = A + C + uniform\ (2.0, 5.0) \times C$$

Note that the C corresponds to the worst-case computation time estimate and not the actual computation time.

- All transactions have *soft-deadline*, and when a transaction's (*value*) $V \leq 1/100$, the transaction is assumed to have lost its validity and therefore is aborted. For $\hat{I}$ and $\hat{C}$, respectively, being the maximum importance level and the maximum computation-time required by any transaction within the entire system, the *diminishing speed* (Δ) of a transaction's value = $\hat{I} / \hat{C}$, and thus, the values of all transactions diminish at the same speed.
- When a transaction is preempted an artificial delay = $\hat{C} / 100$ is introduced in order to simulate the overhead of context switching.
- The simulated load is 80% to 200% controlled by an *exponential* distribution for the transactions' arrival, which in turn is controlled by the *average actual* computation-time ($\hat{C}_a$) of all transactions divided by the desired load (σ). Thus, the submission rate $\equiv \lambda = \hat{C}_a / \sigma$. Note that an exponential distribution allows for *bursty* arrivals, which may result in some transactions missing their deadlines even under normal load.
- The simulation corresponds to the average behavior of 5,000 transactions at each simulated load.

The performance of the simulated techniques is measured according to the following metrics; the interested reader may refer to (Aldarmi and Burns, 1999-b) for other performance metrics.

- *CPU Wastage % = (total time spent on aborted tasks + total time spent on preemption) $\times$ 100 ÷ total time spent on all transactions.*
- *Value-Sum % = total value collected $\times$ 100 ÷ total value of all transactions submitted to the system.*
- *Preemption = total number of preemption $\times$ 100 ÷ total number of transactions submitted to the system.*
- *Restart % = total number of restarted transactions $\times$ 100 ÷ total number of transactions submitted to the system.*
- *Wasted Locks % = total number of lost locks $\times$ 100 ÷ total number of locks acquired by all transactions.*

6. COMPARATIVE STUDY

In this section, we contrast the performance of function (4.3) against several *CPU* scheduling schemes; namely:

- *Earliest Deadline First (EDF)* (Liu and Laylan, 1973) – the transaction with the closest deadline receives the highest priority,
- *Scheduling by Criticalness and Deadline First (CDF)*; i.e., function (2.1),
- *Value-Density (VD)*,
- *Dynamic-Value-Density (DVD)*. Note that we use $(C - \tilde{C})$ in computing the remaining computation time; i.e., $\overline{C}$. However, since the actual computation time may differ from the worst-case estimate in *RTDB* environments as well as our simulator, *VD* and *DVD* might make the wrong decision(s). This scenario exhibits the limitation of *VD* as well as *DVD*.

Figures (2 to 6) show the overall performance of the scheduler. In general, the figures seem to classify the five scheduling schemes being simulated into three groups: *EDF* being in its own group, while *CDF* and *VD* seem to be comparable as in one group, and finally, *DVD* and *ATCC* being somewhat comparable as in a separate group.

Figure (2) shows the total amount of *CPU* time that is wasted on preemption and aborted, partially processed, transactions. The figure shows that *EDF* wastes as much as 38% under a load of 200%. Meanwhile, *VD* and *CDF* waste as much as 10-15% under a load of 80-200%, respectively. On the other hand, while *DVD* waste up to 3% of the *CPU* capacity, *ATCC* wastes as little as 0.19% (practically 0%). The efficiency of *ATCC* in utilizing the *CPU* over the other schemes is also reflected in the rest of the system's performance as depicted in figures 3 to 6.

Figure (3) shows the total amount of value that the scheduler is able to collect from the completed set of transactions. The figure clearly shows that *ATCC* not only is capable of sustaining higher performance under overload situations, but also outperforms the remaining simulated schemes under normal operating conditions as well! Thus, an efficient *CPU* scheduling can be achieved without accurate knowledge of computation time estimates.

Figures (4 and 5), respectively, show the preemption rate and the percentage of transactions that are restarted due to conflict resolution. Notice the strong correlation between the two figures! If a transaction is preempted while it holds some resources, then the problem of *priority-inversion* (Abbott and Garcia-Molina, 1988) becomes an issue, which causes aborting and possibly restarting several conflicting transactions. *Aborting* and *redoing* overhead could accumulate as data-contention rises, which could in-

crease further as preemption rises. Regardless of the concurrency control mechanism being employed (pessimistic or optimistic), a higher preemption rate could significantly contribute to higher amounts of conflict within the system, which translates into performance degradation. Thus, lowering the preemption rate is very important in an *RTDB* system due to the fact that even if data-contention is high between the current set of transactions within the scheduler's queue, due to the nature of the transactions, it is naturally reduced if preemption is reduced. Tailoring the underlying *CPU* scheduling scheme in a manner that minimizes the preemption rate can significantly reduce concurrency control degradation. *ATCC* is a preemptive scheme that attempts to derive and assign the scheduling priorities in a manner that actually construct a sequence of transactions that minimizes preemption while achieving a relatively high level(s) of performance as shown in figures 2 to 6.

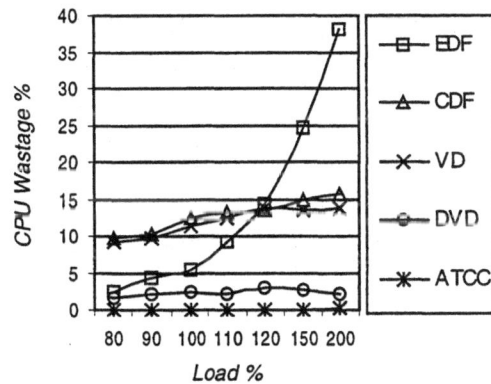

Figure (2)

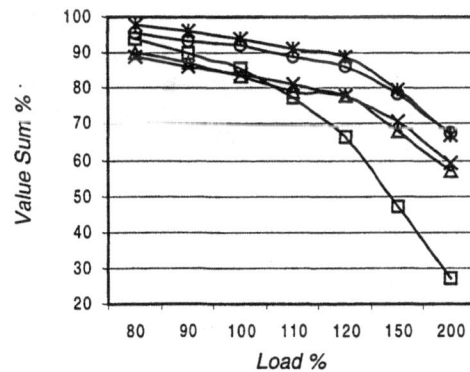

Figure (3)

Finally, figure (6) shows the percentage of locks that is lost (wasted) due to both, restarts as well as aborting partially processed transactions. In contrast to the wasted *CPU* time, figure (6) can be viewed as the wasted database-manger's time. Thus, figures (5 and 6) show that *ATCC* assigns the scheduling priorities in a manner that not only minimizes restarts, but also attempts to prevent wasting any acquired locks and thereby not wasting any of the database-manager's time. Thus, many existing studies conducted in scheduling real-time transactions attempt to minimize restarts and wasted locks by devising new concurrency control protocols. However, figures 5 and 6 show that the answer to reducing restarts and wasted-locks does not highly depend on concurrency control protocols, but mainly on the *CPU* scheduler and the manner in which it derives and assigns the scheduling priorities!

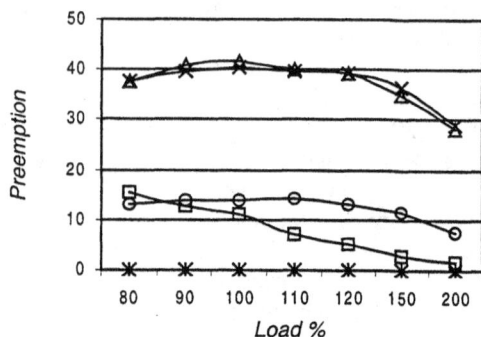

Figure (4)

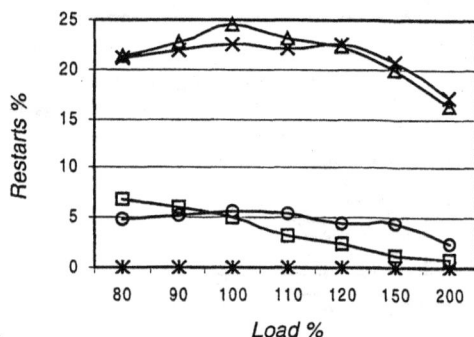

Figure (5)

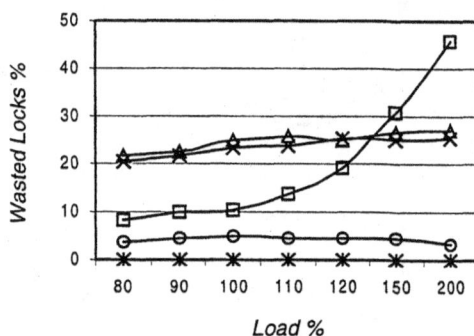

Figure (6)

7. CONCLUSION

In this paper, we constructed a new *CPU* scheduler (*ATCC*) along with the necessary underlying timing model (*ITF*), both of which render an *RTDB* system's scheduler the ability to accommodate *precise* as well as *imprecise* computation-time estimates. The introduced schemes were shown to substantially reduce the amount of wasted system's resources; and thus, redirect such resources towards sustaining a higher level of performance. Hence, providing an *RTDB*

system with a time cognitive behavior complying with the requirements and specifications of time-critical systems. This paper clearly shows that transaction *CPU* scheduling in an *RTDB* system is not hindered by the absence of accurate knowledge of the execution time. In addition, it showed that it is not the concurrency control module, but rather, it is the *CPU* scheduler that is mainly responsible for wasting the system's resources. Hence, with the techniques introduced in this paper, concurrency control is no longer a bottleneck in the performance of *RTDB* systems.

8. REFERENCES

R. Abbott, and H. Garcia-Molina (1988). Scheduling Real-time Transactions: A Performance Evaluation. In: *Proceedings of the 14th International Conference on Very Large DataBases*, Los Angeles - California.

S. Aldarmi and A. Burns (1998). Time-Cognizant Value Functions for Scheduling Real-Time Systems. *Technical Report YCS-306*, Department of Computer Science. University of York.

S. Aldarmi and A. Burns (1999a). Dynamic Value-Density for Scheduling Real-Time Systems. *Technical Report YCS-310*, Department of Computer Science – University of York, U.K. (To appear in *the 11th Euromicro Conference on Real-Time Systems* – York, U.K., 1999).

S. Aldarmi and A. Burns (1999b). Dynamic CPU Scheduling with Imprecise Knowledge of Computation-Time. *Technical Report YCS-314*, Department of Computer Science – University of York, U.K.

J. Huang, J. Stankovic, D. Towsley, and K. Ramamritham (1989). Experimental Evaluation of Real-time Transaction Processing. In: *Proceedings of the 10th Real-time Systems Symposium*, pp. 144-153.

E. D. Jensen, C. D. Locke, H. Tokuda (1985). A Time-Driven Scheduling Model for Real-time Operating Systems. In: *Proceedings of IEEE Real-time Systems Symposium*.

Y. Kim and S. H. Son (1996). Supporting Predictability in Real-Time Database Systems. *IEEE Real-Time Technology and Application Symposium (RTAS' 96)*, Boston, MA.

C. L. Liu, and J. W. Layland (1973). Scheduling Algorithms for Multiprogramming in Hard-Real Time Environments. *Journal of the ACM*, Vol. 20, No. 1.

C. D. Locke (1986). Best-effort Decision Making for Real-time Scheduling. Ph.D. thesis, Computer Science Department, Carnegie Mellon University.

H. Schwetman (1994). CSIM Reference Manual.

SCHEDULABILITY ANALYSIS FOR REAL-TIME PROCESSES WITH AGE CONSTRAINTS

Dieter Zöbel*

* Universität Koblenz-Landau, D-56075 Koblenz

Abstract: Real-time systems not only require the semantical correctness of their operations but also the availability of the computational results within some predefined time intervals. Typical applications are composed of processes which are responsible to execute time-bound computations. Depending on the application specific context processes have to complete due to time constraints. Most frequently are periodic time constraints for which various schedulability tests have been developed. Recently age constraints have shown to be adequate for a certain scope of applications, particularly for those based on real-time data bases. Instead of constructing specific schedules a transformation from processes with age constraints to periodic processes is introduced and a least upper bound to test their schedulability is derived. *Copyright © 1999 IFAC*

Keywords: Real-time systems, real-time databases, scheduling, time constraints, transformations, least upper bounds for schedulability

1. INTRODUCTION

There is an increasing demand for computers to control real-world processes. A principle feature of such a system lies in its composition as a set of software processes which have to keep track with the evolving states of the real-world processes. Computational results though they are semantically correct are valuable only for a narrow interval of time. The same results incorporated into the real-world at some other instant of time may cause catastrophic situations or at least costly damages. An example of such a system is the anti-lock brake system (ABS) for cars where computational processes sense the velocity of the wheels and set the pressure of the brake shoes. The right actuator command which arrives too late endangers human lives.

In this sense a system constitutes a real-time system if there are temporal constraints imposed by real-world processes which have to be managed by computational processes. As a basic model the sensor data which becomes ready at time r must cause a control reaction before some dead-line d. Henceforth, the execution time Δe of a corresponding computational process must fit between the sensing action and the control action. As a generalizing definition we have (Zöbel and Albrecht, 1995)

$$P[r + \Delta e \leq d|B] = p_c \qquad (1)$$

where p_c is a probability constant. The value 1 for p_c characterizes a hard real-time system where timeliness is guaranteed. But even such a system may fail e.g. due to some technical fault. Therefore predicate B accumulates all those cases for which a hard real-time system operates well.

A typical real-time application consists of many computational processes which are responsible for certain control functions. These processes operate on common resources, particularly on the same processor. Various scheduling techniques are applied to assert timeliness for a set of time constrained processes on a single processor. These techniques can be categorized with respect to the following criteria:

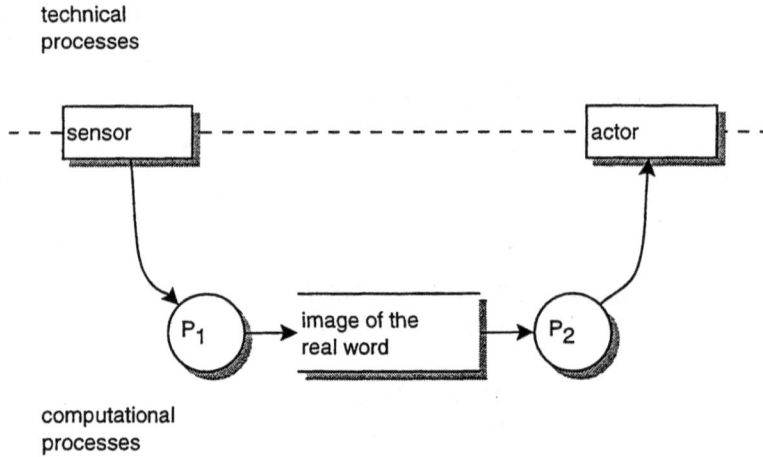

technical
processes

computational
processes

Fig. 1. A database-like approach to real-time systems

- the scheduling of preemptive and non-preemptive processes,
- the construction of schedules before (static, off-line) or during (dynamic, on-line) the runtime of the system,
- the mapping of time intervals to processes (explicit schedule) or the mapping of processes to priorities (implicit schedule).

Typically the components of the entire real-time system are known in advance. Hence, it is possible to apply static scheduling techniques. Furthermore preemptive processes which are mapped to fixed priorities are of major relevance for the analysis, the implementation, and the execution of reliable real-time systems.

The most important criteria for mapping processes to priorities are based on rates, in rate monotonic scheduling (RMS), or on deadlines, in deadline monotonic scheduling (DMS). Both are implicit scheduling techniques and assume that processes repeatedly become ready after some fixed interval of time: the period Δp. For RMS it is additionally assumed that the deadline for the execution of a process coincides with the next ready time of this process: $d^j = r^{j+1}$. Assigning priorities in RMS follows the rule: the shorter the period (the higher the rate), the higher the priority of this process.

Liu and Layland (Liu and Layland, 73) derived a series of important theorems for scheduling real-time processes, particularly for RMS-ruled priority assignments. One theorem says that RMS is optimal among all fixed priority assignments in the sense that no other assignment can schedule a set of processes which cannot be scheduled by RMS. Still more important is the theorem (see eqn. 3) which is based on the definition of the utilization caused by a set P_{per} of n periodic processes:

$$U(P_{per}) = \sum_{i=1}^{n} \frac{\Delta e_i}{\Delta p_i} \qquad (2)$$

Liu and Layland derived a schedulability test based on a least upper bound $U_{lub}(RMS, n)$ such that under RMS there exists an implicit schedule if

$$U(P_{per}) \leq U_{lub}(RMS, n)$$
$$= n \left(\sqrt[n]{2} - 1 \right) \qquad (3)$$

with

$$\lim_{n \longrightarrow \infty} U_{lub}(RMS, n) = \ln 2 \qquad (4)$$
$$\approx 0.693 \qquad (5)$$

Notice that these conditions are only sufficient for schedulability. That is to say, for a set of processes the condition $U(P_{per}) \leq U_{lub}(RMS, n)$ may fail, but the set of processes is still schedulable. Necessary and sufficient conditions for schedulability have been derived by Audsley, Burns, Richardson, Tindell and Wellings (Audsley et al., 1992) and by Buttazzo (Buttazzo, 1997). However, these techniques are more expensive than the simple comparison of equation (3).

The best utilization bound for periodic processes was also derived by Liu and Layland (Liu and Layland, 73). Under the strategy earliest deadline first (EDF) the test

$$U_{lub}(EDF, n) \leq 1 \qquad (6)$$

provides a necessary and sufficient condition for the schedulability of any set of processes. However, this is no implicit scheduling technique. Instead an explicit static schedule or explicit dynamic schedule has to be constructed which governs the execution of strips of processes (Sha and Goodenough, 1990).

2. PROCESSES WITH AGE CONSTRAINTS

A different type of time constraint forces that subsequent executions of processes must not exceed time bounds given by the application context. One type of contraint emerges from the requirement that computational processes should operate on an internal image of the real-world processes. An age constraint is specified by a time bound Δa for the freshness of such an image (see figure 1).

First lets have a look on how an image of a real world process is produced. A corresponding computational process has to be invoked and it lasts the execution time Δe to get this image into a real-time image database, e.g. to read the pixel image of a moving object from a frame grabber card and to detect the objects's actual position. In the sequel this image may be part of the input for other computational processes operating on this data, e.g. to compute the velocity and acceleration of the moving object. The characterizing condition for an age constraint is that for the subsequent process working on this image it must not be older than Δa with respect to the evolving state of the real-world process. This condition can be translated in the following way: The time difference between the start s^j of taking the image and the end of taking the subsequent image c^{j+1} must not exceed Δa:

$$c^{j+1} - s^j \leq \Delta a \qquad (7)$$

Thus consulting the database at some time instant t_I will always refer to an image which is fresh in the sense that it shows the corresponding real-world process at a time not before $t_I - \Delta a$ (see figure 2, case (I)).

Another constellation for the applicability of age constraints is given with the widely used technique of polling. Let E be an event at some unpredictable instant t_E of time. The registration of the event and its notification in the database may last Δe units of time. Events are responsible for changs of states. So a natural requirement is that a dedicated process catches the new state before the expiration of some time bound Δa (see figure 2, case (E)).

3. THE MODEL

Let $P_{age} = \{P_1, ..., P_n\}$ be a set of processes with age constraints. For any such process P_i, $i \in \{1, ..., n\}$, there exists a tuple $(\Delta e_i, \Delta a_i)$ indicating its maximum execution time Δe_i and its age constraint Δa_i. In an analogous fashion a set $P_{per} = \{P_1, ..., P_n\}$ of periodic processes is characterized by the tuples $(\Delta e_i, \Delta p_i)$ with Δp_i

indicating the (minimal[1]) period of process P_i. As a generalization of the latter in a set $P_{dead} = \{P_1, ..., P_n\}$ any process P_i is characterized by the triple $(\Delta e_i, \Delta p_i, \Delta d_i)$ where Δd_i indicates that the j-th execution of process P_i may start at time r_i^j and must be completed not later than $r_i^j + \Delta d_i$. In general this may be equal to, earlier than or even later than $r_i^j + \Delta p_i$.

An explicit schedule for a set $P = \{P_1, ..., P_n\}$ of processes is a function s mapping time to process indices

$$s : \mathbb{R} \longrightarrow \{0, 1, ..., n\} \qquad (8)$$

where $s(t) = 0$ denotes that at time t the processor is occupied by some idle process. A schedule is said to be *feasible* if all processes can start and complete according to their time bounds. We apply the term *schedulable* if there exists an algorithm to produce a feasible schedule (Buttazzo, 1997).

For practical reasons an explicit schedule will always be represented by some finite prefix which is executed repeatedly. However, explicit schedules may become rather long and have to be recomputed when the execution times, periods or age constraints change.

With respect to processes with age constraints a schedule is feasible if for any two consecutive executions of some process P_i the following two conditions hold:

$$c_i^{j+1} - s_i^j \leq \Delta a_i \qquad (9)$$
$$c_i^j \leq s_i^{j+1} \qquad (10)$$

The minimal fraction of processor time contributed to the utilization by a process in a time interval Δa is $2\Delta e/\Delta a$ or for n consecutive overlapping age constraint intervals:

$$\frac{(n+1)\Delta e}{n\Delta a - (n-1)\Delta e} \qquad (11)$$

For $n \longrightarrow \infty$ this quotient converges to $\Delta e/(\Delta a - \Delta e)$ which is the least possible a process contributes to the utilization of a processor. Hence, the utilization for a set P_{age} (or better the *load* imposed) of processes with age constraints is defined by:

$$U(P_{age}) = \sum_{i=1}^{n} \frac{\Delta e_i}{\Delta a_i - \Delta e_i} \qquad (12)$$

The time difference $\Delta a_i - \Delta e_i$ may be regarded as a pseudo-period which marks the maximal distance between two consecutive executions

---

[1] For all theorems applicable to RMS it suffices that the minimal difference $r_i^{j+1} - r_i^j$ is greater or equal Δp_i.

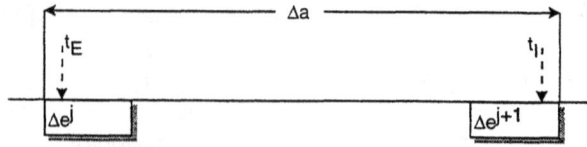

Fig. 2. Two cases are represented: **(I)** the image is dating from a time not before $t_I - \Delta a$ and **(E)** a changing state is caught not later than $t_E + \Delta a$

still satisfying the age constraint of process P_i (Ramamritham, 1993).

4. TRANSFORMING PROCESSES WITH AGE CONSTRAINTS TO PERIODIC PROCESSES

A simple transformation which maps processes with age constraints to periodic ones was proposed by Mok (Mok, 1983). In detail a process with an age constraint denoted by the tuple $(\Delta e_i, \Delta a_i)$ is mapped to a periodic one with the tuple $(\Delta e_i, \Delta a_i/2)$. Generalizing we say that by transformation T a set P_{age} of processes with age constraints is mapped to a set P_{per} of periodic processes. The new period $\Delta p_i = \Delta a_i/2$ guarantees that one complete execution Δe_i of process P_i fits in the interval Δp_i and it follows immediately that both conditions (8) and (9) for processes with age constraints hold. Of course, this transformation only makes sense, if any process leaves time for the execution of other processes:

$$\Delta a > 2\Delta e \qquad (13)$$

For the application of schedulability criteria like those of Liu and Layland (Liu and Layland, 73) it would be very practical for a set of n processes with age constraints P_{age} to test their schedulability by consulting a formula based on a least upper bound (similar to formula (3)):

$$U(P_{age}) \leq U_{lub}(T, A, n) \qquad (14)$$

The notation $U_{lub}(T, A, n)$ represents a least upper bound for the load imposed by a set of n processes with age constraints which are mapped to periodic processes by transformation T. Clearly the least upper bound depends on the strategy A which rules how after transformation T the periodic processes are executed. As the most important strategies we investigate RMS and EDF.

Least upper bounds, as introduced here, are characterized by some *bounding set PLUB* of schedulable processes which

(a) on one hand have a utilization as low as possible,

(b) on the other hand would violate schedulability even for a minimal increase of computation time by some of them.

For EDF any set P_{per} of processes with utilization equal to 1 is a bounding set $PLUB_{per}$ of processes. Instead for RMS, only those sets $PLUB_{per}$ with utilization $U_{lub}(RMS, n)$ are bounding sets if certain relations between execution times and periods hold. For instance such a bounding set $PLUB_{per}$ for two periodic processes is given by:

	Δe_i	Δp_i
P_1	$2 - \sqrt{2}$	$\sqrt{2}$
P_2	$\sqrt{2} - 1$	1

Notice that the quotient $\Delta e_1/\Delta p_1 = \Delta e_2/\Delta p_2 = \sqrt{2} - 1$ has a constant value.

In the sequel let α_i denote the quotient $\Delta e_i/\Delta a_i$, $i \in P_{age}$. In general under the strategy RMS a set $PLUB_{per}$ of n periodic processes is a bounding set only if for all $i \in PLUB_{per}$:

$$\alpha_i = \frac{\sqrt[n]{2} - 1}{2} \qquad (15)$$

With respect to the load imposed by a set P_{age} of processes with age constraints based on the formula (see (11))

$$U(P_{age}) = \sum_{i=1}^{n} \frac{\alpha_i}{1 - \alpha_i} \qquad (16)$$

we now minimize $U(P_{age})$ under the following conditions:

$$0 \leq \alpha_i < 1/2 \qquad i \in P_{age} \qquad (17)$$

$$2 \sum_{i=1}^{n} \alpha_i = U_{lub}(A, n) \qquad (18)$$

The first condition (16) corresponds to formula (12). The second condition (17) refers to the utilization of the set P_{age} of processes with age constraints after transformation to periodic ones. Notice that the α_i depend on each other. So without loss of generality we eliminate α_1:

$$\alpha_1 = \frac{U_{lub}(A, n)}{2} - \sum_{i=2}^{n} \alpha_i \qquad (19)$$

When we are looking for minimal load with respect to formula (15), we have to minimize:

$$U(P_{age}) = \frac{U_{lub}(A, n)/2 - \sum_{i=2}^{n} \alpha_i}{1 - (U_{lub}(A, n)/2 - \sum_{i=2}^{n} \alpha_i)} \qquad (20)$$

$$+\sum_{i=2}^{n}\frac{\alpha_i}{1-\alpha_i} \qquad (21)$$

The derivation after some α_j, $j \neq 1$, leads to the equation:

$$\frac{\delta(U(P_{age}))}{\delta\alpha_j} = \frac{-1}{\left(1-\frac{U_{lub}(A,n)}{2}+\sum_{i=2}^{n}\alpha_i\right)^2} \qquad (22)$$

$$+\frac{1}{(1-\alpha_j)^2} \qquad (23)$$

Setting the left side to zero leaves the following:

$$(1-\alpha_j)^2 = \left(1-U_{lub}(A,n)/2+\sum_{i=2}^{n}\alpha_i\right)^2 \qquad (24)$$

This equation has only one practical solution for α_j:

$$2\alpha_j = U_{lub}(A,n)/2 - \sum_{i=2,i\neq j}^{n}\alpha_i \qquad (25)$$

Due to the arbitrary election of α_1 we find for any $i \in P_{age}$:

$$\alpha_i = \frac{U_{lub}(A,n)}{2n} \qquad (26)$$

At this point of discussion we have the result that a least upper bound $U_{lub}(T,A,n)$ for processes with age constraints exists only if all parameters α_i are equal.

Various steps are still missing to show the existence of a least upper bound $U_{lub}(T,A,n)$ for processes with age constraints and to derive its value. One result on this way is the observation that for a bounding set $PLUB_{per}$ of periodic processes we find a unique set P_{age} of processes with age constraints for which holds: $T(P_{age}) = PLUB_{per}$. This follows immediately from the isomorphic property of T. It remains to show that this set P_{age} is already the bounding set $PLUB_{age}$ of processes with age constraints corresponding to the least upper bound $U_{lub}(T,A,n)$. In this case for any $j \in PLUB_{age}$ the equation (23) holds. For a correct proof we have to show the properties (a) and (b) for bounding sets of processes.

For the proof of property (b) it suffices to show that any increase in computation time Δe_i, $i \in P_{age}$, and as a consequence any increase in α_i both values $U_{lub}(A,n)$ and $U_{lub}(T,A,n)$ also increase. Furthermore this violates the schedulability of the transformed set of periodic processes. This is obvious for $PLUB_{per}$. To see this for $PLUB_{age}$ we increase the execution time of some process P_i resulting in the new set P'_{age} with $\alpha'_i > \alpha_i$. This violates the schedulability of the transformed set of periodic processes and because of

$$\frac{\alpha'_i}{1-\alpha'_i} > \frac{\alpha_i}{1-\alpha_i} \qquad (27)$$

increases the load imposed by P'_{age} over the least upper bound:

$$U(P'_{age}) \not\leq U(PLUB_{age}) = U_{lub}(T,A,n) \qquad (28)$$

So far we have shown that any increase of load in $PLUB_{age}$ leads to a violation of schedulability in the transformed set of periodic processes.

Furthermore it is easy to see that property (a) is valid, because we know by equation (23) that the values $\alpha_i = U_{lub}(A,n)/2n$ minimize the load of processes with age constraints.

For RMS the set of processes with age constraints which by transformation T map to $PLUB_{per}$ must satisfy equation (14) with respect to its values for α_i, $i \in P_{age}$. With $U_{lub}(RMS,n) = n(\sqrt[n]{2} - 1)$ the same values for α_i satisfy equation (23) and constitute a bounding set $PLUB_{age}$ of processes with age constraints satisfying the equation: $T(PLUB_{age}) = PLUB_{per}$.

For EDF any set of periodic processes P_{per} with $U(P_{per}) = 1$ is a bounding set, not imposing any further condition on the values for α_i, $i \in P_{age}$. Instead exactly for all sets of periodic processes with $U(P_{per}) = 1$ and $\alpha_i = 1/2n$ the original set P_{age} with $T(PLUB_{age}) = PLUB_{per}$ has a load which is minimal. It follows that there are other sets of processes P'_{per} with $U(P'_{per}) = 1$ and $T(P'_{age}) = P'_{per}$ but $U(P'_{age}) > U(PLUB_{age})$. That is to say, under EDF P'_{per} may be a bounding set of periodic processes, but P'_{age} is no bounding set of processes with age constraints.

To compute the value of the least upper bound we have to substitute any α_j with the value $U_{lub}(A,n)/2n$. So we get for both RMS and EDF:

$$U_{lub}(T,A,n) = \sum_{i\in PLUB_{age}}\frac{\alpha_i}{1-\alpha_i}$$

$$= \sum_{i\in PLUB_{age}}\frac{U_{lub}(A,n)/2n}{1-U_{lub}(A,n)/2n}$$

$$= \frac{nU_{lub}(A,n)}{2n-U_{lub}(A,n)} \qquad (29)$$

5. EVALUATION

Analytical results like those of Liu and Layland (Liu and Layland, 73) and the one presented here are of an important practical relevance for the early phases of software development. The entire parameters of processes are concentrated in one real number and the validity of a single

225

test may provide a sufficient condition for the schedulability of a given set of processes.

With respect to processes with age constraints we implicitly derived two results, one for the strategy RMS and one for EDF. For the latter we have to test a set P_{age} of n processes with age constraints due to the criterion [2]:

$$U(P_{age}) \leq U_{lub}(T, EDF, n) = \frac{n}{2n - 1} \quad (30)$$

If this condition is valid, the set of processes with age constraints mapped by T to periodic ones can be scheduled, either statically or dynamically, but always in terms of an explicit schedule. If the condition above fails, there exists a set of processes P'_{age} with $U(P_{age}) = U(P'_{age})$ which violates timeliness when transformed to periodic processes. This is the case for the following relation between execution times and age constraints of all n processes $i \in P'_{age}$:

$$\Delta e_i / \Delta a_i = 1/2n$$

For RMS we have to test the following criterion:

$$U(P_{age}) \leq U_{lub}(T, RMS, n) \quad (31)$$

$$= \frac{n \left(\sqrt[n]{2} - 1 \right)}{2 - \left(\sqrt[n]{2} - 1 \right)} \quad (32)$$

If this condition is true we do not need – in contrast to the EDF-strategy – to construct an explicit schedule. Corresponding to the shortness of the age contraints Δa_i a process P_i is assigned a fixed priority.

The negative aspect of this approach based on transformation T lies in the low level of load which is safely schedulable due to the least upper bounds $U_{lub}(T, EDF, n)$ and $U_{lub}(T, RMS, n)$. Here some characteristic values:

n	2	5	10	∞
$U_{lub}(T, EDF, n)$	0.667	0.556	0.526	0.5
$U_{lub}(T, RMS, n)$	0.522	0.401	0.372	≈ 0.347

This negative result has already been anticipated by simulation suits (Albrecht and Zöbel, 1997) and (Albrecht, 1998). The average load achievable by mapping processes with age constraints was at about 0.52. As a consequence of this observation another strategy – already rudimentarily described by Mok (Mok, 1983) – was elaborated in detail by Albrecht (Albrecht, 1998): the execution of the transformed periodic processes under the strategy DMS. Therefore a transformation T' is needed mapping execution times and age constraints to triples consisting of execution times

Δe, periods Δp and deadlines Δd. These parameters must satisfy the following condition: $\Delta p + \Delta d = \Delta a$. There are various heuristics to determine period and deadline. The heuristic applied by Albrecht is to order the processes according to the shortness of their pseudo-periods and than following this order assign periods to processes which are as long as possible and deadlines which are as short as possible. Based on this transformation he could observe a remarkable increase of utilization to at about 0.65 in his simulation suits. However, no analytical results for this new transformation T', particularly no least upper bounds, have been derived so far.

6. REFERENCES

Albrecht, W. and D. Zöbel (1997). Integrating fixed priority and static scheduling to maintain external consistency. In: *Real-Time Database and Information Systems, Research Advances* (A. Bestavros and V. Fay-Wolfe, Eds.). pp. 89–102. Kluwer Academic Publishers. Proceedings of the 2ed International Workshop on Real-Time DataBases, Burlington, Vermont, USA.

Albrecht, Wolfgang (1998). Echtzeitplanung bei Altersanforderungen. PhD thesis. Universität Koblenz-Landau, Fachbereich Informatik.

Audsley, N., A. Burns, M. Richardson, K. Tindell and A. Wellings (1992). Absolute and relative temporal constraints in hard real-time databases. In: *Proc. of IEEE Euromicro Workshop on Real Time Systems*.

Buttazzo, G. C. (1997). *Hard Real-Time Computing Systems: Predictable Scheduling, Algorithms and Applications*. Kluwer Academic Publishers.

Liu, C. L. and James W. Layland (73). Scheduling algorithms for multiprogramming in a hard-real-time environment. *Journal of the ACM* **20**(1), 46–61.

Mok, A. K. (1983). Fundamental design problems of distributed systems for the hard-real-time environment. PhD thesis. Massachusetts Institute of Technology.

Ramamritham, K. (1993). Real-time databases. *Distributed and Parallel Databases* **1**, 199–226.

Sha, L. and J. B. Goodenough (1990). Real-time scheduling theory and Ada. *IEEE Transactions on Computers* pp. 53–62.

Zöbel, Dieter and Wolfgang Albrecht (1995). *Echtzeitsysteme - Grundlagen und Techniken*. Internat. Thomson Publ.. Bonn, Albany.

---

[2] Notice that in contrast to (Liu and Layland, 73) (see formula (5)) this criterion is only sufficient for schedulability.

ON THE SCHEDULABILITY ANALYSIS FOR DISTRIBUTED REAL-TIME SYSTEMS

Shuhua Wang and Georg Färber *

* Laboratory for Process Control and Real–Time Systems
Prof. Dr.–Ing. G. Färber
Technische Universität München, Germany

Abstract: In distributed hard real-time systems, tasks not only have timing constraints but also often have precedence constraints caused by communication among them. In this paper a new schedulability analysis algorithm for distributed hard real-time systems is proposed in which both precedence constraints and communication costs are considered and represented by offsets and modified deadlines. To obtain a tight upper bound for the worst case response time, the concepts of *local critical instant* and *local worst case response time* are introduced. They are used to calculate *global worst case response time* in the system. The proposed schedulability analysis is compared with other schedulability analyses using test cases. The comparison shows that the proposed analysis is more accurate than the compared analyses. *Copyright © 1999 IFAC*

Keywords: Schedulability analysis, distributed systems, hard real-time, precedence constraints

1. INTRODUCTION

A real-time system must be both functionally and temporally correct. To check the hard real-time conditions, schedulability analysis (or test) should be applied.

Fixed priority preemptive scheduling methods are efficient ways of constructing and analyzing schedules for hard real-time systems. Among them, deadline monotonic scheduling is more suitable to be used in parallel environments with communicating tasks, since precedence constraints caused by communication could be taken into consideration by modifying the original deadlines of tasks.

The majority of the schedulability analyses performed to date has assumed a *critical instant* which was introduced by Liu and Layland (Liu and Layland, 1973). They stated that for task set in which all tasks are independent there is a critical instant, i.e., all tasks are simultaneously released. If the schedulability analysis is carried out for the critical instant and all tasks are assumed

to execute by their worst case execution times, then the test is both sufficient and necessary. However, in distributed systems, tasks often have precedence constraints caused by communication among them. This means that all tasks can not be simultaneously released. Hence the schedulability analysis based on the critical instant becomes pessimistic.

There are several approaches in which precedence constraints are considered (Altenbernd, 1995; Tindell, 1994; Bate and Burns, 1997; Wang and Färber, 1998). An exact analysis was presented in (Audsley *et al.*, 1993*b*). Since the exact analysis is computationally intractable, a sufficient but not necessary analysis was developed (Tindell, 1994). Recently, a non-preemptive fixed priority schedulability analysis with offset was proposed (Bate and Burns, 1997). The limitation of these analyses is that they are based on uniprocessor systems.

For multiprocessor systems, an analysis using minimal-maximal offset intervals was presented (Altenbernd, 1995). In that paper, the computation of the worst case response time is divided into two parts: transaction response time and interference of other transactions. However, the communication cost between tasks residing on different processors is not considered. Communication cost has influence on both predecessor task and successor task and makes their deadlines harder to be met. An improved analysis (Wang and Färber, 1998) was proposed by the authors of this paper. In that paper the communication cost is considered and the calculation of transaction response time is more accurate. However the calculation of interference of other transactions is not changed which is based on a less accurate algorithm. Thus the analysis is also pessimistic in some cases.

In this paper, the worst case response time is analyzed from another point of view. The new algorithm is based on *busy period analysis* (Lehoczky *et al.*, 1989; Tindell *et al.*, 1994). Simulation results show that it is more accurate than the algorithm most used in the literature.

The rest of the paper is organized as follows: Section 2 briefly describes the basic assumptions which will be used in the proposed analysis. Then the new schedulability analysis is presented in section 3. Section 4 gives evaluation results in which the proposed schedulability analysis is compared with other approaches using test cases. Finally, Section 5 summarizes the result of this paper.

2. BASIC ASSUMPTIONS

Let's consider a set of $\Gamma = \{\tau_i : 1 \leq i \leq n\}$ of periodic or sporadic hard real-time tasks. Each task $\tau_i \in \Gamma$ has a worst case execution time c_i, a relative deadline D_i and a minimum time interval T_i between two consecutive invocations.

The task set and the intertask precedence constraints are represented by a directed acyclic graph which is also called task graph $TG = (N, A)$. N is a set of nodes representing the tasks in the set Γ. A is a set of directed arcs representing the precedence constraints between the tasks in Γ, that is, if $\tau_j \prec \tau_i$ then $(\tau_j, \tau_i) \in A$. Additionally, the task graph contains also the information about task allocation. An example of task graphs is shown in Fig.1.

In addition to task parameters and task graph, the general assumptions which will be used in the new analysis are follows:

- Tasks are statically allocated on a number of homogeneous processors.
- Communication between two tasks residing on the same processor is done via accessing shared memory and its cost can be negligible.

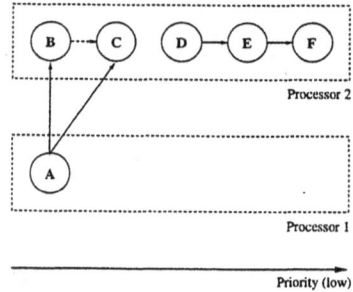

Fig. 1. An example of the task graph

- Communication between two tasks residing on different processors is done via a shared bus system with token passing protocol.
- Scheduling is done locally for each processor by deadline monotonic scheduling policy. Preemptive due to higher priority tasks is allowed.
- Operating system overhead and release jitter are zero.
- For all tasks, $D_i \leq T_i$.

3. THE NEW SCHEDULABILITY ANALYSIS

In distributed hard real-time systems, tasks often have precedence constraints caused by communication among them. This means that all tasks can not be simultaneously released. Hence the schedulability analysis based on critical instant becomes pessimistic.

To solve the above problem, a new schedulability analysis is proposed in which both precedence constraints and communication costs are considered and represented by offsets and modified deadlines. The new approach can be seen as an extension of the original busy period analysis which is first proposed by Lehoczky (Lehoczky *et al.*, 1989) and later extended by Audsley (Audsley *et al.*, 1993a) and Tindell (Tindell *et al.*, 1994).

The new schedulability analysis consists of two steps: priority assignment and worst case response time computation. If for all tasks in the task set, their worst case response times are less than or equal to their modified deadlines, the task set is said to be schedulable.

3.1 *Priority Assignment*

The priority assignment is performed by using the task graph to compute appropriate deadlines. The original predecessor's deadline D_i does not take into account the necessary execution time following τ_i and the communication cost. To consider the influence of precedence constraints and communication cost, the original deadline should be modified by:

$$d_i := \begin{cases} Min\{d_j - c_j - cd_{ij}\}, & \forall j : (\tau_i, \tau_j) \in A \\ D_i, & otherwise \end{cases} \quad (1)$$

where cd_{ij} denotes the communication cost between tasks τ_i and τ_j.

It is assumed that the communication between two tasks residing on different processors is done via a shared bus system with token passing protocol. In the worst case, cd_{ij} is given by (Wang, 1999):

$$cd_{ij} = \begin{cases} 0, & \tau_i, \tau_j \text{ are on the same processor} \\ \sum_{p=1}^{m}(\sum_{l=1}^{n(p)} \frac{M_{l,p}}{S} + TO), & otherwise \end{cases} \quad (2)$$

where $n(p)$ is the number of tasks on processor p, $M_{l,p}$ is the total size of messages sent on the bus from the lth task residing on processor p, m is the number of processors in the system, TO is the time taken to transmit the token and S is the speed of the bus.

After the original deadlines are modified, tasks are assigned priorities with respect to these modified deadlines.

3.2 Worst Case Response Time Computation

To simplify the computation, the concept of *Job* is introduced as follows:

Definition 1 (Job)
A job is a collection of related tasks which have precedence constraints among them.

For example, in Fig. 1 task A, B and C constitute a job, task D, E and F constitute another job. During allocation, tasks in the same job may be allocated to different processors due to the consideration on fault tolerance or I/O requirements. In correspondence to this, the concepts of the *global input task* and the *local input task* are introduced:

Definition 2 (Global Input Task)
A global input task is the input task of a job.

Definition 3 (Local Input Task)
A local input task of a job for a specific processor is the first allocated task of that job on that processor.

In the task set in Fig.1, task A and D are global input tasks, task B is a local input task (with respect to processor 2).

With the above concepts, the *global offset* O_{1i} and the *local offset* O_{2i} are given by:

$$O_{1i} := \begin{cases} Max\{O_{1j} + c_j + cd_{ji}\}, & \forall j : (\tau_j, \tau_i) \in A \\ 0, & otherwise \end{cases} \quad (3)$$

$$O_{2i} := \begin{cases} Max\{O_{2j} + c_j\}, & \forall j : (\tau_j, \tau_i) \in A \\ 0, & otherwise \end{cases} \quad (4)$$

where cd_{ji} is the communication cost between task τ_j and τ_i and its definition is given in equation (2).

According to the above equations, global offset O_{1i} is the offset with respect to the global input task, and the local offset O_{2i} is the offset with respect to the local input task. The meaning of the offset is the minimum time interval between the release of task τ_i and the input task of the job to which the task τ_i belongs.

In addition, the communication cost has no influence on the local offset due to the assumption that the communication between two tasks residing on the same processor is done via accessing shared memory and its cost is negligible.

As an example, we examine the task set in Fig.1. For task B:

$$O_{1B} = c_A + cd_{AB} = 2$$
$$O_{2B} = 0$$

For tasks with precedence constraints, all tasks can not be released simultaneously. This means that the original concept *critical instant* proposed by Liu and Layland is no longer valid. To cope with precedence constraints, we introduce the concept of the *local critical instant*. It is the extension of *critical instant* described as follows:

Definition 4 (Local Critical Instant)
Local critical instant is the time when all local input tasks on a processor are released simultaneously.

Following this, the concepts of *global worst case response time* and *local worst case response time* are further introduced:

Definition 5 (Global Worst Case Response Time)
Global worst case response time R_{1i} is the worst case response time measured with regard to the beginning of global input task of the job to which task τ_i belongs.

Definition 6 (Local Worst Case Response Time)
Local worst case response time R_{2i} is the worst case response time measured with regard to the beginning of local input task of the job to which task τ_i belongs.

The relation between global worst case response time and local worst case response time is:

$$R_{1i} = R_{2i} + O_{1i} - O_{2i} \qquad (5)$$

Fig.2 shows part of the execution of the example system in Fig. 1. For illustration propose, we consider only the execution of task A on processor 1, task B, C and D on processor 2.

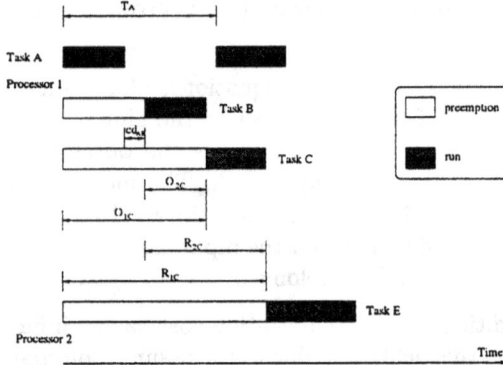

Fig. 2. Global and local worst case response times

The local worst case response time is considered first. Because of the precedence constraints, tasks in the same job can not be released simultaneously. They have minimum time intervals, i.e., the local offset O_{2i}, between their release. Thus the largest number of invocations of a higher priority task τ_j falling in the level i busy period of length w_i changes from $\left\lceil \frac{w_i}{T_j} \right\rceil$ to:

$$K_{j,w_i} = \left\lceil \frac{w_i - O_{2j}}{T_j} \right\rceil \qquad (6)$$

Therefore the length of the level i busy period should be modified as follows:

$$w_i = c_i + \sum_{j \in hp(i)} \left\lceil \frac{w_i - O_{2j}}{T_j} \right\rceil c_j \qquad (7)$$

In addition, for tasks within the same job, they can not interfere with each other due to their precedence constraints. Or in other words, the local offset O_{2i} can be seen as the interference of the higher priority tasks within the same job. To obtain a tight bound on the response time, this important fact must be taken into consideration. Thus when task blocking time is also included, the local worst case response time is given by:

$$R_{2i} = O_{2i} + c_i + b_i + \sum_{j \in hp^J(i)} \left\lceil \frac{R_{2i} - O_{2j}}{T_j} \right\rceil_0 c_j \qquad (8)$$

where b_i is the maximum blocking time that task τ_i could be delayed by lower priority task. The notation $\lceil X \rceil_0$ denotes a modified ceiling function that return zero if $X \leq 0$.

It is important to note the difference between the task set $hp(i)$ and $hp^J(i)$. The task set $hp(i)$ includes all higher priority tasks than task τ_i.

The task set $hp^J(i)$ is the subset of $hp(i)$ which includes only the higher priority tasks of different jobs as τ_i.

To calculate the local response time , the following recurrence relation can be used:

$$r_i^{(n+1)} = O_{2i} + c_i + b_i + \sum_{j \in hp^J(i)} \left\lceil \frac{r_i^n - O_{2j}}{T_j} \right\rceil_0 c_j \quad (9)$$

The initial condition of recurrence relation is $r_i^0 = O_{2i} + c_i + b_i$. A minimum solution is found when $r_i^{(n+1)} = r_i^n$, then $R_{2i} = r_i^n$. If $r_i^{(n+1)} > d_i$, then the task is not schedulable. Similar proof to the convergence of the recurrence is given by Audsley (Audsley et $al.$, 1993a).

With the local worst case response time R_{2i}, it is easy to determine the global worst case response time R_{1i} according to equation (5).

4. TEST CASE EVALUATION

The proposed analysis is first compared with the most used analysis in the literature (Tindell et $al.$, 1994) but in which no precedence constraints are considered. Then it is compared with the analysis proposed in (Wang and Färber, 1998). The comparison is performed by evaluation the difference between analyses and simulation (i.e., the real behavior of the system) results for test cases.

The J. Santo's example and the K. Tindell's example in (Santos et $al.$, 1997) are completely specified with execution times, original deadlines, periods, allocations, precedence constraints and communication requirements for each task. Therefore they can be taken as test cases to evaluate the new schedulability analysis. For clarity we will use Tind1 and Tind2 to represent the solution from K. Tindell and the solution 1 from J. Santo for Tindell's example respectively. More detailed information about these test cases can be found in (Santos et $al.$, 1997).

The above test cases are evaluated under the assumption that the speed of the shared bus is 10 Mb/s and token passing protocol is used to arbitrate between processors when they access the shared bus.

In case of neglecting the influence of precedence constraints, the worst case response time can be computed according to:

$$r_i = c_i + b_i + \sum_{j \in hp(i)} \left\lceil \frac{r_i}{T_j} \right\rceil c_j \qquad (10)$$

The evaluation results are shown in Fig. 3, Fig. 4 and Fig. 5. In these figures " Algorithm 2" refers to the analysis proposed in this paper. "No Precedence Constraint" refers to the analysis according to equation (10).

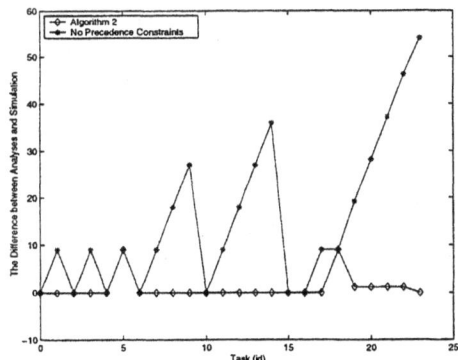

Fig. 3. The results of the J. Santo's example

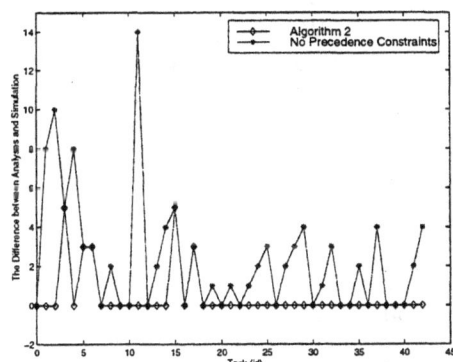

Fig. 4. The results of the test case Tind1

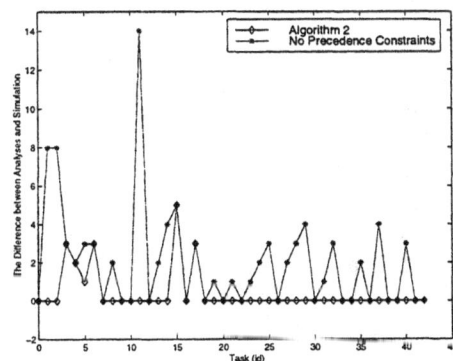

Fig. 5. The results of the test case Tind2

From these figures, it can be seen that the results of the proposed analysis are almost the same as the simulation results. This means that the proposed analysis is fairly accurate. On the contrary, the results of the analysis in which precedence constraints are not considered are very different from the simulation results. This means that in case of existing precedence constraints, the analysis neglecting these precedence constraints will be relatively pessimistic.

In addition to the above comparison, the new analysis is also compared with the analysis proposed in (Wang and Färber, 1998). Although the precedence constraints are considered in that paper, the computation of the worst case response time is pessimistic in some cases. More details about that algorithm please see the original paper.

The evaluation results of the two analyses for different test cases are shown in Fig. 6, Fig. 7 and Fig. 8. In these figures "Algorithm 1" refers to the analysis proposed in (Wang and Färber, 1998). "Algorithm 2" refers to the analysis proposed in this paper.

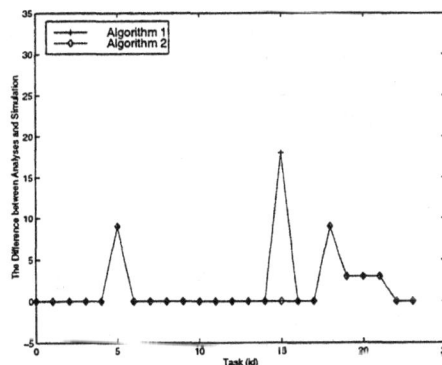

Fig. 6. The results of the J. Santo's example

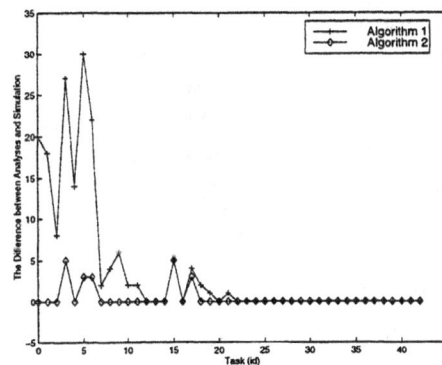

Fig. 7. The results of the test case Tind1

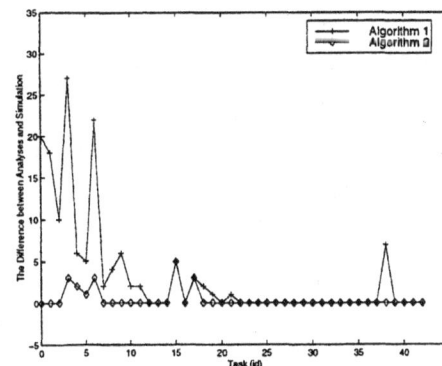

Fig. 8. The results of the test case Tind2

From these figures it can be seen that in most cases the differences between the results of two analyses and simulation are very small. This means that both algorithms are relative accurate. On the other hand, from Fig. 7 and Fig. 8 it can be seen that the most differences between the two algorithms are job 0 which consists of task 0, 1, 2, 3, 4, 5 and 6. Because these tasks have the lower priorities than all the other tasks and are the only tasks in the transaction with period of 60 time unit (see (Santos *et al.*, 1997)). Therefore, the interference caused by higher priority tasks consists of only the interference of other transactions. However, the calculation of this type interference is just the less exact part of the worst case response time computation in algorithm 1. This is the reason why algorithm 1 is less accurate than algorithm 2 in these cases. The same tendency can also be seen in Fig. 6.

5. CONCLUSION

In this paper, a new schedulability analysis algorithm for distributed hard real-time systems is proposed. The influence of both precedence constraints and communication costs is considered. It is represented by the offsets and modified deadlines.

To obtain a tight upper bound of the worst case response time, the concept of *local critical instant* is introduced. It is the extension of the Liu and Layland's original concept of *critical instant*. Then the concepts of *local offset* and *global offset* are further introduced. They are used in the computation of *local worst case response time* and *global worst case response time*. The computation is based on busy period analysis.

The simulation results show that when the precedence constraints between tasks are ignored, the analysis results will be relatively pessimistic. Therefore in a good schedulability analysis the precedence constraints should be considered. The proposed schedulability analysis is more accurate than those in (Tindell, 1994; Wang and Färber, 1998). Another useful characteristic of the proposed algorithm is the consideration of the influence of communication costs which makes deadlines harder to be met and is not considered in (Altenbernd, 1995; Tindell, 1994; Bate and Burns, 1997).

6. REFERENCES

Altenbernd, P. (1995). Deadline–monotonic software scheduling for the co–synthesis of parallel hard real–time systems. In: *Proceedings of the European Design and Test Conference.* pp. 190–195.

Audsley, N. C., A. Burns, M. F. Richardson, K. Tindell and A. J. Wellings (1993*a*). Applying new scheduling theory to static priority pre–emptive scheduling. *Software Engineering Journal* **8**(5), 184–292.

Audsley, N. C., K. Tindell and A. Burns (1993*b*). The end of the line for static cyclic scheduling. In: *Proceedings of the 5th Euromicro Workshop on Real–Time Systems.* pp. 36–41.

Bate, I. and A. Burns (1997). Schedulability analysis of fixed priority real-time systems with offsets. In: *Proceedings of the Ninth Euromicro Workshop on Real-Time Systems.* IEEE Computer Society. Toledo, Spain. pp. 153–160.

Lehoczky, J. P., L. Sha and Y. Ding (1989). The rate monotonic scheduling algorithm–exact charaterization and average case behaviour. In: *Proceedings of IEEE Real-Time Systems Symposium.* pp. 166–171.

Liu, C. L. and J. W. Layland (1973). Scheduling algorithms for multiprogramming in a hard real-time environment. *ACM* **20**(1), 46–61.

Santos, J., E. Ferro, J. Orozco and R. Cayssials (1997). A heuristic approach to the multitask-multiprocessor assignment problem using the empty–slots method and rate monotonic scheduling. *Real-Time Systems* **13**(2), 167–199.

Tindell, K. (1994). Adding time–offsets to schedulability analysis. Technical Report YCS221. Department of Computer Science. University of York.

Tindell, K., A. Burns and A. J. Wellings (1994). An extendible approach for analyzing fixed priority hard real–time tasks. *Real–Time Systems* **6**(2), 133–152.

Wang, Shuhua (1999). Specification, Allocation and Schedulability Analysis for Fixed-Priority Hard Real-Time Systems. PhD thesis. Laboratory for Process Control and Real–Time Systems, Technische Universität München. Munich, Germany.

Wang, Shuhua and Georg Färber (1998). Schedulability analysis for communicating tasks on a multiprocessor system. In: *Proceedings of the 23th IFAC/IFIP Workshop on Real Time Programming WRTP'98.* Shantou, China. pp. 25–30.

A FLEXIBLE MODEL OF TIME CONSTRAINTS FOR CONTROL AND MULTIMEDIA REAL-TIME SYSTEMS

José M. López, Daniel García

University of Oviedo
Department of Computer Science
Campus de Viesques, 33204, Gijón, Spain
{chechu,daniel}@atc01.etsiig.uniovi.es

Abstract: Real-time constraints of periodic tasks are usually expressed as hard deadlines. Nevertheless, in many systems, such as multimedia, packet communication and some control systems, these constraints are too restrictive. In this case, a better model of real-time constraints for these periodic tasks, called *occasionally skippable tasks*, permits some deadlines to be missed provided that most deadlines are met. A general model for occasionally skippable tasks is developed, which includes real-time constraints ranging gradually from soft deadlines to hard deadlines. An efficient feasibility test is proposed. This feasibility test is the basis of an algorithm which provides the optimal assignment of static priorities to occasionally skippable tasks. *Copyright © 1999 IFAC*

Keywords: Real-Time scheduling, occasionally skippable tasks, busy period, static priorities, feasibility test

1. INTRODUCTION

This paper deals with uniprocessor systems in which all tasks are periodic, independent and scheduled by a static priority pre-emptive scheduler. Scheduling costs are assumed to be negligible. The classic model of a periodic task is defined by a period, computation time, offset (time at which the first task instance is released) and a hard deadline. In this model, missing a deadline is not acceptable. However, there are applications, such as multimedia, packet communication and some control systems, in which missing a deadline is acceptable provided that most deadlines are met. Thus, the objective is to provide a defined quality of service.

Periodic tasks that allow some deadlines to be missed are called occasionally skippable [2]. A simple model for occasionally skippable tasks consists of defining a skip parameter s in the range $1 \leq s \leq \infty$, indicating in periods the minimum distance between two missed deadlines. With $s = 1$ all the deadlines of the task can be missed, and so the task deadline is a soft deadline. With $s = 2$, if a task instance misses its deadline, then the deadline of the next task instance must be met. With $s = \infty$ all the deadlines must be met, and so the task deadline is a hard deadline. This simple model of real-time constraints for occasionally skippable periodic tasks consists of a deadline and a skip parameter. Nevertheless, this model can be improved to cover more complex situations. For example, consider a video system allowing the loss of five consecutive video frames (because of five consecutive missed deadlines), provided that the fifty previous consecutive frames were processed in time (because of fifty consecutive met deadlines).

In order to deal with complex occasionally skippable tasks, the feasibility function $F_i(f_i^1, f_i^2, \ldots, f_i^j, \ldots)$ for any occasionally skippable task τ_i is defined.

233

$(f_i^1, f_i^2, \ldots, f_i^j, \ldots)$ is the sequence of feasibilities of the task τ_i for all its instances from $t = 0$ to $t = \infty$. Every instance of a task is associated to a feasibility value which is 1 if the task instance is feasible (meets the deadline) and 0 if the task instance is not feasible (misses the deadline). For example, the sequence of feasibilities $(0, 0, 1, 1, \ldots)$ indicates that the first and second instances of the task from $t = 0$ are missed, while the second and third are met.

The feasibility function of a task receives its sequence of feasibilities as input. It generates the value 1 if the sequence of feasibilities is acceptable, and the value 0 if the sequence of feasibilities is unacceptable.

The general model for the real-time constraints of occasionally skippable tasks is defined by a deadline and a feasibility function. For example, the simple model of real time constraints of a task τ_i defined by the skip parameter s_i can be expressed as the feasibility function

$$F_i = \begin{cases} 0 & \text{if } \exists j, k \mid f_i^j = f_i^k = 0 \text{ and } k - j < s_i \\ 1 & \text{otherwise} \end{cases}$$

In this paper a periodic task set $\Gamma = \{\tau_i(T_i, C_i, O_i, D_i, F_i), \ i = 1, \ldots, n\}$ is considered. Where T_i is the period, C_i the computation time, O_i the offset, D_i the deadline and F_i the feasibility function of the the task τ_i. For the sake of simplicity the infeasible task instances are considered to be completely processed. Thus, they are processed, even after their deadlines have been missed.

This paper proposes an exact and efficient algorithm which solves the feasibility test for skippable periodic tasks. This test was considered to be NP-hard [2], however in practice, it can be solved in pseudo-polynomial time.

The feasibility test is used to calculate an optimal assignment of static priorities for occasionally skippable tasks. It is optimal in the sense that it provides a priority assignment which makes the periodic task set schedulable if one priority assignment which makes it schedulable exists.

2. THE LOAD FUNCTION

In this section the concept of load function is defined.

Definition 1. Let $\Gamma = \{\tau_i(T_i, C_i, O_i, D_i, F_i), \ i = 1, \ldots, n\}$ be a periodic task set. The load function, $L(t)$, of the periodic task set at time t is defined as

$$L(t) = L_a[0, t] - L_p[0, t], \quad t \geq 0 \qquad (1)$$

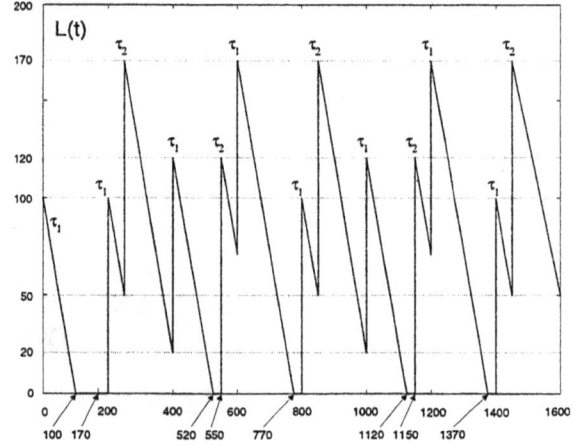

Fig. 1. Load function for the set of tasks $\{\tau_1, \tau_2\}$ of periods $\{200, 300\}$, computation times $\{100, 120\}$ and offsets $\{0, 250\}$.

where

$$L_a[0, t] = \sum_{i=1}^{n} \left\lceil \frac{t - O_i}{T_i} \right\rceil C_i \qquad (2)$$

is the *load activated* in the interval $[0, t)$, and

$$L_p[t_0, t_1] = \int_{t_0}^{t_1} \alpha_p(t) dt \quad, 0 \leq t_0 \leq t_1 \qquad (3)$$

is the *load processed* in the interval $[t_0, t_1)$.

$$\alpha_p(t) = \begin{cases} 1 & \text{if } L(t) > 0 \\ 0 & \text{if } L(t) = 0 \end{cases} \qquad (4)$$

is the *processing rate* at time t. Figure 1 depicts an example of load function for a set of periodic tasks.

$L_p[t_0, t_1]$ represents the load processed in the interval $[t_0, t_1)$, at the rate of one load unit per time unit, when the load is greater than zero. The scheduler is assumed to grant the processor to a task instance whenever there are task instances that have not yet been completely processed.

To every time t, the load function assigns a value equal to the total time required to finish the processing of all task instances not completely processed at t. Hence, the load function is independent of the scheduling algorithm.

The interval $[t_0, t_0)$ is an empty interval, and so $L_a[t_0, t_0] = L_p[t_0, t_0] = L(0) = 0$. Below, some properties of the load function are given

(1) $L(t) \geq 0, \quad t \geq 0$
(2) $L(t_1) = L(t_0) + L_a[t_0, t_1] - L_p[t_0, t_1]$,
   $0 \leq t_0 \leq t_1 \leq t_2$
(3) $L(t_0) \leq L(t_0 + P), \quad t_0 \geq 0$

where P is the least common multiple of the periods, also called hyperperiod.

Properties 1 and 2 are deduced directly from the definition. From Property 2 says that the load function at a given time t_1 depends only on the load function at a previous time t_0, and the sequence of task instances from (at and after) that previous time. The sequence of task instances is periodic, of a period P at any time, for any set of periodic tasks. For example, the sequence of task instances from (at and after) $t = 0$, $t = 600$ and $t = 1200$ is the same for the task set shown in Figure 1. Starting from these times, task instances are found with the same computation times and deadlines, released at the same relative instants with regard to the starting time. The load at time $t = 0$ is zero (the minimum possible), therefore, the load at time $t = P$ is higher than or equal to the load at time $t = 0$. The sequence of task instances is the same from $t = 0$ and from $t = P$, and so the load at time $t_0 + P$ cannot be lower than the load at time $t = t_0$. This demonstrates Property 3.

Figure 1 shows two different states of the load function: an initial transient state, and a final steady state. From $t = 0$ to $t = 170$ the load is said to be in a transient state. After $t = 170$ the load becomes periodic of a period 600, and is said to be in a steady state of load. Next, the periodicity of the load function will be proved.

Theorem 1. Let $\Gamma = \{\tau_i(T_i, C_i, O_i, D_i, F_i),\ i = 1, \ldots, n\}$ be a periodic task set. Let P be the hyperperiod.
If $\sum_{i=1}^{n} C_i/T_i < 1$ then the load becomes periodic of a period P, from a time previous to P.

PROOF. Firstly, it is proved by contradiction that there is at least one time t_p, with $0 \le t_p < P$, verifying that $L(t_p + P) = 0$.

If $L(t) > 0\ \forall t \in [P, 2P)$ then applying Property 2 $L(2P) = L(P) + P\sum_{i=1}^{n} C_i/T_i - P < L(P)$. This is a contradiction, as applying Property 3 with $t_0 = P$ makes $L(2P) \ge L(P)$. Therefore, a $t_p \in [0, P)$ verifying that $L(t_p+P) = 0$ exists. The load at t_p verifies $L(t_p) \le L(t_p + P)$ as deduced from Property 3. The load is no negative applying Property 1, and so $L(t_p) = L(t_p + P) = 0$. The sequence of task instances repeats periodically with a period P from t_p. Applying Property 2 the load repeats periodically from $t_p < P$. □

The periodicity of the load has been proved when the utilization factor is less than one. Let t_z be the earliest time in $[0, P)$ verifying $L(t_z + P) = 0$. It can be easily demonstrated that t_z is the first instant from which the load is periodic. For example, $P = lcm\{200, 300\} = 600$ for the set of tasks given in Figure 1. In $[600, 1200)$, the first time the load function is zero is $t_z + P = 770$. Therefore, the earliest time at which the system is in steady state of load is $t_z = 170$.

3. BUSY PERIODS AND RESPONSE TIMES

In this section an algorithm to obtain the response time of the task instances released in the first period $[t_z, t_z + P)$, of the load function is proposed. Once these response times are obtained, the array of feasibilities associated to any task is calculated directly by comparison with the task deadline.

If the offsets were zero, the deadlines less than or equal to the periods and the task set were feasible, then the first instance of every task would be that of maximum time response.

An algorithm which calculates the response time of any task instance, applicable to periodic task sets of arbitrary offsets and deadlines less than or equal to the periods is proposed in [1]. It is based on the calculation of the interference due to higher priority tasks. The interference is calculated using a set of tuples $\{C_j, t\}$ for each task instance to test the feasibility, where C_j is the interference of a higher priority task at time t. Each set of tuples requires ordering before calculating the response time, which reduces the efficiency of the algorithm.

This paper proposes a simple and efficient algorithm to obtain the response time of any task instance, applicable to task sets of arbitrary offsets and deadlines less than or equal to the periods. The algorithm is based on the load function and the concept of busy periods, defined below. Throughout this paper it is considered that the infeasible task instances are completely processed, even after missing their deadlines.

The load function has an initial transient state and a final steady state. In order to calculate the maximum response time of the task instances released in the steady state, it is necessary and sufficient to consider the response times of the task instances released in the first period of the steady state, that is to say, in the interval $[t_z, t_z + P)$. However, what happens with the response time of the task instances released in $[0, t_z)$?

After assigning the priorities, the static priority pre-emptive schedulers run the task instance of highest priority at any time. Instances coming from different tasks have different priorities. If there are two or more instances of the same task ready for execution, the processor is granted to the oldest one.

There are the same task instances (the same relative release times and the same parameters) from $t = 0$ and $t = P$. There are no task instances released before $t = 0$ that can compete with the task instances released at or after $t = 0$. However, there may be task instances that can compete with those released at or after $t = P$. Therefore, the response time of the task instances

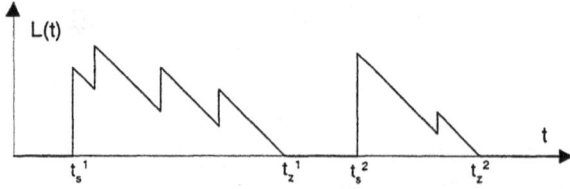

Fig. 2. Beginning and end of busy periods.

from $t = 0$ can never be higher than that of their counterparts from $t = P$.

The conclusion is that the maximum response times occur in the steady state.

The array of feasibilities corresponding to any task is equivalent to the array of feasibilities obtained by repeating the array of feasibilities within the interval $[t_z, t_z + P)$.

The busy periods are the time intervals in which the processor is working [3]. The busy periods can be defined in terms of the load function, as the intervals in which the load function is greater than zero. For example, Figure 1 contains the busy periods $[0, 100)$, $[200, 520)$, $[550, 770)$, $[800, 1120)$ and $[1150, 1370)$.

Figure 2 shows the beginning and end of several consecutive busy periods of a load function. The beginning t_s^k of the k-th busy period can be obtained from the end of the previous busy period t_z^{k-1}, by using (5)

$$t_s^k = \min_{1 \leq i \leq n} \left\{ O_i + \left\lceil \frac{t_z^{k-1} - O_i}{T_i} \right\rceil T_i \right\} \quad (5)$$

The term between brackets is the first time at which the task τ_i is released at or after the end of the previous busy period. The beginning of the first busy period can be obtained more easily, as $t_s^1 = \min_{1 \leq i \leq n} \{O_i\}$. The end of any busy period, t_z^k, is the first time after the beginning of the busy period that makes the load zero. Therefore, all the task instances released in $[t_s^k, t_z^k)$ have been completely processed at t_z^k. The number of instances of task τ_i in $[t_s^k, t_z^k)$, called m_i^k, can be calculated from (6)

$$m_i^k = \left\lceil \frac{t_z^k - O_i}{T_i} \right\rceil - \left\lceil \frac{t_s^k - O_i}{T_i} \right\rceil \quad (6)$$

The first term of (6) is the number of τ_i instances in $[0, t_z^k)$, whereas the second term is the number in $[0, t_s^k)$. In addition, t_s^k can be expressed as $t_z^k = t_s^k + \sum_{i=1}^n m_i^k C_i$. The values t_z^k and m_i^k must be obtained by iteration. The iteration starts with the value $t_z^k = t_s^k + C_s^k$, where C_s^k is the sum of computation times for all task instances released at t_s^k.

Section 5 contains an example in which the busy periods are calculated.

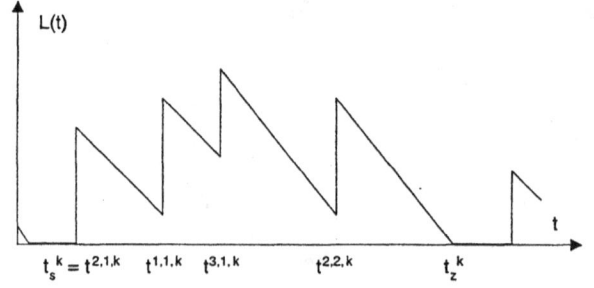

Fig. 3. Feasibility of periodic tasks in a busy period.

Below, an algorithm to obtain the busy periods included in $[0, t_z + P)$ is proposed.

calculation of busy periods in $[0, t_z + P)$
  $k := 1; t_s^k := \min_{1 \leq i \leq n} \{O_i\}$
  repeat
    new $t_z^k := t_s^k + C_s^k$
    repeat
      $t_z^k :=$ new t_s^k
      for each task τ_i
        $m_i^k := \left\lceil \frac{t_z^k - O_i}{T_i} \right\rceil - \left\lceil \frac{t_s^k - O_i}{T_i} \right\rceil$
      end of for
      new $t_z^k := t_s^k + \sum_{i=1}^n m_i^k C_i$
    while new $t_z^k > t_z^k$
    $k := k + 1$
    $t_s^k := \min_{1 \leq i \leq n} \left\{ O_i + \left\lceil \frac{t_z^{k-1} - O_i}{T_i} \right\rceil T_i \right\}$
  while $t_s^k < P$
end of calculation of busy periods in $[0, t_z + P)$

Once the busy periods included in $[0, t_z + P)$ are obtained, the busy periods included in $[t_z, t_z + P)$ are extracted. The response time of any task instance released in $[t_z, t_z + P)$ is calculated efficiently using these busy periods. To explain the algorithm, Figure 3, which depicts a general busy period of value $[t_s^k, t_z^k)$ is used. In the figure, $t^{i,j,k}$ represents the time at which the jth instance in $[t_s^k, t_z^k)$ of task τ_i is released . If τ_1 had the lowest priority, the response time of the instance released at $t^{1,1,k}$ would be $(t_z^k - t^{1,1,k})$. If τ_3 had the lowest priority, the response time of the instance released at $t^{3,1,k}$ would be $(t_z^k - t^{3,1,k})$. If τ_2 had the lowest priority, the response time of the instance released at $t^{2,2,k}$ would be $(t_z^k - t^{2,2,k})$. The calculation of the response time of the instance released at $t^{2,1,k}$ is more complex, since the second instance of τ_2 is released before the processing of the first one is finished. As deadlines less than or equal to the periods are considered, the instance released at $t^{2,1,k}$ would miss its deadline.

In general, let $\Gamma = \{\tau_i(T_i, C_i, O_i, D_i, F_i), \ i = 1, \ldots, n\}$ be a periodic task set, with $D_i \leq T_i \ \forall i = 1, \ldots, n$. To analyze the feasibility of instances of the lowest priority task τ_l, released in the busy period $[t_s^k, t_z^k)$, the first step is to calculate the

236

number of τ_l instances, m_l^k, in that busy period. If $m_l^k = 0$ there is nothing to calculate. If $m_l^k = 1$ then the response time of the single task instance of τ_l in the busy period takes the value

$$R_l^{1,k} = t_z^k - O_l - \left\lceil \frac{t_s^k - O_l}{T_l} \right\rceil T_l$$

This task instance is feasible if, and only if $R_l^{1,k} \leq D_l$. The term $O_l + \left\lceil \frac{t-O_l}{T_l} \right\rceil T_l$ is the instant at which τ_l is released in the busy period $[t_s^k, t_z^k)$. If $m_l^k > 1$, the first $(m_l^k - 1)$ task instances of τ_l in the busy period are not feasible. The response time of the last τ_l instance in the busy period $[t_s^k, t_z^k)$ takes the value

$$R_l^{m_l^k,k} = t_z^k - O_l - \left\lceil \frac{t_s^k - O_l}{T_l} \right\rceil T_l - (m_l^k - 1)T_l \tag{7}$$

This instance is feasible if, and only if $R_l^{m_l^k,k} \leq D_l$.

After this, the task τ_l is removed from the task set, and the same process is used for the resultant task set until a task set with a single periodic task is obtained.

4. OPTIMAL STATIC PRIORITY ASSIGNMENT

The assignment of static priorities to a set of n tasks consists of giving a priority within $\{1, \ldots, n\}$ to each task, where n is the lowest priority.

If offsets are zero and deadlines less than or equal to the periods, the algorithm deadline monotonic provides the optimal assignment of static priorities to non-skippable periodic tasks [4].

An algorithm of complexity $O((n^2 + n)E)$ that provides the optimal static priorities, where E is the complexity of the feasibility test for the lowest priority task is proposed in [1]. In each step of the this algorithm a task which is feasible with the lowest priority of the task set is tried. If such a task does not exist, the task is not feasible. If it does, then it is assigned the lowest priority and removed from the task set. This process is repeated until all the priorities have been assigned.

In order to determine if a task is feasible with the lowest priority, it is necessary to obtain the array of feasibilities in the interval $[t_z, t_z + P)$, and apply the feasibility function of the tasks to its array of feasibilities. If two or more tasks are found to be feasible with the lowest priority in one step of the algorithm, they are also feasible with a higher priority.

Below, an optimal algorithm which incorporates this observation is proposed.

optimal assignment of static priorities
  n := number of tasks
  $task\ set$:= FEASIBLE
  while $n \geq 1$ and $taskset$ = FEASIBLE
    obtain the value of F_i $\forall i = 1, \ldots, n$
    $task\ set$:= NO FEASIBLE
    for i := n to 1 decreasing 1
      if F_i = FEASIBLE
        assign priority n to τ_i
        remove τ_i from the task set ($n := n - 1$)
        $task\ set$:= FEASIBLE
      end of if
    end of for
  end of while
end of the optimal assignment of static priorities

5. EXAMPLE

The problem is to assign static priorities $\{p_1, p_2, p_3\}$ to the occasionally skippable tasks $\{\tau_1, \tau_2, \tau_3\}$ of periods $\{10, 20, 5\}$, computation times $\{1, 5, 3\}$, offsets $\{1, 0, 4\}$, deadlines $\{8, 20, 4\}$, and feasibility functions:

- $F_1 = 0$ if there are at least 2 consecutive missed deadlines.
- $F_2 = 0$ if there is at least a missed deadline.
- $F_3 = 0$ if there is at least a missed deadline.

$\sum_{i=1}^n \frac{C_i}{T_i} = (1/10 + 5/20 + 3/5) = 0.95 \leq 1$ and so the load reaches the steady state and may be feasible. The hyperperiod is $P = lcm\{10, 20, 5\} = 20$. Firstly, the interval $[t_z, t_z + P)$ is calculated. Prior to calculate t_z it is necessary to obtain the busy periods.

$$t_s^1 = \min_{1 \leq i \leq n} \{O_i\} = min\{1, 0, 4\} = 0$$
$$t_z^1 = t_s^1 + C_s^1 = 0 + 5 = 5$$
$$m_1^1 = \left\lceil \frac{t_z^1 - O_1}{T_1} \right\rceil - \left\lceil \frac{t_s^1 - O_1}{T_1} \right\rceil = 1$$
$$m_2^1 = \left\lceil \frac{t_z^1 - O_2}{T_2} \right\rceil - \left\lceil \frac{t_s^1 - O_2}{T_2} \right\rceil = 1$$
$$m_3^1 = \left\lceil \frac{t_z^1 - O_3}{T_3} \right\rceil - \left\lceil \frac{t_s^3 - O_3}{T_3} \right\rceil = 1$$
$$t_z^1 = t_s^1 + m_1^1 C_1 + m_2^1 C_2 + m_3^1 C_3 = 9$$

Repeating the same calculations, but now using $t_z^1 = 9$ gives the same values of m_1^1, m_2^1 and m_3^1. Therefore, $t_z^1 = 9$, $m_1^1 = 1$, $m_2^1 = 1$ and $m_3^1 = 1$. Following, the time t_s^2 at which the next busy period starts is obtained.

$$O_1 + \left\lceil \frac{t_z^1 - O_1}{T_1} \right\rceil T_1 = 1 + \left\lceil \frac{9-1}{10} \right\rceil 10 = 11$$
$$O_2 + \left\lceil \frac{t_z^1 - O_2}{T_2} \right\rceil T_2 = 0 + \left\lceil \frac{9-0}{20} \right\rceil 20 = 20$$
$$O_3 + \left\lceil \frac{t_z^1 - O_3}{T_3} \right\rceil T_3 = 4 + \left\lceil \frac{9-4}{5} \right\rceil 5 = 9$$
$$t_s^2 = min\{11, 20, 9\} = 9$$

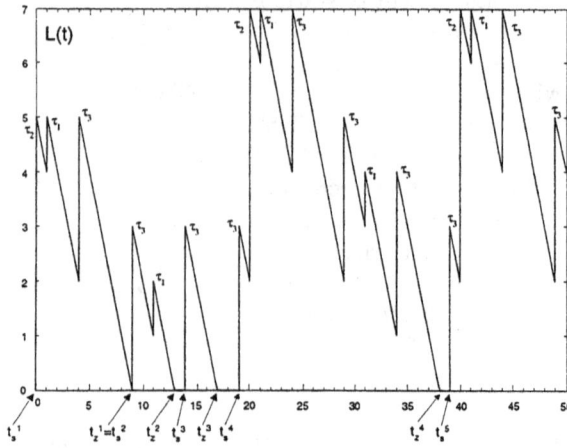

Fig. 4. Load function for the example of static priority assignment.

Operating in the same way produces $[t_s^2, t_z^2) = [9, 13)$, $[t_s^3, t_z^3) = [14, 17)$, $[t_s^4, t_z^4) = [19, 38)$, $m_1^2 = 1$, $m_2^2 = 0$, $m_1^3 = 1$, $m_1^3 = 0$, $m_2^3 = 0$, $m_3^3 = 1$, $m_1^4 = 2$, $m_2^4 = 1$ and $m_3^4 = 4$.

The next busy period begins at $t_s^5 = 39 > P = 20$. Therefore, this busy period and the ones following it can be neglected. Therefore, $t_z + P = 38 \Rightarrow t_z = 18$. To obtain the feasibility array, it is sufficient to take into account the task instances released in the busy periods included in $[18, 38)$.

Figure 4 depicts the load function, which can be used to confirm the previous results.

$m_1^4 = 2$, so there are 2 instances of τ_1 in the busy period $[19, 38)$. Thus, $f_1^1 = 0$. The response time of the last instance of τ_1 in the busy period $[19, 38)$, if τ_1 would receive the lowest priority, is obtained by applying 7

$$R_1^{2,4} = t_z^4 - O_1 - \left\lceil \frac{t_s^4 - O_1}{T_1} \right\rceil T_1 - (m_1^4 - 1)T_1 = 7$$

Therefore, $R_1^{2,4} = 7 \le D_1 = 8 \Rightarrow f_1^2 = 1$.

$m_2^4 = 1$ and so the response time of the single task instance of τ_2 in the busy period $[19, 38)$, if τ_2 would receive the lowest priority, is

$$R_2^{1,4} = t_z^4 - O_2 - \left\lceil \frac{t_s^4 - O_2}{T_2} \right\rceil T_2 = 18$$

and $R_2^{1,4} = 18 \le D_2 = 20 \Rightarrow f_2^1 = 1$.

$m_3^4 = 4$. Therefore the first three instances of τ_3 in $[t_z, t_z + P)$ are not feasible and so $f_3^1 = f_3^2 = f_3^3 = 0$. The response time of the last instance of τ_3 in the busy period $[19, 38)$ if τ_3 would receive the lowest priority, takes the value

$$R_3^{4,4} = t_z^4 - O_3 - \left\lceil \frac{t_s^4 - O_3}{T_3} \right\rceil T_3 = 4$$

Therefore $R_3^{4,4} = 4 \le D_3 = 5 \Rightarrow f_3^4 = 1$.

$[19, 38)$ is the only busy period included in $[t_z, t_z + P)$. Hence, the array of feasibilities for the tasks is

obtained by repeating the feasibilities calculated for the busy period $[19, 38)$.

$$\{f_1^1, \dots, f_1^j, \dots\} = \{0, 1, 0, 1, 0, 1, 0, 1, 0, 1, 0, 1 \dots\}$$
$$\{f_2^1, \dots, f_2^j, \dots\} = \{1, 1, 1, 1, 1, 1, 1, 1, 1, 1, 1, 1 \dots\}$$
$$\{f_3^1, \dots, f_3^j, \dots\} = \{0, 0, 0, 1, 0, 0, 0, 1, 0, 0, 0, 1 \dots\}$$

Using the previous feasibility arrays the value of the feasibility functions $F_1 = 1$, $F_2 = 1$ and $F_3 = 0$ are obtained. Thus, tasks τ_1 and τ_2 are feasible when they receive the lowest priority, that is to say, priority 3. Priority 3 is assigned to task τ_1 and priority 2 to task τ_2 (τ_2 is obviously feasible when it receives a higher priority, such as 2). After this, tasks τ_1 and τ_2 are removed from the task set, resulting in a task set which consists only of task τ_3, which is also feasible with priority 1 (the highest priority).

Therefore, the task set is feasible with the priorities $p_1 = 3$, $p_2 = 2$ and $p_3 = 1$.

6. CONCLUSIONS AND FUTURE WORK

A new model of real-time constraints for periodic tasks has been proposed. This model is able to specify real-time constraints ranging gradually from soft deadlines to hard deadlines. A necessary and sufficient feasibility test has been proposed, which is used to obtain an optimal assignment of static priorities for skippable periodic tasks.

The feasibility test will be improved to account for precedence relations, shared objects, and scheduling overloads. In addition, a feasibility test for dynamic scheduling will be developed.

7. REFERENCES

[1] Audsley, N.C. 1991. Optimal Priority Assignment and Feasibility of Static Priority Tasks with Arbitrary Start Times. Technical report YCS 164, Department of Computer Science. University of York, England.

[2] Koren, G. and Shasha D. 1995. Skip-Over: Algorithms and Complexity for Overloaded Systems that Allow Skips. Proc. of the 16th IEEE Real-Time Systems Symposium. Pisa, Italy, pp. 110-117.

[3] Lehoczky J.P. 1990. Fixed Priority Scheduling of Periodic Task Sets With Arbitrary Deadlines. Proc. of the 11th IEEE Real-Time Systems Symposium. Lake Buena Vista, FL, USA, pp. 201-209.

[4] Leung, J.Y.T., and Whitehead, J. 1982. On the Complexity of Fixed-Priority Scheduling of Periodic, Real-Time tasks. Performance Evaluation 2:237-250.

AUTHOR INDEX

www.ingramcontent.com/pod-product-compliance
Lightning Source LLC
Chambersburg PA
CBHW072058220326
41598CB00068BA/4452